ELECTRONIC INSTRUMENTS

FOURTH EDITION

ELECTRONIC INSTRUMENTS
INSTRUMENTATION TRAINING COURSE

DALE R. PATRICK, PROFESSOR

Department of Industrial Education and Technology
College of Applied Arts and Technology
Eastern Kentucky University

PRENTICE HALL, Englewood Cliffs, NJ 07632

Library of Congress Cataloging-in-Publication Data

PATRICK, DALE R.
  Electronic instruments : instrumentation training course / Dale R.
Patrick.—4th ed.
      p.   cm.
  Includes index.
  ISBN 0-13-251208-4
  1. Engineering instruments.   2. Electronic instruments.
  I. Title.
  TA165.P33   1992
  621.381′54—dc20   91-31939
                                                               CIP

Editorial/production supervision and
   interior design: **Marcia Krefetz**
Cover design: **Ben Santora**
Manufacturing buyer: **Ed O'Dougherty**
Prepress buyer: **Ilene Levy**
Managing editor: **Mary Carnis**
Supervising editor: **Barbara Cassel**
Publisher: **Susan Willig**
Editorial assistant: **Shirley Cholopak**
Production assistant: **Bunnie Neuman**

© 1992 by Prentice-Hall, Inc.
A Simon & Schuster Company
Englewood Cliffs, New Jersey 07632

Printed in the United States of America
10 9 8 7 6 5 4 3 2 1

ISBN 0-13-251208-4

Prentice-Hall International (UK) Limited, *London*
Prentice-Hall of Australia Pty. Limited, *Sydney*
Prentice-Hall Canada Inc., *Toronto*
Prentice-Hall Hispanoamericana, S.A., *Mexico*
Prentice-Hall of India Private Limited, *New Delhi*
Prentice-Hall of Japan, Inc., *Tokyo*
Simon & Schuster Asia Pte. Ltd., *Singapore*
Editora Prentice-Hall do Brasil, Ltda., *Rio de Janeiro*

CONTENTS

CHAPTER 9 ELECTRONIC TRANSMITTERS 159

CHAPTER 10 pH INSTRUMENTS 172

CHAPTER 11 FLOWMETERS AND CONVERTERS 190

CHAPTER 12 RECORDERS AND INDICATORS 204

PREFACE

Many technicians who are competent with mechanical/pneumatic instruments are at a loss when confronted with electronic instrumentation. There are perhaps two good reasons for this. First, we shy away from electronics because we fear it is just too complicated. Second, unlike the workings of mechanical equipment, which can be observed, electricity cannot be seen and an elaborate symbolism has resulted, which discourages the novice. Much of the symbolism is highly mathematical, which is a further deterrent. However, in this book the only symbolism that must be mastered is the notation of electronic circuitry. No prior study of mathematics or theory is needed.

The course of study is based on functional training methods, which include:

- Direct study of actual industrial instruments
- Extensive use of the oscilloscope
- Development of principles and theory only when needed to understand a piece of equipment and circuits.

With the knowledge obtained from reading this book, the industrial technician should be able to perform routine measurements using electronic instruments that were formerly considered only as precision laboratory equipment.

Dale Patrick

ELECTRONIC INSTRUMENTS

CHAPTER 1

MEASUREMENT CIRCUITS AND PRIMARY DEVICES

OBJECTIVES

Upon completion of this chapter, you will be able to:

1. Identify common electronic symbols used in schematic diagrams.
2. Explain how electricity is generated electrochemically and with a thermocouple.
3. Demonstrate Ohm's law with the use of a slide-wire resistor.
4. Explain the operation of a thermocouple.
5. Show how a thermocouple is used to measure an unknown temperature.
6. Explain the operation of a galvanometer.
7. Show how a potentiometer is used to measure electrical voltage values.
8. Explain the operation of a Wheatstone bridge.

IMPORTANT TERMS

In the study of electronic instruments one frequently encounters a number of new and somewhat unusual terms. These terms play an important role in the presentation of this material. As a rule, it is very helpful to review these terms before proceeding with the chapter.

Bimetallic strip: Two dissimilar metal strips commonly bonded together into a single unit. This unit will change its shape when subjected to heat.

Bourdon element: A pressure-sensing element made of a hollow tube that is closed at one end. Pressure applied to the open end causes the structure to change its shape.

Cold junction: The reference junction of a thermocouple measuring element. This junction is generally maintained at a temperature of 0 degrees Celsius or 32 degrees Fahrenheit when measuring temperature.

Constantan: A copper–nickel alloy used as the negative lead of a thermocouple.

Coulomb: A unit of electrical charge that represents a large number of electrons.

Current: The movement of electrical charge; the flow of electrons through an electrical circuit.

Electromotive force (EMF): The "pressure," or force, that causes electrical current to flow.

Electron: A small particle which is part of an atom that is said to have a negative (−) electrical charge. An electron causes the transfer of electrical energy from one place to another.

Electronic: A branch of science that deals with the flow of small electrical quantities such as electrons and ions.

Filled temperature system: A temperature-measuring assembly that consists of a sensing bulb, capillary tube, and pressure element. The assembly is filled with a liquid or gas. Changes in temperature will cause the internal pressure of the unit to vary, which in turn produces a physical change in the pressure element.

Galvanometer: An electromagnetic instrument used to measure small values of current through the deflection of an indicating hand.

Neutron: A particle in the nucleus or center of an atom that has no electrical charge and is considered to be neutral.

Nucleus: The center part of an atom, made of protons or positive charges and neutrons that have no electrical charge.

Ohm's law: A law that explains the relationship of voltage, current, and resistance in an electrical circuit.

Potentiometer: A variable resistor used to control an electrical circuit. An unknown voltage-measuring circuit that determines values by comparing them with a known reference to produce a null condition.

Proton: A particle in the center of an atom that has a positive (+) electrical charge.

Resistance: An opposition to the flow of electrical current in a circuit. The fundamental unit of measurement is the ohm.

Slide wire: An electrical resistance that has a contacting slider that permits adjustment of its value.

Standard cell: A voltage cell that serves as a standard or reference in the measurement of an electromotive force.

Thermocouple: A junction of two dissimilar metals that will produce a voltage that is proportional to the temperature applied.

Variable: A process condition such as temperature, flow, or level that is susceptible to change and can be measured, altered, or controlled.

Wheatstone bridge: A network of four resistors, a voltage source, and a galvanometer that makes precision resistance measurements by comparing an unknown resistance with a known or standard resistance value.

INTRODUCTION

The development of electronic instruments that test, measure, and control industrial processes has gone through a rather impressive growth pattern. It is now common practice for industrial technicians to perform routine measurements using electronic instruments that were formerly considered precision pieces of expensive laboratory equipment. As a result of this, industry in general is very concerned about training its personnel to become competent instrument users in order to take full advantage of the expanding field of electronic instrument technology.

INDUSTRIAL ELECTRONIC SYMBOLS

In the electronics field, a number of symbols are commonly used as a means of representing electrical components in diagrammed form. These symbols serve as the basis for schematic diagrams, which range from

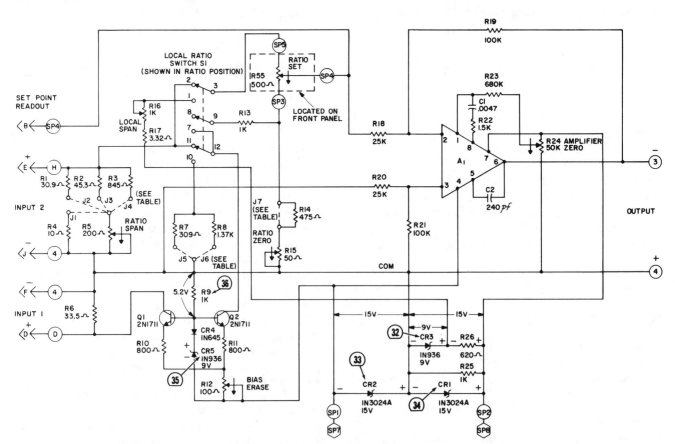

FIGURE 1-1   Example of an electronic diagram. (Courtesy of The Foxboro Co.)

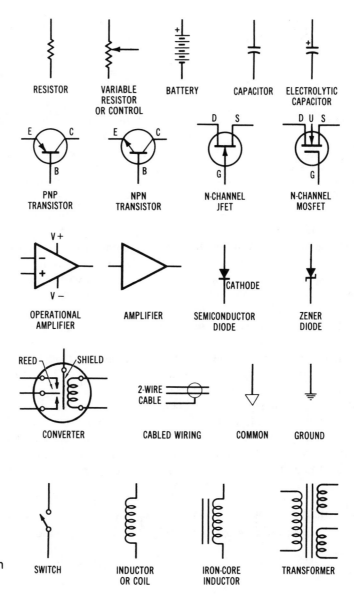

FIGURE 1-2 Some common electronic symbols.

simple to very complex circuits. As a rule, only a few basic symbols are in common use. These symbols are, however, extremely important because they are in much common use. It is essential that these symbols, and the components they represent, be understood in order to effectively interpret the circuitry of a schematic diagram.

We will employ electronic symbols in schematic diagrams throughout this book. A diagram of an amplifier of a controller is shown in Fig. 1-1 as an example. As you will note in this diagram, the symbols are numbered R_{10}, C_2, Q_1, and so on. This numbering generally serves as a reference for replacements.

A number of important component symbols are shown in Fig. 1-2 for reference. These represent some of the more common symbols used in this book. A more extensive list of industrial electronic symbols is provided in the Appendix at the back of the book.

Refer to this list when you encounter an unfamiliar symbol.

ELECTRON FLOW

In the study of electronics the generation of electricity is an extremely important concept. We consider only the three methods of generation that are particularly applicable to the industrial field: thermocouples, electrochemical cells, and mechanical generators. It is well known that when two wires of different metals are joined together and placed in a closed loop, as in Fig. 1-3, a current will occur. *Current* simply refers to the flow or movement of electrons in a closed loop.

An electron is one of the basic particles of an *atom* of matter. An atom of any one of the 105 natural elements, when divided into its basic particles, con-

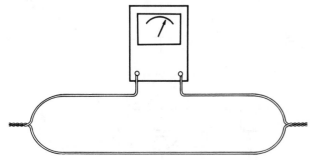

FIGURE 1-3   Simple closed circuit.

tains a number of *electrons*, *protons*, and *neutrons*. Figure 1-4 shows a representative drawing of an atom of carbon. Note that it contains six protons, represented by plus signs; six electrons, represented by minus signs; and six neutrons, represented by the letter N. The center of the atom is called the *nucleus*. It houses the protons and neutrons. Electrons revolve around the nucleus, which tends to attract the electrons by the positive charge of the protons. Atoms are electrically neutral when the number of electrons equals the number of protons in the nucleus. Atoms in general are electrically neutral unless they have been acted upon by some outside source of energy.

As far as we know today, the physical makeup of all atoms is basically the same as the carbon atom, the only difference being the number of particles. Hy-

drogen, for example, has 1 electron, 1 proton, and no neutrons. Helium has 2 electrons, 2 protons, and 2 neutrons. Copper, which is a very common electrical conductor, has 29 electrons, 29 protons, and 34 neutrons.

It is a natural property of metals that if two different metals are joined together, the natural equilibrium of the atomic structure is disturbed in such a way that the electrons have a tendency to flow from one of the metals into the other. If, for example, copper is joined to constantan, which is an alloy of different metals, the electrons of the constantan tend to leave it and flow into the copper. If this is to happen, electrons in the copper must move to make room for the constantan electrons. They do this by flowing through the wires connected to the thermocouple and back into the constantan wire, as shown in Fig. 1-5. In this way, a current, or movement of electrons, occurs.

Current is an important electrical unit that is measured in amperes. An *ampere* is defined as a flow of 1 coulomb per second. A coulomb is a unit of electrical quantity that refers to a 6 280 000 000 000 000 000 electrons. This large number may be written as 6.28 $\times 10^{18}$.

In practical *electronic* circuits, 1 ampere (A) of current is considered to be quite large. A *milliampere* (mA), which is 1/1000 (0.001) of an ampere, is a somewhat more reasonable value of current. An expression of an extremely small value of current is the *microampere* (μA). This value is 1/1 000 000 (0.000 001) of an ampere. Microamperes of current frequently appear in solid-state circuits and as the output of sensor circuits.

To produce an electron flow, an electrical force must be applied. In the case of the thermocouple just described, the dissimilar wire and the temperature difference between the two junctions of the wires will provide such a force. The force itself is called *electromotive force* and is abbreviated EMF. The basic unit of EMF is the *volt* (V).

As might be expected, all materials have a tendency to oppose the flow of electrons to some extent.

Measurement Circuits and Primary Devices

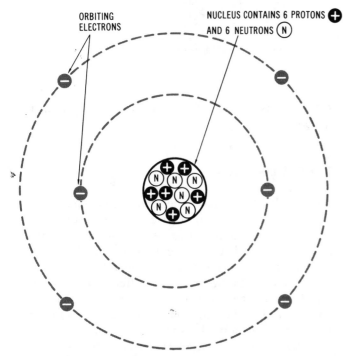

FIGURE 1-4   Representative drawing of an atom of carbon.

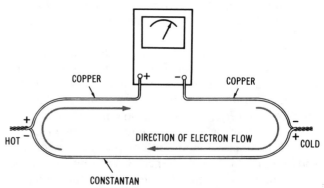

FIGURE 1-5   Electron movement in a thermocouple circuit.

This opposition, in general, is called *resistance*. The resistance of a material is primarily dependent on the structure of the atoms in its physical makeup. You may recall that current is based on the movement of free electrons in a material. If the material does not have an abundance of free electrons in its atomic structure, it tends to resist the flow of electrons. In practice, the resistance of a conductor is based on the atomic structure of the material, its length, cross-sectional area, and temperature. The fundamental unit of resistance is the *ohm*. The Greek capital letter omega (Ω) is used as a symbol to represent ohms of resistance. Resistance is expressed in thousands of ohms by the letter combination $k\Omega$, and in millions of ohms by the letter combination $M\Omega$.

Some materials are such poor conductors of electricity that they are often called *insulators*. Glass, porcelain, dry wood, paper, many of the plastics, and rubber are examples of common electrical insulating materials. Essentially, insulating materials do not contain an abundance of free electrons in their atomic structure.

In general, metals, carbon, ionized gases, or ionized molecules in a chemical solution make good electrical conductors. In practice, copper is the most common conductor of electricity. Silver is a better conductor, but is more costly and rarely used.

A fundamental relationship exists between the electrical units of resistance, current, and voltage. In this relationship, when 1 volt of EMF is applied to 1 ohm of resistance, a resulting current of 1 ampere will occur. Another way of expressing the relationship of resistance, current, and voltage is to say that it takes 1 volt of EMF to force 1 ampere of current through 1 ohm of resistance. This basic relationship is known as *Ohm's law*, which can be expressed mathematically as

$$ I = \frac{V}{R} $$

where I is the symbol for current, V the symbol for voltage, and R the symbol for resistance. Ohm's law can also be expressed mathematically as

$$ R = \frac{V}{I} $$

and

$$ V = I \times R $$

Notice that this law states (among other things) that current times resistance is equivalent to voltage. It says that if a current exists, a voltage also exists because all materials have some electrical resistance. In subsequent work we will frequently describe circuits where a desired voltage is obtained by causing a particular current to pass through a selected resistance. In this way it is possible to "generate" a particular voltage.

Voltage, resistance, and current are easily measured with ordinary test equipment. Voltage is often considered as a measurement of the potential difference that exists between two points. It is simply not possible to have a voltage at a single test point. It is incorrect to speak of the "voltage in a wire" or "the voltage at a terminal." When these expressions are used, what is being said is that the "voltage of a wire compared to the voltage of a second wire is of a certain value." Or, perhaps, "the voltage of a wire compared to the voltage of earth's ground is a particular value." Similarly, the expression "the voltage at a terminal" is short for saying "the voltage between terminals A and B" is a particular value.

The measurement of current, on the other hand, is a measurement of a condition at one point. Or, more exactly, it is a measurement of the number of electrons flowing past a particular point of a conductor. To make a current measurement, it is necessary to open the circuit and to cause the current to pass through the measuring instrument. In practice, this is called connecting the meter in series with the circuit under test. Notice that the current measurement of Fig. 1-5 can be made at any place in the circuit since the current is the same throughout the circuit.

The measurement of resistance is accomplished by causing a current (originating in the measuring instrument) to pass through the device in whose resistance we are interested. To make a resistance measurement, the device whose resistance is to be measured must be removed from its associated circuitry.

ELECTROCHEMICAL ACTION: THE STANDARD CELL

There are many varieties of electrochemical cells, but their operating principles are the same. The essential components of a cell are two electrodes made of dissimilar materials immersed in an electrolyte. An *electrolyte* is a solution that readily enables electrons to "skip" from molecule to molecule. The electrodes are usually metals or some material such as carbon that has metallic characteristics. A common electrode combination is zinc and carbon, using sulfuric acid solution as the electrolyte. A variation of this combination is found in the so-called *dry cell*, which is not truly dry, since the fluid electrolyte, sal ammoniac (NH_4Cl), is made into water paste. The electrodes are the zinc outer casing and a carbon rod in the middle. The paste mixture is packed around the central carbon electrode, and the assembly is sealed with tar.

A third combination is that found in the so-called *mercury cell*—a "dry" type. One electrode is zinc (−) and the other is mercuric oxide (+). The electrolyte is a solution of potassium hydroxide.

The *lead–acid cell*, used in the common automobile battery, is composed of a lead electrode (−) and a lead oxide, PbO_2 (+) electrode, with sulfuric acid solution as the electrolyte. Some additional combinations are nickel and cadmium with potassium hydroxide, and nickel oxide (+) and iron with an electrolyte of potassium hydroxide. The latter is sometimes called the *Edison cell*.

Some cells can be recharged; that is, electrical power can be put back into them. Others cannot be recharged.

The *standard cell* (or *Weston cell*) is unusual in some of its properties because, unlike some cells, if it is used properly, the voltage across its electrodes remains essentially constant. In fact, it remains so constant that the voltage can be used as a standard for comparison with other voltages—hence the reason for calling this special cell a *standard* cell.

The Weston standard cell uses mercury as one electrode and a cadmium amalgam (mixture of cadmium and mercury) for the second electrode. The electrolyte is a cadmium sulfate solution. Cadmium sulfate is formed by reacting cadmium with sulfuric acid. Sulfuric acid is a compound of the elements sulfur, oxygen, and hydrogen. Using chemical notation, sulfuric acid is written as H_2SO_4, where H stands for hydrogen, S for sulfur, and O for oxygen and the numbers indicate the number of atoms of each element in the molecule. When added to water, this compound tends to dissociate into hydrogen ions and sulfate ions. An ion is an atom with too many or too few electrons. This results in the atom having a charge. If the hydrogen ion and sulfate ion rejoin, they neutralize each other. An electron carries a negative charge; hence the particle lacking an electron becomes positively charged. This positively charged particle makes a good "stepping stone" for migrating electrons. In fact, the positively charged particle actively attracts electrons in an effort to get balanced.

If cadmium is added to hydrogen sulfate (sulfuric acid) it will be found that the cadmium and sulfate form a stronger compound than the hydrogen and sulfate, with the result that the cadmium drives off the hydrogen, which bubbles off as a gas, to form cadmium sulfate.

Like the hydrogen sulfate just discussed, cadmium sulfate ionizes and becomes a "stepping stone" for the migration of electrons.

The construction of a Weston standard cell is shown in Fig. 1-6. The cell is made of glass in an H configuration. A cadmium amalgam is placed in one leg, forming one electrode. Mercury is placed in the

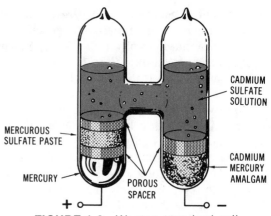

FIGURE 1-6   Weston standard cell.

other leg, forming the second electrode. Solid cadmium will "dissolve" into mercury (this is called an *amalgam*). In doing so the mercury tends to reduce the attraction between the cadmium atoms. A paste of mercury sulfate and mercury is placed over the mercury. The two electrodes are covered with a solution of cadmium sulfate, the electrolyte.

When the cell operates, the cadmium ionizes and goes into the cadmium solution. As the cadmium ionizes, it liberates electrons, which pile up on the platinum lead of the mercury–cadmium electrode. The electrons are negatively charged; therefore, the lead becomes negative and is labelled with the negative sign (−). If a path is provided, the electrons will flow from the cadmium–mercury electrode to the mercury electrode. At the mercury electrode the electrons will be taken up by the mercury ions (mercurous ions) formed in the mercury electrode, deionizing the mercury and returning it to its original state. In this way a flow of electrons is caused. The pressure or electromotive force behind this flow depends on the electrode materials used. In the standard cell this EMF (or voltage) is 1.0183 V, provided that very small currents are drawn for very small periods of time. In the ordinary dry cell using zinc and carbon electrodes, the voltage is approximately 1.5 V. The basic voltage depends only on the electrode materials, as is the case with thermocouples. The size of the cell does not change the voltage—only the current-delivering capacity. If several cells are joined in series to form a battery, the voltage of the battery is the sum of the individual cell voltages.

Two methods for generating current have been considered. The thermocouple is an example of the solid-state method, wherein thermal energy (heat) is converted to electricity. The electrochemical cell is an example of converting chemical energy into electrical energy. There are other methods of generating current, particularly the mechanical-to-electrical method. Each of these three methods is reversible; that is, electrical energy can be converted back to thermal, chemical, or mechanical energy, for example, as in the charging of an automobile battery (electrical to chemical) and the operation of the electric motor (electrical to mechanical).

THE SLIDE WIRE AND OHM'S LAW

In the foregoing discussion, two methods of generating current have been discussed. Also, current (electron flow) was considered and it was noted that various materials resist the flow of electrons according to the composition of the material. Now the properties of electrical resistance will be considered in some detail. The slide wire will be studied as an example of some of the properties and will be considered as a demonstration of Ohm's law.

Let us examine the circuit of Fig. 1-7. The cell is an ordinary No. 6 dry cell, with a voltage of 1.5 V between the terminals. This voltage will cause electrons to flow through the wires and the resistance. The electrons will leave the negative (−) terminal and flow through the wire and back to the positive (+) terminal, establishing a current. The amount of current is determined by how much resistance there is to the flow of electrons. The factors that affect the resistance are the kind of material used to make the wire, the length of the wire, the diameter of the wire, and, depending on the materials, the temperature of the wire.

For a given material, as we might suspect, the smaller the diameter, the greater the resistance, and the longer the wire, the greater the resistance. Some materials offer relatively small resistance. Such materials are said to be *good conductors*. Silver is the best conductor; copper is next, followed by gold and aluminum. Carbon is a relatively poor conductor, having a resistivity about 2000 times that of copper. The resistivity of iron is about five times that of copper. Nichrome, an alloy made to have a high resistivity, is about 65 times more resistive than copper. For specific examples, 630 feet (192 meters) of 12-gauge copper wire has a resistance of 1 ohm. If the wire were iron, only 110 ft (33.5 m) would be required to obtain 1 Ω of resistance.

Almost all conductors change resistance with temperature. This property is most useful in some cases, but a great nuisance in others. For example, the varying resistance property of some conductors makes a convenient and reliable method of temperature measurement, but if a fixed resistance is required, special alloys are needed. These alloys must be designed so that a temperature change results in a very small change in resistance—preferably no change. Manganin is an alloy widely used when variations in resistance due to temperature are undesirable. Manganin is composed of 4% nickel, 84% copper, and 12% manganese. Unlike most metals and alloys, the resistance of manganese decreases slightly as the temperature increases. The resistance of copper changes about 100 times the amount that manganin does for the same temperature change, but its resistance increases with temperature. In addition to its temperature stability, manganin has a relatively high resistivity, making it possible to obtain high resistance values with comparatively small lengths of wire.

Suppose that we rebuild the circuit of Fig. 1-7, as shown in Fig. 1-8. In this figure, a resistance wire has been selected so that 10 ft (3 m) has a resistance of 10 Ω. The battery has a potential difference of 1 V between its terminals Therefore, the wire has 1 V applied across it. That is, between the ends of the wire, as they are attached to the battery, there is a difference in voltage equal to 1 V. The 1 V causes electrons to flow through the wire. Suppose that we attach the negative (−) probe of a sensitive voltmeter (one whose resistance is so high that 1 V cannot force any current through it) to the negative (−) terminal of the battery and then touch the negative terminal of the battery with the positive (+) probe of the meter. The meter will show 0 V. Now suppose that we move the positive probe of the meter to the positive terminal of the bat-

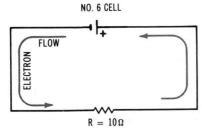

FIGURE 1-7 Closed circuit with cell and resistance.

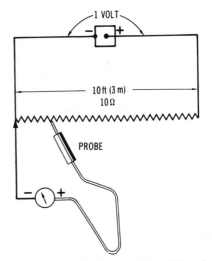

FIGURE 1-8 Principle of the slide wire.

tery. The meter will read 1 V. Return the positive probe to the negative terminal and slowly draw it along the wire away from the negative terminal toward the positive terminal. The meter reading will gradually increase as the probe moves along the wire until the probe arrives at the positive terminal, where the meter will read 1 V. Suppose that we start at the negative terminal and draw the probe along the wire until the meter reads 0.5 V and mark that point on the slide wire. If we measure the distance from that point to the negative terminal, it will measure exactly 5 ft (1.5 m). Suppose that we continue moving the probe until the meter reads 0.75 V and measure the distance. It will be exactly equal to $\frac{3}{4}$ of 10 ft, or $7\frac{1}{2}$ ft (2.3 m). Once this was done, we would no longer need the voltmeter because we would know that at the 2-ft (0.61-m) mark, for example, there would be a voltage difference of 0.2 V between the negative ($-$) terminal and that mark; at the 6-ft (1.83-m) mark the voltage difference between that point and the negative terminal would be 0.6 V. Or, if we choose to compare the 6-ft (1.83 m) mark to the positive terminal, the voltage would be 0.4 V. Suppose that we are interested in the voltage between the 2-ft (0.61-m) mark and the 6-ft (1.83-m) mark. This we know would be 0.4 V.

When we marked the wire using a voltmeter, we made it possible to know what the voltages are across various points on the wire by knowing only where the marks were on the wire. Briefly stated, we calibrated the wire so that even though we measure in feet (or meters) we can speak of voltages.

Let us change the battery of Fig. 1-8 from a 1-V battery to a 2-V battery. The voltage drop across the wire would be 2 V. The 5-ft (1.5-m) mark would have a voltage of 1 V when referenced to the negative terminal. Select any other mark, say 7 ft (2.13 m), and compare the voltage difference between it and the negative terminal. It would equal 0.7 of the applied voltage (0.7 × 2, or 1.4 V).

Instead of using the 10-Ω resistance wire, let us replace it with a 20-Ω wire of the same length. The 2-V battery remains the same. Again, as was the case in the previous examples, the voltage drops across the wire are related to the lengths of wire involved. The maximum voltage drop will equal the 2 V supplied. At a point 5 ft (1.5 m) from the negative terminal, the voltage referenced to the negative terminal will equal 1 V:

$$\frac{5}{10} \times 2 = 1$$

It is recognized that if the voltage on a given circuit is changed, something else must change. The voltage is what drives the current, and if the voltage is doubled and the circuit stays the same, the current will double. But consider the case where the resistance of the wire is doubled and the applied voltage stays at 2 V. Something must happen to the current. If the resistance is doubled, the current finds it twice as difficult to get through the wire; therefore, if the driving force (voltage) stays the same, it will be able to drive only half as much current through the wire.

The relationship between voltage, current, and resistance can be expressed by saying that the voltage between any two points of a circuit is equivalent to the resistance of the wire between the same two points multiplied by the current through the points. Or, in mathematical notation,

$$V = IR$$

where V is in volts, I in amperes, and R in ohms.

Let us see what current is in the circuit of Fig. 1-7. $V = 1.5$ V and $R = 10$ Ω; therefore, $1.5 = 1 \times 10$. Dividing both sides of the equation by 10, we get $1.5/10 = I$; therefore, $0.15 = I$. The current is 0.15 A. In Figure 1-8, $V = 1$ V and $R = 10$ Ω, so

$$I = \frac{V}{R} = \frac{1}{10} = 0.1 \text{ A}$$

In the example where $V = 2$ V and the resistance is 10 Ω,

$$I = \frac{V}{R} = \frac{2}{10} = 0.2 \text{ A}$$

In the example where $V = 2$ V and the resistance is 20 Ω,

$$I = \frac{2}{20} = 0.1 \text{ A}$$

Examine the equation $V = IR$ and, keeping this in mind, reexamine Fig. 1-8. Notice that when a known voltage is applied to a known resistance, the equation permits us to predict how much current there will be. If we probe along this resistance, we select various resistances. When these are multiplied by the current, we are able to predict the voltages involved; and if we go one step further and calibrate this resistance with a series of marks obtained by measuring the voltages between the marks and the negative terminal, we can predict what the voltage is by simply referring to the position on the wire.

The configuration just discussed is an example of a slide wire as used in potentiometer-type measuring instruments and in other instruments. The slide wire is a practical application of Ohm's law. In later work, the slide-wire notion will be expanded into a voltage-measuring instrument.

THERMOCOUPLE PRINCIPLES

The thermocouple is, perhaps, the only practical industrial method for measuring temperatures between 500 and 1500°Celsius. The filled system is not designed for these high temperatures. The resistance thermometer must be specially designed if it is to be used in those ranges. For temperatures less than 500°C, the thermocouple is often used, notwithstanding the fact that certain thermocouple installations cost more than a filled system would for the same job.

One of the distinct advantages of the thermocouple is that its voltage output can readily be transmitted over large distances. A second advantage is that a thermocouple can be fabricated in about 10 minutes in almost any instrument shop. The thermocouple itself is relatively inexpensive. The recording or indicating instruments used with a thermocouple may be of the null-balance type or of the deflection type. The use of null-balance instruments usually results in a higher installation cost than that for a filled system.

EMF Property of Metals

One of the natural properties of metals is that if two different metals come in contact with each other, an electromotive force (EMF) is developed. The contact point is a property of all metals. The amount of EMF developed depends on two things: the metals involved and the temperature at the junction. It is possible to take any two metals, join them, and heat the junction through a range of temperatures. A table of equivalent values can be tabulated by noting the temperature and recording the EMF developed. Such data, called *conversion tables*, have been developed for certain combinations of metals. The EMF is expressed in millivolts.

These tables are used to convert EMF back to temperature. This is the more important aspect since the purpose of the thermocouple is to measure temperature, yet the output of the thermocouple is EMF. This is in some ways analogous to recording flow when the output is differential pressure.

Thermocouple Materials

Certain standard metals and pairs have been adopted for thermocouple construction. These are:

    Chromel–alumel
    Iron–constantan
    Copper–constantan
    Platinum–rhodium–platinum

There are other combinations, but the ones listed are the most widely used. These metals are all alloys, that is, mixtures of metals. The manufacture of these alloys is carefully controlled so that a specific EMF will be developed for each temperature.

The desirable properties for thermocouple materials are that the metals be relatively inexpensive and develop substantial EMF for small temperature changes. In addition, the alloys should be stable over long periods of time, resistant to chemical attack, and easily manufactured to specifications. No thermocouple materials are perfect or even particularly good on all requirements. In general, the ones mentioned are adequate in the practical plant situation.

Additional Properties

In addition to the property of developing a voltage, each thermocouple pair has a definite polarity. That is, the electrons will flow (if permitted to) in a direction that depends on the materials used. The polarities of the thermocouple materials listed previously are:

(+)	(−)
Chromel	Alumel
Iron	Constantan
Copper	Constantan
Platinum	Rhodium–platinum

It is extremely important that polarity be observed when making connections. There is only one method of determining polarity, and that is by knowing which wire is which. The wires can be identified by testing them with a permanent magnet. Alumel is slightly magnetic, iron is strongly magnetic, constantan not at all. Copper can be identified by its color.

It is important to remember that whenever and wherever two dissimilar wires come in contact, an EMF is generated. It is a function of the temperature at the junction and the metals involved. The thermocouple is a utilization of this phenomenon, and certain alloys and combinations have been adopted as standard. Tables are available relating the temperature of the junction and the EMF developed for the standard combinations so that it is possible to determine the temperature if the EMF is known, or to determine the EMF if the temperature is known. Figure 1-9 shows the temperature–voltage curves for the thermocouples mentioned. Observe that the curves are not straight. This means that temperature charts and scales are not linear. Note that they all meet at 32°F (0°C). The slope of the platinum–rhodium–platinum curve shows that the EMF output for a given change in temperature is small compared to that of iron–constantan or chromel–alumel.

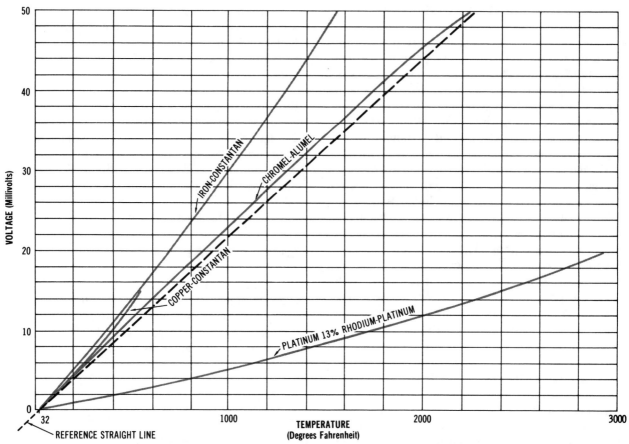

FIGURE 1-9   Temperature/voltage curves for some thermocouple materials.

THE REFERENCE JUNCTION

In the discussion on thermocouple principles, it was emphasized that whenever two dissimilar wires are joined, a voltage will be generated. If we are to measure this voltage, it follows that an electrical circuit be completed (i.e., a closed path must be completed through the measuring instrument). In this way a continuous path is furnished for the flow of electrons. A moment's thought will show that if the circuit is closed, the two dissimilar wires are rejoined. This rejoining of the wires is accomplished in a somewhat devious manner in that the dissimilar thermocouple wires are joined through the measuring instrument, as shown in Fig. 1-10. Nevertheless, a second junction of dissimilar wires is formed. This second junction is the reference junction or "cold" junction. The designation *reference junction* will be used in our discussions. Thus it can be seen that in every thermocouple circuit there are at least two junctions of dissimilar wires. The problem then becomes one of determining how to make a temperature measurement with such a system.

Properties of the Reference Junction

It will be recalled that the voltage generated by a junction of two dissimilar metals is determined by the metals involved and the temperature of the junction. It has been stated that every circuit necessarily has two junctions. This is so because if two different wires are joined at one end, the other ends must also be joined to form a circuit. Unfortunately, the second junction (reference junction) in many cases is not physically obvious and may appear in different locations in different circuits. To further complicate the matter, there are several junctions of dissimilar metals that do not form reference junctions. Consider the circuit in Fig. 1-11.

Terminal A is a junction of iron to copper, and terminal B is also a junction of iron to copper. If terminals A and B are at the same temperature, the voltage generated at terminal A will be equal to the voltage generated at terminal B. Notice that if we proceed around the circuit in the direction indicated by the arrow, the voltage at (say) terminal B increases by 1

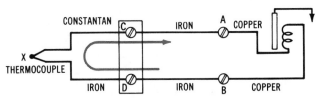

FIGURE 1-12 Basic thermocouple action.

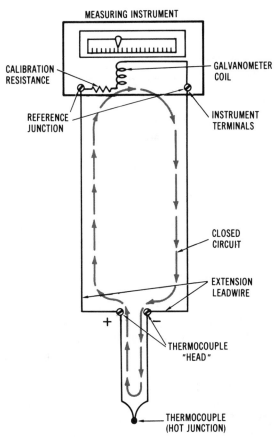

FIGURE 1-10 Closed path in a thermocouple instrument.

mV, but the voltage at terminal A decreases by 1 mV. In other words, they oppose each other, and since they are equal, they cancel each other out. As far as the contribution of terminals A and B to the voltage in the circuit is concerned, they could just as well be directly joined together.

Now let us add a thermocouple to the circuit, as in Fig. 1-12. Starting at terminal D and proceeding in the direction shown by the arrow, observe that at terminal D, iron is joined to iron; hence no voltage is generated. At point X (the thermocouple), iron is joined to constantan and a voltage is generated. Pro-

ceeding to terminal C, constantan is joined to iron, and at this terminal a voltage is generated. In our previous discussion it has been shown that as far as voltages are concerned, terminals A and B could be joined. If this is done, iron is joined to iron; therefore, terminals A and B may be disregarded, leaving only point X (the thermocouple) and terminal C, both of which are junctions of iron and constantan. Notice, however, that at the thermocouple, iron is joined to constantan (when proceeding in the direction of the arrow) and at terminal C, constantan is joined to iron. This is the reference junction. The consequence is that the voltages at the two junctions oppose each other. If the temperatures at these two points are the same, the voltages at the two junctions will cancel each other. The only time that there will be a net gain in voltage is whenever there is a difference in voltage between the thermocouple and the reference junction. It is this difference in voltage that the measuring instrument "sees." In this case, terminal C is the second or reference junction. It is absolutely necessary that the temperature of the reference junction be known, as well as the net voltage of the circuit, before the thermocouple temperature can be determined. The galvanometer can indicate only the millivolts of the circuit. Various means for determining the reference junction temperature are available and will be discussed later. First, however, the location of the reference junction will be considered.

Reference Junction Location

An examination of Fig. 1-12 will show two terminals, A and B, where iron and copper are joined. These have been shown to "cancel out." The actual circuits, however, are arranged somewhat differently in that the iron wire between C and A and D and B is replaced with copper wire; or a second possibility is that an iron wire may be used between D and B and a constantan wire between C and A. The overall operation is the same in either case and is the same as the circuit of Fig. 1-12. The difference between the two methods is one of location of the reference junction. In the case where copper is used, the reference junction is located at the terminals of the thermocouple, as in Fig. 1-13. For the case of the iron and constantan wire, the reference

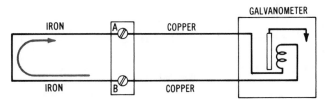

FIGURE 1-11 Junctions of dissimilar metals.

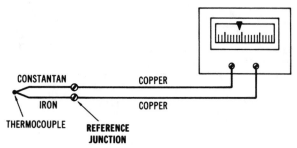

FIGURE 1-13 Reference junction at thermocouple terminals.

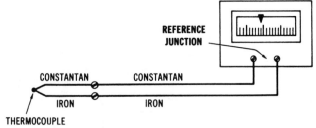

FIGURE 1-14 Reference junction at instrument terminals.

junction is located at the instrument terminals, as in Fig. 1-14. The reference junction is always located at the junctions in the complete circuit where two different pairs of wire are found. For example, in Fig. 1-13 the reference junction is located where the pairs are constantan–copper and iron–copper. In Fig. 1-14 the same pairs are to be found at the instrument where the constantan and iron are joined with the copper of the measuring instrument. Thus the use of constantan and iron wires resulted in the reference junction being located at the measuring instrument. The location of the junction at the instrument makes it possible to take into account the reference junction contribution to the net voltage in the thermocouple circuit.

The special iron–constantan wire is called *extension lead wire*. Lead wire is available in combination for use on the various kinds of thermocouples. It should be noted that the lead wire is not necessarily made up of the same material as the couples. For example, it is not uncommon to find a lead wire of iron–constantan used on a chromel–alumel couple. This usage may seem to be contrary to what has been said. It is possible to mix these wires because the voltage generated by the two different combinations is substantially the same if the temperatures are less than 150°F (65.56°C). An examination of the curves in Fig. 1-9 will show that the iron–constantan and chromel–alumel curves coincide at a low temperature.

Reference Junction and Conversion Tables

Conversion tables are available relating the output of the several thermocouple types. It has been pointed out that the output of a thermocouple of a given material depends on the difference in temperature between the couple and the reference junction; therefore, each table is based on a known reference junction temperature. The most common reference temperatures are 0°C and 32°F. The reason for two tables is one of convenience. The two tables make it unnecessary to convert from the Fahrenheit scale to the Celsius scale, or vice versa. If the actual temperature of the reference

junction is different from the reference temperature of the table, a correction must be made.

Methods for Compensating (Correcting) for Reference Junction Temperature

It will be remembered that the net voltage in a thermocouple circuit is determined by the temperature of two junctions: the measuring junction and the reference junction. The voltage generated is determined by the difference in temperature between these two junctions. This difference is expressed in millivolts and must be converted to degrees. In other words, the signal is calibrated in terms of temperature. However, before this can be done, the reference junction temperature must be taken into consideration. The methods for doing this are called *reference junction compensation*. The methods might be classified as follows: (1) paper and pencil, (2) manual, and (3) automatic. Manual and automatic compensation will be discussed with potentiometers.

The paper-and-pencil method consists of finding the millivolt equivalent of the reference junction temperature. In order to do this the reference junction

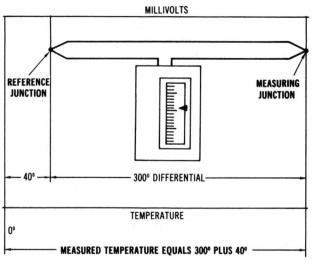

FIGURE 1-15 Compensation for reference junction temperature.

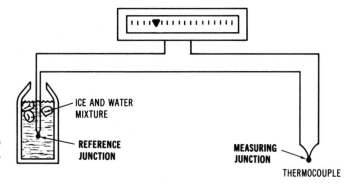

FIGURE 1-16 Laboratory setup for controlling reference junction temperature.

temperature must be measured with a glass-stem thermometer. This temperature value can be located in the proper table and the millivolt equivalent can be determined. This must then be added to the differential millivolt reading of the measuring instrument (see Fig. 1-15). This problem is somewhat analogous to the problem of absolute pressure versus gauge pressure. However, there are further complications in that the millivolt reading has to be converted to degrees. It might be helpful to think in terms of millivolts alone for the time being, recognizing that given a millivoltage, it can be converted to degrees.

In addition to the three methods suggested, there is a fourth possibility that deserves some comment. This method consists of holding the reference junction at a known temperature. One of the most convenient ways of doing this in the laboratory is to immerse the reference junction in a water/ice bath, as in Fig. 1-16. The temperature of this bath is 32°F or 0°C.

To summarize, every temperature measurement made with a thermocouple is necessarily a measurement of the difference in two temperatures. One of these temperatures is the temperature of the reference junction. The second is the temperature of the measuring junction.

The reference junction temperature must be known before the temperature of the measuring junction can be determined. This is accomplished in different ways. The paper-and-pencil method and the constant-temperature method have been considered; other methods are discussed in detail in following paragraphs.

MEASURING TEMPERATURE WITH THERMOCOUPLES

In preceding paragraphs the principle of thermocouples, the problem of the reference junction, and a method for making thermocouples have been discussed. Now we discuss the two basic instruments used with thermocouples and consider some of the problems in the practical use of thermocouples.

Thermocouple Measuring Instruments

The measuring instruments used with thermocouples fall into two general categories: (1) deflectional and (2) null. On the surface it might appear that all measuring instruments are *deflectional* in that the indicator or pointer moves over the scale. However, a study of the operation of the instruments will show a clear distinction in the two types. The deflectional type is characterized by an arrangement of the millivolt detector so that there is a change in position of the detector for each change in measured voltage. For example, consider the galvanometer mentioned previously. As the current through the coil increases, the angular rotation of the galvanometer increases. In this way the galvanometer can be used directly to measure the voltage of a thermocouple circuit.

The *null* method is somewhat more complicated but has many advantages. The essential characteristic of a null-type measurement is that the detector assumes the same position regardless of input; that is, the position of the detector after the new input is balanced is the same as the position prior to the new input. With appropriate circuitry, this is readily accomplished.

The null and deflectional concepts are discussed later in more detail. The various specific instruments will be classified as either null or deflectional. At this time, an example of the two different concepts can be found in weight scales. Consider the spring scale that junk dealers and fishermen may use. These scales are essentially a spring fixed at one end. The object to be weighed is attached to the movable end. The greater the weight of the object, the more the spring will be stretched. A pointer is fastened to the movable end of the spring. The spring movement is calibrated in pounds and ounces. This scale is a deflectional instrument.

Now consider the laboratory balance. Prior to making any measurements, the scale is balanced. That is, an indicator is brought to a midpoint on a scale. A determination of weight is made in the following manner: The unknown weight is added to one pan, deflecting the scale. Weights of known magnitude are added to the other pan until the indicator is brought back to the midpoint. The unknown weight is determined by knowing the total weight of the balancing weights added to the scale.

Notice that the scale pointer is at the same point as it was before the weight was added. This will be recognized to be the essential requirement of null measurements. The scale, as used here, provides a method for comparing the weight of an unknown body to a known weight.

Thermocouple Installations

The thermocouple in most cases is placed within a closed protecting tube called a *thermowell*. Thermowells are available in numerous sizes and materials. Their selection is based on:

1. The physical and chemical properties of the materials being measured
2. The temperature range
3. The dimensional requirements

For example, extremely high temperatures may require that a ceramic well be used; corrosive acids may require stainless steel; a deep tank may call for a well 15 to 20 ft (4.6 to 6 m) long.

Thermocouples may be welded (or soldered) in the bottom of the well. This cuts down the amount of time it takes for the temperature change to reach the thermocouple. The disadvantage of a welded (or soldered) thermocouple is that visual inspection of the thermocouple is impossible. Removal or replacement of the well cannot be made unless the process conditions permit the removal of the well. There is no practical way to check the thermocouple with a second thermocouple. In general, it would be wise to avoid thermocouples that are welded or soldered in their wells. However, where fast response is absolutely necessary, a thermocouple that is welded into the well must be used.

Lead wire is run from the junction block of the thermocouple to the recorder or indicators. If possible, the lead wire run should be a single piece of wire without any splices. If splices are necessary, great care should be exercised in making them. Mechanical connectors are not recommended. A satisfactory method is to separate the wires for several inches, clean away the insulation for approximately $2\frac{1}{2}$ in. (6.5 cm), then clean the wires thoroughly. Twist the wires to form a pigtail about 2 in. (5 cm) long. Fold the pigtail in halves and solder the connection. Tape the joint with plastic tape. Use great care to ensure a vaportight insulation. In general, four to five times the amount of tape that

might be used on an electrical connection is required on a thermocouple connection. Even the best-made splices are potential trouble spots and are not satisfactory in conduit that becomes moisture soaked.

The reason for the elaborate precaution with thermocouple wiring arises out of the fact that the system carries milliamperes and millivolts. The slightest leakage between the wire and the conduit may "leak out" significant amounts of current, causing errors. If the leak is from wire to wire, the effect is that of a junction which tends to short out the thermocouple.

The installation of thermocouple wiring is a much more exacting job than the installation of electrical wiring. Techniques that are satisfactory on electrical wiring may cause all sorts of problems on thermocouple wiring.

Servicing Thermocouple Systems

A breakdown of typical service calls on thermocouple systems follows:

Thermocouple failure (due to wetness, damp corrosion, age, damaged insulators, broken junction blocks, etc.) would represent about	80%
Extension lead wire failure (caused by faulty splices, polarity errors, wet or damaged insulation, especially at the thermocouple)	10%
Recording or indicating instruments	5%
Others	5%
	100%

The type of service depends to some extent on whether a null- or a deflection-type instrument is being used. In general, the null instrument is more tolerant of poor (high-resistance) connections, whereas the deflectional type is more tolerant of wet or damp systems.

A thermocouple system in good repair will be satisfactory on either instrument, subject to the requirement of the deflectional instrument that the resistance of the lead wire must not exceed a certain fixed amount.

GALVANOMETERS

An important instrument component is the galvanometer. Galvanometers are the basic components of many power and current meters. Also, they are widely used in conjunction with the thermocouple to measure temperature. In our work we discuss two galvanometer types. The types are determined by the method of supporting the galvanometer coil. In the first type the coil is supported by pivots running in a "bearing." This type is based on the D'Arsonval meter movement. In the second type the coil is suspended by straps that twist as the coil rotates.

Galvanometers may also be classified according to usage. One classification is composed of the deflectional galvanometer; the second is the galvanometer as a null detector. However, the galvanometers themselves are the same in principle and operation. The difference comes about in the circuitry associated with the specific galvanometer. In general, the suspension galvanometer is used as a null detector, and the pivot-bearing (D'Arsonval-type) galvanometer is used as a deflectional galvanometer. There are many galvanometer varieties in addition to those mentioned. For the most part, our discussion will be limited to the D'Arsonval-type, direct-current galvanometer. Details of the D'Arsonval meter movement are shown in Fig. 1-17.

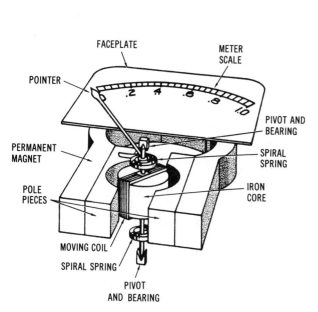

(A) Basic components of meter.

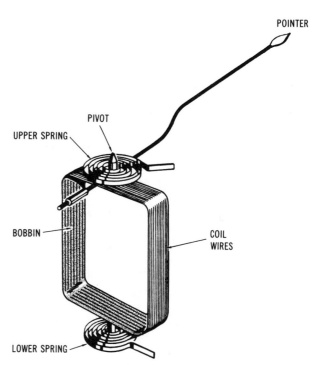

(B) Details of moving-coil assembly.

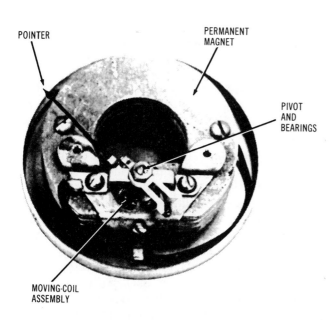

(C) Photograph of movement.

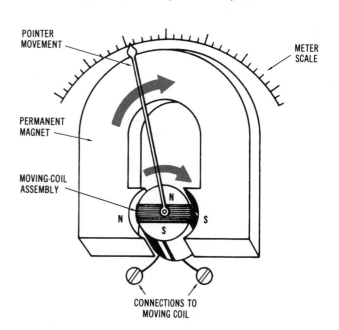

(D) Principle of operation.

FIGURE 1-17   D'Arsonval-type galvonometer.

Components

A D'Arsonval-type galvanometer consists basically of (see Fig. 1-18):

1. A movable coil and pointer
2. A permanent magnet with "pole pieces"
3. Spiral balancing springs called "hair springs"
4. A scale
5. An RJ indicator (added when the galvanometer is used as a pyrometer)

Arrangements of the D'Arsonval-Type, Pivot-Bearing Galvanometer

The coil is supported on bearings and mounted so that it rotates between the pole pieces. The hair springs tend to oppose the rotation. The electrical path is from one terminal through the upper spring, through the moving coil, through the lower spring, and out to the opposite terminal.

Principle of Operation

When a voltage is applied to the galvanometer, a current will result. The current through the moving coil causes a magnetic field to build up around the coil. This coil field reacts with the magnetic field produced by the permanent magnet. The coil field tends to align with the permanent field, which causes the coil to rotate. This rotation is resisted by the hair springs.

The greater the current through the coil (which means the greater the applied voltage), the stronger the coil field and the greater the attraction between the coil field and the magnet field. The greater the attraction, the larger the force available to rotate the coil.

This larger rotation is opposed by a stronger hair-spring force, since the hair spring is under a larger deformation. Finally, the spring balances the force of the magnetic fields and the galvanometer pointer comes to rest.

As a consequence of this chain of events, the input voltage has been converted to a pointer rotation. If the pointer rotation is read with a scale calibrated in volts, a measurement of the input voltage is obtained.

Suspension Galvanometer

The coil is frequently suspended on thin straps. These "suspensions" do the same job as the hair springs and the pivots and bearings of the D'Arsonval meter. The resistance to twisting of suspension is the opposing force that balances the forces of the magnetic fields. Usually, a suspended galvanometer is used in a null-balance type of measurement.

Service and Calibration

A galvanometer is an extremely low-power device, and it is imperative that it be as free of friction as is possible. Therefore, the bearings and pivots must be in top condition and free from all traces of dirt and oil. The clearance between pole pieces and coil must be free of any dust, dirt, or lint. To test for cleanliness, use the "tap" test. With a constant EMF applied to the galvanometer, watch the pointer closely and gently tap the galvanometer. If the pointer shifts position, there is a *tap error*. Such an error is a mechanical stoppage which might be due to lint, dirt, metallic chips, or any other foreign body lodged between the pole pieces and the coil.

Other possible causes of tap error are blunt pivots and cracked bearings. The pivots may be checked for sharpness by sliding them lightly across the thumbnail. If they are in good condition, they will scratch the nail. The bearings may be examined with a jeweler's eyepiece for cleanliness and cracks. The convolutions of the hair springs should not touch each other.

The pointer is a lightweight aluminum tube that slides onto the pointer base. Shellac is used to attach the pointer to the pointer base. The pivot bases are attached to the coil by shellac. Care must be exercised so that the bases are not dislodged. If they are, a small pencil-type soldering iron is used to melt the shellac.

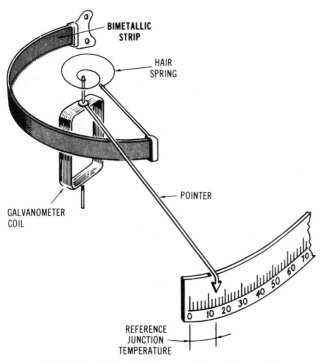

FIGURE 1-18 Reference junction compensation with bimetallic strip.

The pivot base is then set in position and the shellac is allowed to harden.

Reference Junction Compensation

A review of thermocouple theory will show that whenever a temperature measurement is made with a thermocouple, two temperatures are involved—the temperature of the reference junction and the temperature of the thermocouple. Therefore, it is necessary that some means be developed to measure the cold-junction temperature.

Suppose that the galvanometer was caused to rotate upscale as the reference junction temperature increased. The effect of such an arrangement would be to cause the pointer to move as the temperature of the reference junction changed. In a sense, the instrument would indicate ambient temperatures. Such devices are readily available in the form of bimetallic strips.

Bimetallic strips consist of two flat strips of different metals fastened back to back. The combination is formed into a helix. Since the strip is composed of two metals, the rate of expansion of the different metals causes the helix to unwind. The movable end of the helix is fastened to the hair spring, and its motion is passed to the galvanometer coil, causing the coil to rotate (see Fig. 1-18). This corrective action can be compared to the action of the compensating Bourdon element on filled temperature systems.

BASIC POTENTIOMETER

One of the few basic electrical measuring circuits is the potentiometer circuit. A very large and important segment of electrical measuring instruments involves the potentiometer. The basis for many of the temperature measurements made with a thermocouple is the potentiometer. The following discussions are on the potentiometer as used with the thermocouple. It is important, however, to recognize that the potentiometer is also used in measurements of pH, in infrared gas analyzers, and for electrical measurements such as current, voltage, or any variable that can be converted to an electrical signal: for example, speed, r/min, thickness, and density. It would not be an overstatement to say that any and all process variables may be measured using the potentiometer.

Components

The basic potentiometer circuit consists of:

1. A source of voltage.
2. A uniform resistance so constructed that it is pos-

sible to run a contact over the length of the resistance. This resistance is called a slide wire.

3. A connection to the movable contactor.

NOTE: In general electronics work, the three-terminal variable resistance is commonly called a potentiometer. However, in our work we shall consider the variable resistance together with all of its associated circuitry as a potentiometer. For example, we shall describe the instrument that records temperatures measured by thermocouples as a recording potentiometer.

Arrangements

The voltage source, usually a battery, is connected across the slide wire. The signal to be measured is connected to one end of the slide wire and the contactor. A basic potentiometer circuit is shown in Fig. 1-19.

Principle of Operation

If the voltage appearing between the contactor and one end of the slide wire is measured, it will be found that the value of that voltage is always related to the position of the contactor on the slide wire. For example, if the contactor is at the halfway position, the voltage is equal to one-half of the applied voltage. If the contactor is at the three-fourths position, the voltage appearing at the contactor is three-fourths of the applied voltage. This relationship is true for all positions. Therefore, it is possible to determine, without measuring, what the voltage is at the contactor if the position of the contactor on the slide wire is known and if the magnitude of the voltage being applied to the slide wire is known. Thus we have an instrument that

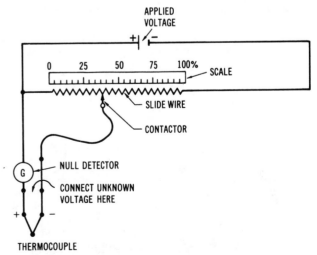

FIGURE 1-19 Basic potentiometer circuit.

can be used to measure an unknown voltage by comparing the unknown voltage to the known voltage generated by the potentiometer. To do this we require a device that will indicate when the unknown voltage is being matched by the voltage generated by the potentiometer. Such a device is called a *null detector*. A galvanometer is a common null detector. Other common null detectors are electronic amplifiers and relays.

The null detector is used in the following way. One side of the null detector is connected to the unknown voltage. The other side of the null detector is connected to the zero end of the slide wire. The remaining connection of the unknown voltage goes to the contactor, thereby completing the circuit.

If there is an imbalance between the voltage appearing at the contactor and the unknown voltage, a current will pass through the null detector, causing it to indicate that an imbalance exists. It will also indicate the direction of the current. Since the direction of the current is determined by whether the unknown voltage is greater or smaller than the contactor voltage, we have a method for determining in which direction to move the contactor. Suppose that the unknown voltage is greater than the contactor voltage. To balance the null detector we must increase the contactor voltage. We do this by sliding the contactor to the right on the slide wire until there is no imbalance across the null detector.

The balancing action of the potentiometer can be likened to the action of a pair of balance scales as shown in Fig. 1-20. An unknown weight is placed on one pan of the scales and enough scale weights are placed on the other pan to reach a balance. The value of the unknown weight is equal to the value of scale weights required for balance.

Examine the position of the contactor as indicated by the scale. Say that it is 75% of scale. Then the unknown voltage is equal to 75% of the applied voltage. Thus we have a device that can be used to measure voltage.

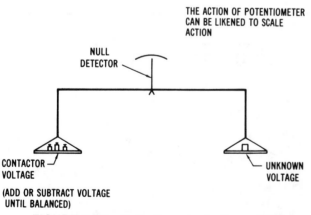

THE ACTION OF POTENTIOMETER CAN BE LIKENED TO SCALE ACTION

NULL DETECTOR

CONTACTOR VOLTAGE

UNKNOWN VOLTAGE

(ADD OR SUBTRACT VOLTAGE UNTIL BALANCED)

FIGURE 1-20 Potentiometer action compared to balance scales.

It is possible to modify the basic potentiometer circuit so that the applied voltage will always be known. When this step is taken, a complete voltage-measuring device will have been obtained.

BASIC POTENTIOMETER WITH STANDARDIZATION

A limitation of the basic potentiometer circuit arises out of the fact that to measure an unknown voltage using the potentiometer, it is necessary to know the magnitude of the voltage that is being applied across the slide wire. The means for making known the applied voltage are referred to as the *standardizing circuit* or, simply, *standardization*.

Components

To incorporate a method for standardizing a potentiometer, the following components are required in addition to those of the basic potentiometer:

1. An adjustable resistance called the *battery rheostat*
2. A resistance called the *standardizing resistance*
3. A *standard cell* (described earlier in the chapter)
4. A switch called the *standardizing switch*

Arrangements

The battery rheostat is put in series with the battery and slide wire. The standardizing resistance is put in series with the slide wire. The standard cell is put in parallel with the standardizing resistance. The standardizing switch is connected so that throwing it will connect the standard cell and disconnect the voltage under test (Fig. 1-21).

Principle of Operation

The standard cell is used in a balancing action similar to that which we have just described. To use the standard cell to "standardize" the circuit, the standardizing switch is thrown, connecting the cell to the galvanometer and disconnecting the thermocouple. This switching applies the voltage of the standard cell across the slide wire and standardizing resistance, thereby comparing the voltage across the slide wire and resistance to the standard voltage of the cell. If these voltages are different, current will pass through the galvanometer, causing it to indicate the imbalance and also the direction of the current.

If the galvanometer is deflected, the battery rheostat is adjusted, thereby changing the current through the slide wire and resistance. This changes the voltage

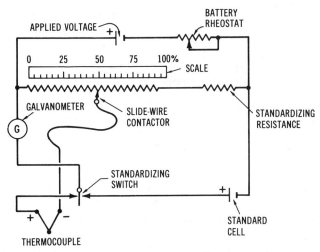

FIGURE 1-21 Basic potentiometer plus standardizing components.

across the slide wire and resistance ($V = IR$). Thus it is possible to balance the standard-cell voltage across the slide wire and resistance.

Suppose that on throwing the standardizing switch, the galvanometer indicated that a current was being delivered from the standard cell into the slide-wire circuit. This means that the standard-cell voltage is greater than the voltage across the slide wire and resistance. In order to increase the voltage across the slide wire and resistance it is necessary to cause more current to be delivered through them. This is accomplished by reducing the resistance of the battery rheostat.

This increase in current will result in an increase

in voltage across the slide wire because of the relationship in which voltage equals the current times the resistance ($V = IR$). When the voltage across the slide wire and resistance equals the standard-cell voltage, the galvanometer will come to zero and the instrument is said to be standardized. Figure 1-22 shows an industrial voltage potentiometer.

CAUTION: Remember that the standard cell has very little current-delivering capacity and therefore must not be connected into the circuit for long periods of time. Two or three seconds is about as long as the standardizing switch should be closed at any one time.

BASIC POTENTIOMETER WITH STANDARDIZING CIRCUIT AND REFERENCE JUNCTION COMPENSATION

It has been shown that a thermocouple is essentially a device that converts degrees of temperature to millivolts, and that the equivalent millivolts are determined by the difference in temperature between the reference junction and the hot junction. This fact makes it necessary that the reference junction temperature be known or that there be a method for compensating for the reference junction temperature. The discussion will show how the reference junction is compensated for when using the potentiometer circuit to measure temperature. Two methods are discussed: manual and automatic.

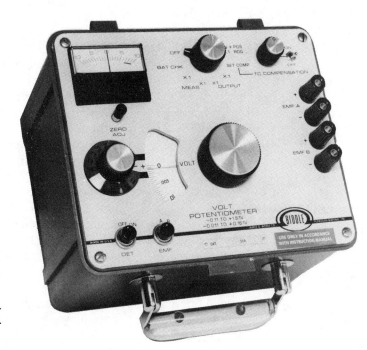

FIGURE 1-22 Industrial potentiometer. (Courtesy of Biddle Instrument Co.)

Components

The components for manual compensation are a fixed resistance and a variable resistance. The variable resistance is knob operated for manual compensation. In the case of automatic compensation the variable resistance is made of special resistance wire.

Arrangements

The variable resistance is put in series with the fixed resistance. The variable and fixed resistances are put in parallel with the slide wire. The tap of the variable resistance is connected to the null detector.

Principle of Operation (Manual)

In the discussion of reference junction compensation it was shown that the voltage output of a thermocouple circuit is determined by the difference in temperature between the hot and reference junctions. Therefore, it is necessary to know the value of the reference junction temperature. This temperature is converted into millivolts and added to the millivolts output of the thermocouple circuit. This can be accomplished electrically (see Fig. 1-23) as follows.

Measure the reference junction temperature with a thermometer. Rotate the reference junction compensation, moving the contactor upscale. This upscale travel is the millivoltage equivalent of the reference junction temperature. Since one side of the null detector is connected to the movable contact, the effect of this adjustment is to move the starting point upscale; thus a new reference point has been established and this new reference is equal to the reference junction temperature.

The thermocouple output is equal to the difference between the reference junction and the hot junction. The reference junction has been shifted upscale by an amount equal to its temperature. Therefore, the hot junction can now be read directly on the scale of the instrument.

Principle of Operation (Automatic)

Suppose that it were possible to get the variable resistance to change as the temperature of the reference junction changes. If this were possible, it would not be necessary to move the tap manually, and an automatic reference junction compensation would be obtained. To get a resistance that will change as the tem-

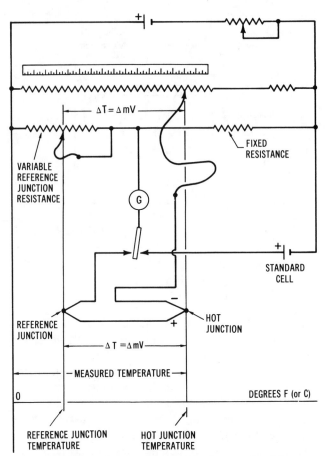

FIGURE 1-23 Basic potentiometer with manual reference junction compensation.

perature changes is not at all difficult. A resistor made of nickel wire has that property. As the temperature changes, so does its resistance. If such a resistance is used instead of the variable resistance, automatic reference junction compensation is obtained.

This reference junction compensator is located so that it will be at the temperature of the reference junction. The reference junction is at the point where the dissimilar wire of the thermocouple is rejoined, which almost invariably is at the terminal strip of the instrument (see Fig. 1-24).

It is interesting to note that this method of compensation results in a Wheatstone bridge arrangement. However, the instrument is still a "potential" meter, or a potentiometer. The introduction of the Wheatstone bridge method of reference junction compensation, unfortunately, does result in confusing the potentiometer with the Wheatstone bridge. All too frequently both instruments are called "bridges," but it should be remembered that the potentiometer is a voltage-measuring instrument, whereas the Wheatstone bridge is a current-measuring instrument; in a certain sense they are opposites.

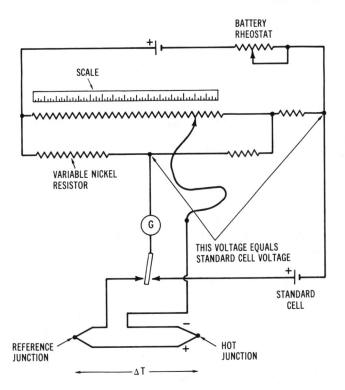

FIGURE 1-24 Basic potentiometer with automatic reference junction compensation.

THE WHEATSTONE BRIDGE

A substantial portion of process variable measurements are made using the Wheatstone bridge. Examples are conductivity, gas analyses, temperature, and strain-gauge measurements. Figure 1-25 shows a typical Wheatstone bridge that is often used to make industrial measurements. This particular instrument measures resistance values from $1 \, \Omega$ to $11.01 \, M\Omega$. Two resistance decades and a slide-wire resistor are built into this instrument. All resistance values and dial indications are displayed on the central window.

The Wheatstone bridge is, in certain respects, the opposite of the potentiometer. The potentiometer has been shown to be a voltage-measuring device. The Wheatstone bridge is a resistance-measuring device. Since it is a voltage-measuring instrument, the potentiometer requires a fixed and known voltage, hence the need for the standardizing circuit. Varying resistances in the external circuit make very little difference. The Wheatstone bridge functions successfully with wide variations of applied voltage, but varying resistances (other than the one being measured) are intolerable. It is important that the basic purpose and operation of each of these two instruments be well understood.

Components

The Wheatstone bridge consists of the following components:

1. One known, fixed resistance
2. One known, variable tapped resistance called a slide wire
3. A null detector
4. A voltage source
5. The unknown resistance that completes the circuit

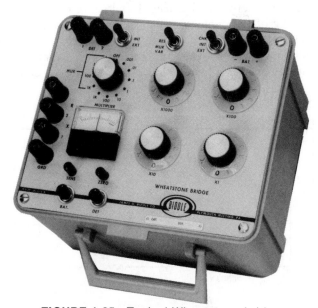

FIGURE 1-25 Typical Wheatstone bridge. (Courtesy of Biddle Instrument Co.)

Arrangements

The unknown resistance is put in series with the fixed resistance, and this combination is put in parallel with the variable tapped resistance, as shown in Fig. 1-26. The tap is connected to the null detector. The other side of the null detector is connected to a point between the unknown resistance and the known fixed resistance. Power is connected to the ends of the tapped resistance.

If the bridge is to be used for temperature measurement, the unknown resistance must change with temperature. If the bridge is to be used for measuring variables other than temperature, then the unknown resistance must be made to change as the measured variable changes. For example, if the bridge is to be used to measure the percentage of SO_2 in air, the unknown resistance must change as the percentage of SO_2 changes.

Operation

If a voltage is applied to the tapped resistance and the parallel branch composed of the unknown resistance, a current will be delivered through the tapped resistance and the parallel branch. The current will distribute itself according to the relative resistance of the two branches.

The null detector is brought to zero by moving the contactor. When the null detector is at zero, the current through it must also be zero. If the current through the null detector is zero, it means that the voltage across the null detector must be zero. The only time that this voltage is equal to zero is when the voltage drops across resistances A and C are equal. The moving of the contactor changes the resistance of C and simultaneously causes a change in the relative amount of current. This combined effect results in the current through C times the resistance of C to equal the current through the unknown resistance times its resistance, or

$$I_C \times C = I_A \times A$$

If there is no current through the null detector, the current through resistance C must also pass through resistance D, and the current through A must also pass through B, or

$$I_C \times D = I_A \times B$$

If the two equations are divided, the following equation is obtained:

$$\frac{C}{D} = \frac{A}{B} \quad \text{and} \quad A = \frac{C}{D} \times B$$

In other words, A, the unknown resistance, equals C divided by D, times B. The values of D and B are known and the value of C is determined by the contact position on the variable resistance. If a scale is used to indicate the contact position, the value of C can be read on the scale. Hence we have a device that measures resistance.

This instrument can be used to measure temperature if a resistor could be made so that its resistance changed with temperature. When a resistor made of platinum or nickel, for example, is enclosed in a pressure-tight bulb, it is possible to measure the temperature of certain materials by immersing the bulb in them. A Wheatstone bridge set up for temperature measurement is shown in Fig. 1-27.

Figure 1-28 shows a commercially available resistance thermometer bridge. This is a portable instrument with an electronic null detector that provides sufficient sensitivity to detect resistance changes of $0.0005\ \Omega$, or values within 0.001% accuracy. Readouts are made on the central readout window.

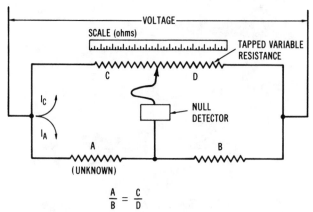

FIGURE 1-26 Wheatstone bridge circuit for resistance measurement.

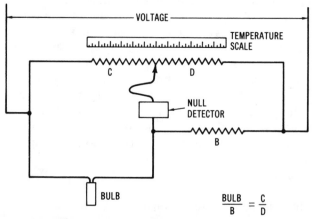

FIGURE 1-27 Wheatstone bridge circuit for temperature measurement.

FIGURE 1-28 Commercial resistance thermometer bridge. (Courtesy of Leeds & Northrup.)

Lead-Wire Compensation

Usually, the point of measurement is at some distance from the place where the temperature is read, making necessary a fairly long run of connecting wire between the bulb and the instrument. This wire has resistance, and what is worse, the resistance changes a slight amount as the temperature of the connecting wire changes. This results in an extraneous change in resistance in addition to the change in the resistance of the bulb. This action is similar to that of a filled system bulb and its connecting capillary. The lead-wire resistance change must be compensated for by running three wires to the bulb, as shown in Fig. 1-29. One wire is on one side of the null detector and a second wire is on the other side. These two wires will change resistance in equal amounts and since this change is equal on both sides of the bridge, they cancel themselves out. The third wire is connected to the null detector and at balance, there is no current through it; hence its change in resistance makes no difference.

Properties of the Wheatstone Bridge

Since, at balance, there is no current through the null detector, the resistance between the contactor and the slide wire does not affect the accuracy of the instrument. On the other hand, any variable resistance in the connections to the instrument or in the external

lead wire and the bulb connections will result in very serious errors. As a consequence, as far as it is possible to do so, all connections should be soldered. All other connections must be clean and tight.

It was shown that the absolute amount of current through each branch of the bridge was not too important, the distribution of current was the important consideration. The absolute amount of current is determined by the voltage supplied to the bridge. If this voltage changes, it does not affect the distribution of current; therefore, the accuracy of the supply voltage is not important. The only effect of a low supply voltage is that it reduces the sensitivity of the instrument.

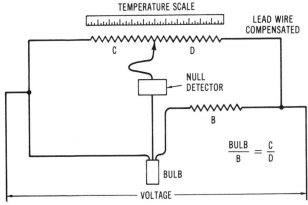

$$\frac{BULB}{B} = \frac{C}{D}$$

FIGURE 1-29 Bridge circuit with lead-wire compensation.

SUMMARY

In the study of electronics, symbols are commonly used in diagrams to indicate electronic components. A few of these symbols are quite important because they are duplicated many times in schematic diagrams.

An electron is one of the basic particles of an atom. The flow or movement of electrons through an electrical conductor constitutes an electric current. One coulomb (6.28×10^{18} electrons) passing through a given point in an electrical conductor in 1 second is representative of 1 ampere of current.

An opposition to electron flow is called resistance. Resistance is measured in ohms and exists in all conductors to some extent.

Voltage is representative of the electromotive force (EMF) required to drive electrons through a conductor to overcome resistance. A voltage drop occurs when electrons flow through a resistance.

Ohm's law shows the relationship between voltage, current, and resistance. It takes 1 volt of electromotive force to drive 1 ampere of current through 1 ohm of resistance.

Thermocouples generate electricity as the result of heat applied to the junction of two dissimilar metals. These devices are used to determine temperature and serve as reference junctions in measuring applications. Thermocouple measuring applications usually employ a reference junction and a measuring junction. The measured temperature is actually a comparison of the temperature difference between these two junctions.

Galvanometers are instruments that produce a pointer deflection on a meter scale as an indication of electric current. This type of instrument employs a moving coil that is placed in a permanent magnetic field and energized by an electric current. An interaction between the coil field and the permanent magnet field causes the coil to move accordingly. A pointer attached to the coil is used to indicate values on a graduated scale.

A potentiometer is a basic electrical circuit that can be used to measure electrical voltage values. Voltage appearing across a slide-wire resistor is always proportional to the position of the contactor. Potentiometers are used by industry to measure many process variables.

Wheatstone bridges are also commonly employed by industry to measure process variables. This type of instrument is essentially a resistance-measuring instrument. When a known voltage is applied to a bridge circuit, an unknown resistance value placed in one leg of the bridge is determined by nulling at the common connection point. Circuits of this type are used to determine unknown resistances that vary in value due to changes in temperature, conductivity, pH levels, and so on.

ACTIVITIES

ELECTRONIC SYMBOL IDENTIFICATION

1. Refer to the schematic diagram of Fig. 1-1.
2. Using the symbols of Fig. 1-2 and the Appendix, identify all of the components of the schematic.

THERMOCOUPLE CIRCUITRY

1. Construct the thermocouple circuit of Fig. 1-13. Use copper–constantan (type T), iron–constantan (type J), or any convenient type of thermocouple in the circuit. The meter should be a galvanometer or a millivoltmeter.
2. Grasp the hot junction in your hand while observing the galvanometer. Does this produce a display on the indicator?
3. Apply heat from a soldering pencil, 100-W lamp, or heating cone while observing the indicator. Describe your findings.

ELECTROCHEMICAL CELL EVALUATION

1. Construct the electrochemical cell evaluation circuit of Fig. 1-7.
2. Measure the voltage across the cell when connected to the resistor and with the resistor disconnected. This shows how the cell responds under a load and with no load. Is there a difference in the measured voltage?
3. If a standard cell is available, run the same test procedure. Use a 100-Ω resistor for the loading operation.
4. Construct some cells by using two dissimilar metals and a saltwater solution. Run the same test procedure using a 100-Ω load resistor. What metal combination produces the greatest cell voltage?

BASIC POTENTIOMETER EVALUATION

1. Construct the basic potentiometer circuit of Fig. 1-19. The applied voltage should be a standard cell, 1-V dc source, or a type D or C flashlight cell. The slide wire should be 100 Ω or a potentiometer of the same size.
2. Connect the thermocouple to the circuit and adjust the slide wire to produce a null indication on the galvanometer. The voltage output of the thermocouple can be obtained from the slide-wire scale or

measured across the potentiometer with a voltmeter. Record your findings.

3. Apply heat to the thermocouple, null the circuit again with the slide wire, and determine the thermocouple voltage.

INDUSTRIAL POTENTIOMETER OPERATION

1. Review the operational manual of the potentiometer.
2. Check the value of the battery of the instrument. Standardize the instrument.
3. Apply an unknown voltage to one of the inputs of the potentiometer.
4. Adjust the slide-wire potentiometer of the instrument to produce a null indication.
5. Read the voltage scale to determine the value of the unknown voltage.
6. Try the same procedure on another unknown voltage.

INDUSTRIAL WHEATSTONE BRIDGE OPERATION

1. Review the operational manual of the bridge being used for this measurement.
2. Connect an unknown resistor value to the appropriate input of the bridge.
3. Set multiplier switches to an approximate resistance value.
4. Turn on the voltage source.
5. Set the galvanometer to the shunt position.
6. Adjust the variable resistor to produce a null condition.
7. Switch the indicator to the direct galvanometer position.
8. Adjust the variable resistor to produce a more accurate null condition.
9. Determine the value of the unknown resistor.
10. Try the same procedure on another unknown resistor value.

QUESTIONS

1. Why are the symbols of a schematic diagram numbered such as R_5, C_3, and Q_1?
2. Describe the physical construction of a thermocouple.
3. What are the basic particles of an atom?
4. What is a coulomb?
5. What is the relationship between an ampere, a milliampere, and a microampere of current?
6. What is Ohm's law?
7. What letter symbols are used for voltage, current, and resistance?
8. Describe the physical construction of an electrolytic cell.
9. Using Ohm's law, solve the following problems.
   (a) $I = 2$ A, $R = 50 \Omega$. Find the value of V.
   (b) $R = 20 \Omega$, $I = 3$ A. Find the value of V.
   (c) $V = 2$ V, $R = 100 \Omega$. Find the value of I.
   (d) $R = 50 \Omega$, $V = 10$ V. Find the value of I.
   (e) $I = 5$ A, $V = 20$ V. Find the value of R.
   (f) $V = 20$ V, $I = 0.5$ A. Find the value of R.
10. Why does the indicating hand of a galvanometer deflect when current passes through the instrument?
11. What are the components of a basic potentiometer?
12. How is a potentiometer used to measure an unknown voltage?
13. What are the components of a basic Wheatstone bridge?
14. What does a Wheatstone bridge measure?
15. What is meant by the term *null*, and how is it used in the operation of a Wheatstone bridge?

CHAPTER 2

POWER SUPPLIES

OBJECTIVES

Upon completion of this chapter, you will be able to:

1. Explain the difference between ac and dc electricity.
2. Explain how ac is generated.
3. Demonstrate how an ac voltmeter is used to measure voltage.
4. Explain the operation of a transformer.
5. Define rectifier terms such as *covalent bonding*, *n-type material*, *p-type material*, *anode*, *cathode*, *holes*, and *biasing*.
6. Explain the characteristic differences between half-wave and full-wave rectification.
7. Given a schematic diagram of a divided power supply, indicate the direction of current flow for each alternation of the ac input that will produce an output.
8. Identify different types of filter circuits.
9. Explain how a zener diode is used to achieve regulation.
10. Identify the components of a three-terminal IC regulator.

IMPORTANT TERMS

In the study of electronic instruments one frequently encounters a number of new and somewhat unusual terms. These terms play an important role in the presentation of this material. As a result, it is very helpful to review these terms before proceeding with the chapter.

Ac: The abbreviation used for alternating current.

Alternating current: Current produced when electrons move first in one direction and then in the opposite direction.

Alternation: One-half of a complete cycle of alternating current.

Alternator: A rotating machine that generates ac.

Anode: The positive terminal or electrode of an electronic device such as a solid-state diode.

Armature: The movable part of a relay, rotating coils of a motor, or the part of a generator into which current is induced.

Bias (forward): Voltage applied across a p-n junction that causes the conduction of majority current carriers.

Bias (reverse): Voltage applied across a p-n junction that causes little or no current flow.

Capacitance: The property of a device to oppose changes in voltage due to energy stored in its electrostatic field.

Capacitive reactance: An opposition to the flow of ac current caused by a capacitor. X_c is used to identify capacitive reactance and is measured in ohms.

Capacitor: A device that has capacitance and is usually made of two metal plate materials separated by a dielectric material.

Cathode: The terminal of an electronic device that is a source for electron emission.

Covalent bonding: Atoms joined together by an electron-sharing process that causes a molecule to be stable.

Current carriers: Electrons or holes that support current flow in solid-state devices.

Dielectric: An insulating material placed between the metal plates of a capacitor.

Dielectric constant: A number that compares the insulating ability of a material to dry air. Dry air has a dielectric constant that is slightly greater than 1.0.

Electrostatic field: The field that is developed around a material due to the energy of an electrical charge.

Hole: A charge area or void where an electron is missing.

IC regulator: An integrated circuit device that achieves voltage regulation.

Lenz's law: The induced voltage of a circuit that opposes the force that originally produced it.

Load: The part of an electrical system that does work or converts electrical energy into another form of energy.

Negative alternation: The half of an ac cycle that appears below the zero reference line.

N-type material: A semiconductor material that has extra electrons, or negative charges, that do not take part in the covalent bonding procedure.

Polarity: The direction of an electrical potential (− or +) or a magnetic field (north or south).

Positive alternation: The half of an ac cycle that appears above the zero reference line.

P-type material: A semiconductor material that has extra holes, or positive charged areas, that do not take part in the covalent bonding arrangement.

Rectification: The process of changing ac into pulsating dc.

Regulation: A constant value holding of some electrical conditions, such as voltage, current, or power.

Ripple: An ac component present in the output of a dc generator, rectifier, or power supply.

Root-mean-square (rms) voltage: The value of ac sine-wave voltage that has the same effect as an equal value of dc voltage.

Time constant (RC): The time required for the voltage across a capacitor in an RC circuit to increase to 63% of its maximum value or decrease to 37% of its maximum. Time equals resistance times capacitance or $t = RC$.

Transformer: An ac power control device that transfers energy from its primary winding to its secondary winding by mutual inductance.

Watt: The basic unit of electrical power. The amount of work accomplished by an electrical device when 1 ampere of current flows under a pressure or force of 1 volt.

Zener diode: A voltage-regulating device that has high impedance at low voltage values and becomes conductive above the breakdown voltage.

INTRODUCTION

In Chapter 1 we discussed some fundamental concepts that are necessary to make an electronic measuring instrument operate. You should now be able to see how basic components fit together to form a simple circuit, and the influence that voltage, current, and resistance have on the circuit, on electrical generation, on measuring devices, and on potentiometer operation.

The continuation of our investigation of electronic instruments calls for the combining of components to perform specific functions. A function, for example, might change the shape of a waveform, amplify a weak signal, or change the nature of an applied voltage from ac to dc. Electronic instruments have a variety of specific functions that must be performed to achieve a particular operation. These functions are basic to all instruments, with only a few minor component modifications. An understanding of these basic functions is an essential part of the operational theory of an electronic instrument.

In this chapter, attention is directed toward the power supply as a functional block of an electronic instrument. This part of the instrument provides a primary source of electrical energy that makes the system operate. In its simplest form the power supply could be a battery or combination of one or more electrochemical cells. As a general rule, only portable instruments derive their operating power from this type of source. Other instruments utilize alternating current (ac) from the electrical power line as a source of electrical energy.

ALTERNATING CURRENT

In industry today the primary source of electric power to operate electronic instruments is derived from the electrical power line or from portable electrical generators. As a rule, this power is in the form of alternating current (ac). As you will see in this chapter and those that follow, electronic devices such as transistors, integrated circuits, and vacuum tubes generally require dc electrical power for their operation. This means that before most electronic devices can be operated, ac power must be changed or converted into dc power. This process is commonly called *rectification*.

To understand rectification, we must first have some understanding of the nature of alternating current. In Chapter 1 we discussed the nature of direct current (dc). You may recall that dc resulted from a constant electromotive force or voltage being applied to a closed circuit. This force caused electrons to move through the conducting material in a single direction.

Alternating current differs from direct current due to the direction of electron flow. In ac, electrons flow first in one direction for a short time, then reverse direction and flow in the opposite direction for a short time. The flow of electrons in one direction and then in the other is called a *cycle* of ac. The number of cycles that occur in 1 second of time is called the *frequency* of ac. Frequency is expressed in *hertz* (Hz), which means "cycles per second." In the United States, the standard power-line frequency is 60 Hz.

A clearer understanding of alternating current can be obtained by investigating one method by which an alternating current is generated. When a conductor is passed through a magnetic field, electrons are caused to flow in that conductor. Notice the three conditions that must be met to induce a current electromagnetically: the presence of a magnetic field, a conducting material, and relative motion between the two. If the poles of a permanent magnet are placed in the position shown in Fig. 2-1, the magnetic field is directed from the north magnetic pole toward the south magnetic pole. When a conductor is moved downward through the field, a current is induced into the conductor away from the observer. The symbol ⊗ indicates a current away from the observer, and the symbol ⊙ indicates a current toward the observer.

The preceding discussion is summed up by *Lenz's law*, which states: "When a conductor cuts through magnetic lines of force, a current is induced into the conductor so as to oppose the motion." This is an experimental law and can only be stated here as demonstrated by experiment. The law states that a current is induced in such a direction as to oppose the motion of the conductor. This opposition occurs because of the magnetic field produced by the current in the conductor itself. Since the electrons are flowing in the conductor away from the observer (Fig. 2-1), magnetic lines of force are produced about the conductor in a counterclockwise direction. It can be seen in the illustration that this means the lines of force below the conductor are in the same direction as the magnetic field of the permanent magnet, while the lines of force above the conductor are in the opposite direction. The effect of this is to strengthen the total magnetic field below the conductor while weakening the total magnetic field above the conductor. Thus the current induced in the conductor produces a magnetic field about the conductor that interacts with the field of the permanent magnet in such a way as to produce a force opposite to the original force that induced the current.

The direction of the induced current described by Lenz's law can be easily remembered by a device called the *left-hand rule*, shown in Fig. 2-2. If the thumb and the first two fingers of the left hand are placed at right angles to one another, the first finger should point in the direction of the magnetic field, the thumb in the direction of movement of the conductor, and the second finger in the direction of the induced electron flow. In the illustration of Fig. 2-1, the direction of electron flow away from the observer is indicated by the left-hand rule.

The method of producing an alternating current that will be discussed here involves an ac generator, often called an *alternator*. The alternator is composed

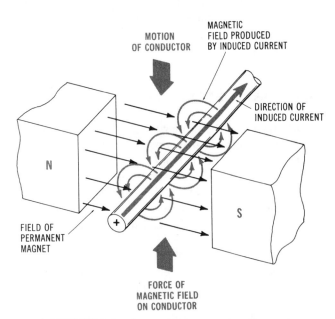

FIGURE 2-1 A conductor moving through a magnetic field induces a current in the conductor in a direction opposite the motion of the conductor.

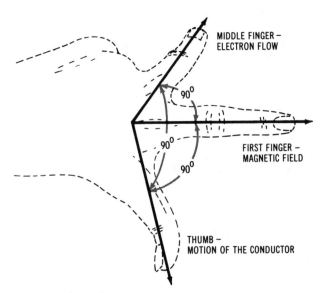

FIGURE 2-2 Left-hand rule for induced current.

of a rotor and a stator, the stator being an integral part of the frame of the alternator. We shall consider an alternator in which the magnetic field is produced by the stator. This means that the stator forms the pole pieces that make up the magnetic field through which the rotor moves. These pole pieces may be either permanent magnets or electromagnets. Electromagnets utilize the principles just described, which state that magnetic lines of force are produced about a conductor through which electrons are flowing. This magnetic field is strengthened by winding the conductor into a coil around a soft-iron core. In the alternator described here, permanent magnets produce the field. The rotor or "armature" is turned or rotated through the stationary magnetic field. The armature is made up of many turns of copper wire wound about a shaft. The armature is then driven by a turbine, a gasoline engine, or some other prime mover. Electrical contact is made with the rotor through brushes that connect to the external circuit.

In the discussion of the ac generator outlined here we shall be concerned with a cross section of just one winding of the conductor. It should be remembered that what is described here takes place in many windings of the armature. The basic points to remember in this discussion are the following: Lenz's law states the direction of the induced current. The conductor must cut magnetic lines of force to induce current. No current is induced if the conductor moves parallel to the lines of force.

Now let us consider the situation in Fig. 2-3 where the conductor is rotated through the magnetic field. In position A, the conductor moves parallel to the magnetic lines of force; hence no current is induced into the conductor, as indicated in the graph at point A. To get to position B, the conductor must move downward through the field. When this occurs, current is induced into the conductor. Remember that the direction of induced current under these conditions is away from the observer. We will call this the positive direction as indicated on the diagram. Eventually the conductor is again moving parallel to the magnetic field in position C. The electron flow at this position has again decreased to zero. Moving toward position D, the conductor is moving upward through the magnetic field and is again cutting through magnetic lines of force. Lenz's law states that induced current is now

moving toward the observer. This you will note is opposite to the direction of the current observed at point B. This direction is labeled negative in the diagram. As the conductor continues to rotate in the clockwise direction, it returns to point A and the current again drops to zero. The alternator then repeats this series of operations with each rotation of the rotor.

The amount of current induced into the conductor at any given time is determined by the number of lines of force cut per unit of time. The points of maximum induced current therefore occur at points B and D. Between points A and B, current is produced, but since the motion is not perpendicular to the field, fewer lines of force are cut than at point B. Less induced current is the result. A gradual rise and fall of current then occurs between the zero and maximum points. The current so generated follows the pattern of the mathematical sine function and is therefore called a *sine wave*. Alternating current usually varies in this manner, although many other waveforms will be encountered.

Let us investigate the characteristics of alternating current. Figure 2-4 is a graphical representation of an ac sine wave. This graph will also represent an ac voltage, since, by Ohm's law, V depends directly on I. The maximum voltage or current occurs at point B. This is called the *amplitude* of this signal. The amplitude can also be measured at points D, F, and H and is sometimes referred to as the peak voltage (V_p), or peak current (I_p). Also, ac signals are measured by their peak-to-peak voltage ($V_{p\text{-}p}$). In a sine wave, $V_{p\text{-}p}$ is twice the value of V_p. Ac voltmeters measure the *rms* or *root-mean-square value* of a sine wave. This is the effective value of an ac sine wave voltage or current. This means that an ac voltage with an rms value of 117 V will do the same effective amount of work as 117 V dc. The rms value of an ac voltage or current is 70.7% of the peak value. Mathematically this is expressed as rms = 0.707 × peak value. The peak value of ac is obviously greater than the rms value. Mathematically, peak value is expressed as 1.414 × 117 = 165.438 V.

In Fig. 2-4, the portion of the sine wave between points A and C is called the *positive alternation*. The portion between points C and E is the *negative alternation*. The complete sine wave, from point A to point

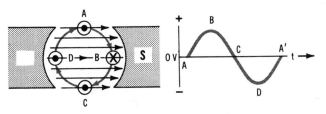

FIGURE 2-3   Generation of a sine wave.

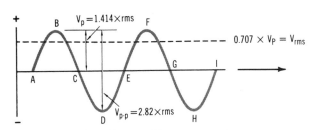

FIGURE 2-4   Sine-wave values.

E, is called one *cycle* of the sine wave. The sine wave repeats itself between points E and I. Two cycles of an ac voltage are shown in Fig. 2-4. The time it takes for an alternating current to complete one cycle is called its *period*. The number of cycles that occur in 1 second is an indication of the frequency in hertz. Ordinary line voltage has a period of $\frac{1}{60}$ of a second when the frequency is 60 Hz.

TRANSFORMER ACTION

Power supplies of an electronic instrument rarely operate with ac obtained directly from the power line. A transformer, for example, is commonly used to step the line voltage down or up to a desired operating value. A schematic representation of a simple transformer is shown in Fig. 2-5.

The winding on the left of the diagram is called the *primary winding*, to which the input voltage is applied. The winding on the right is the *secondary winding*, from which the output voltage is taken. The parallel lines between the two windings represent the transformer core. In a power transformer, this core is usually made of laminated soft iron. Both windings are wound on the same core. It is not uncommon today for transformers to employ alternate primary windings to accommodate different line voltages and to have more than one secondary winding.

The output voltage of a transformer may have the same polarity as the input voltage, or it may have the opposite polarity. This depends on the way the secondary winding is wound. A dot is sometimes used to indicate the two terminals having the same polarity. The output of the "ideal" transformer depends on the turns ratio of the transformer. If the primary winding is composed of 100 turns of wire, and the secondary has 600 turns, the transformer has a 1:6 turns ratio. Since the secondary has the most turns, it is called a step-up transformer. The output voltage will be six times the input. For example, if the input voltage in Fig. 2-5 is 3 V ac, V_{out} equals 18 V ac. When the secondary winding contains fewer turns than the primary, the voltage will be stepped down. Current varies inversely with the voltage. When the voltage is stepped up, the current will be stepped down. The output current also depends on the turns ratio of the transformer; if the voltage is stepped up six times, the output current

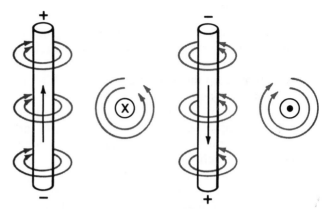

FIGURE 2-6 Magnetic field around a conductor.

will be one-sixth of the primary current. Of course, no transformer is ideal; therefore, V_{out} will be slightly less than the values described.

The question naturally arises as to how this output voltage is developed. It can be demonstrated that when a current passes through a conductor, a magnetic field is built up around it. This field is concentric with the conductor and is in a direction that opposes electron flow. The direction of this field is demonstrated in Fig. 2-6 by the use of the left-hand rule. If the thumb of the left hand is pointed in the direction of electron flow, the natural curve of the fingers follows the direction of the field. When the conductor is wound into a coil, the magnetic field of each turn of the conductor strengthens the field created by its neighboring turns of wire, forming an electromagnet.

As can be seen in Fig. 2-7, if the fingers of the left hand curve in the direction of electron flow around the coil, the thumb points in the direction of the north pole of the electromagnet. When an alternating current is applied to such a coil, the magnetic field is constantly changing. At the beginning of a cycle, the current is zero. No magnetic lines of force are present. As electrons begin to flow in the coil, magnetic lines of force begin to build up around the coil. As the current rises,

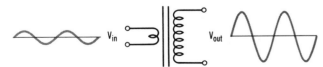

FIGURE 2-5 Transformer action.

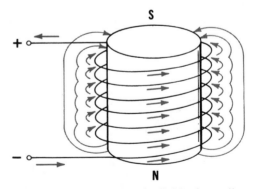

FIGURE 2-7 Magnetic field of a coil.

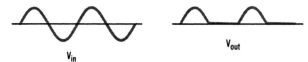

FIGURE 2-8   Half-wave rectification of a sine wave.

THE RECTIFIER

Assume now that a power transformer is used to step down the 117-V ac line voltage to approximately 24 V ac. A power supply that uses this type of transformer must then change the ac voltage into a usable form of dc voltage for operation of a voltage amplifier circuit. The first step in changing alternating current to direct current is rectification. This process involves eliminating one alternation of the sine-wave input so that the output flows only in one direction. Figure 2-8 shows a graphic representation of the rectification process.

In the rectification of a sine wave, you will notice that the output only displays the positive alternation. In this case, the negative alternation has been eliminated by the rectifier. The resulting output of the rectifier is then called *pulsating dc*. In practice, either the positive or negative alternation can be removed by rectification. When the positive alternation appears in the output, the developed dc is positive with respect to a common point or ground. A negative dc output with respect to ground appears when the positive alternation has been removed. Since only half of the ac wave appears in the output, this is called *half-wave rectification*. Both half- and full-wave rectification are discussed in the next section.

In a power supply, rectification is normally

the magnetic field becomes stronger. The stronger magnetic field results in more magnetic lines of force that expand and move away from the center of the coil. The magnetic field then starts collapsing as the current passes its peak. At the zero point, the field has completely disappeared. The current then begins to move in the opposite direction in the coil. The lines of force are now in the opposite direction. The end of the coil that was previously a north magnetic pole now becomes the south pole. The magnetic lines of force again expand about the coil and then collapse.

In the transformer, these expanding and collapsing lines of force cut through the conductor forming the turns in the secondary winding of the transformer. Remember that all of the conditions necessary for inducing a current into the secondary winding of the transformer are present. A voltage is then available at the terminals of the secondary winding. The more turns of wire present in the secondary the higher the voltage developed across the secondary of the transformer.

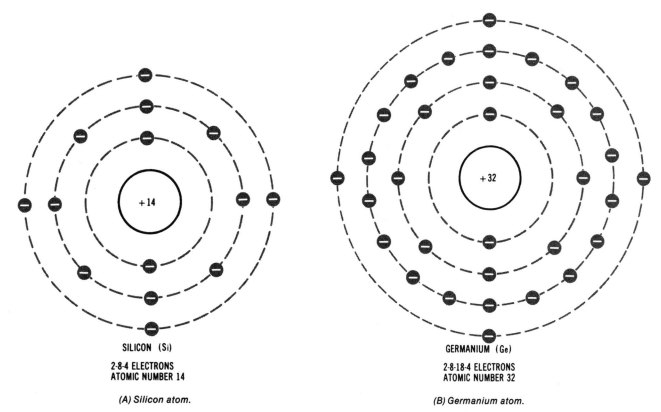

SILICON (Si)

2-8-4 ELECTRONS
ATOMIC NUMBER 14

(A) Silicon atom.

GERMANIUM (Ge)

2-8-18-4 ELECTRONS
ATOMIC NUMBER 32

(B) Germanium atom.

FIGURE 2-9   Structure of silicon and germanium atoms.

achieved by a solid-state *diode*. This type of device contains two electrodes, called the anode and the cathode. A diode has an unusual property that will let electrons flow easily in one direction but not in the other direction. As a result of ac applied to a diode, electrons only flow when the anode is positive and the cathode is negative. Reversing the polarity of voltage applied to a diode will not permit electron flow.

To understand the rectification process and the operation of a diode, we must briefly discuss some of the electrical properties of the elements silicon and germanium. Silicon primarily, or in some cases germanium, is the basis of nearly all solid-state electronic devices today. Figure 2-9 shows the structure of silicon and germanium atoms. Note that the atomic number of silicon (Si) is 14 and germanium (Ge) is 32. This means that these elements have 14 and 32 orbiting electrons, respectively. Note also that they are similar to the extent that both atoms have four electrons in their outer shells.

In a crystal structure of pure silicon, such as shown in Fig. 2-10, the four electrons in the outer shell of each atom tend to join together. In this structure, which shows only the nucleus and the outer-shell electrons, atoms are connected together in what is called *covalent bonding*. This structure forms a strong binding force that tends to hold the atoms together in a manner similar to the way that the nucleus of an atom attracts its own electrons. As a result of this structure, a crystal of silicon does not readily accept new electrons, nor does it permit bonded electrons to be removed or move around. A crystal of pure silicon therefore has rather poor conduction characteristics.

When a pure form of silicon or germanium is formed, it makes a good insulating material. If a minute amount of certain impurities is added to the silicon crystal, its conduction characteristics change significantly. This process is known as doping. Arsenic (As), bismuth (Bi), and antimony (Sb) are typical impurities used to dope silicon.

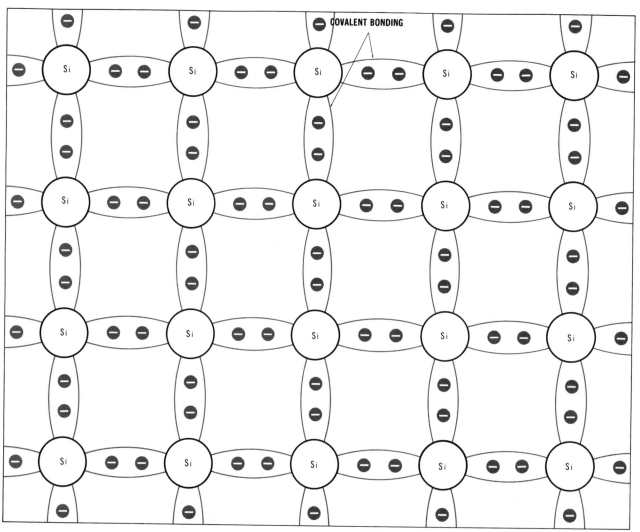

FIGURE 2-10   Crystal structure of pure silicon.

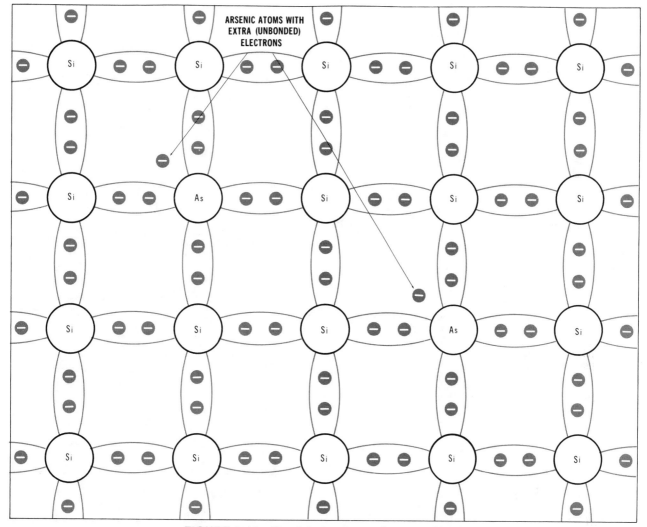

FIGURE 2-11 Crystal structure of n-type silicon.

When an impurity such as arsenic, which has five outer-shell electrons in its atomic structure, is added to a crystal structure of atoms containing four outer-shell electrons, such as silicon, the covalent bonding structure of the silicon is upset. As shown in Fig. 2-11, the structure now has an extra electron for each atom of the doping material. The extra electrons do not take part in the covalent bonding structure. As a result, there are extra electrons in the structure that are free to move around and contribute to the conduction of current. A doped crystal of silicon therefore becomes a conductor when voltage is applied to it. This type of structure is commonly called an *n-type material* because it has extra electrons, or negative charges, that do not take part in the covalent bonding arrangement. An n-type material will conduct current very effectively in either direction.

In the same manner, a *p-type material* is formed by doping silicon with a material that has only three outer-shell electrons, such as gallium (Ga), boron (B),

or indium (In). In this structure, as shown in Fig. 2-12, electron voids, or *holes*, appear in the covalent bonding structure. A hole is often described as an electron deficiency spot, which represents a positively charged area. A p-type material has an abundance of positively charged areas, or electron holes, in its covalent bonding structure and, therefore, becomes a good electrical conductor because it permits the "movement" of holes. Hole flow is in a direction opposite to that of electron flow. A p-type material will also conduct current in either direction very effectively.

When n-type and p-type materials are fused together, a solid-state diode is produced. As noted in Fig. 2-13, the p-type material represents the anode element and the n-type material serves as the cathode.

A silicon diode has the property of conducting current when connected as indicated in Fig. 2-14. In this condition, the negative side of the dc source forces electrons into the n-type material and drives them to

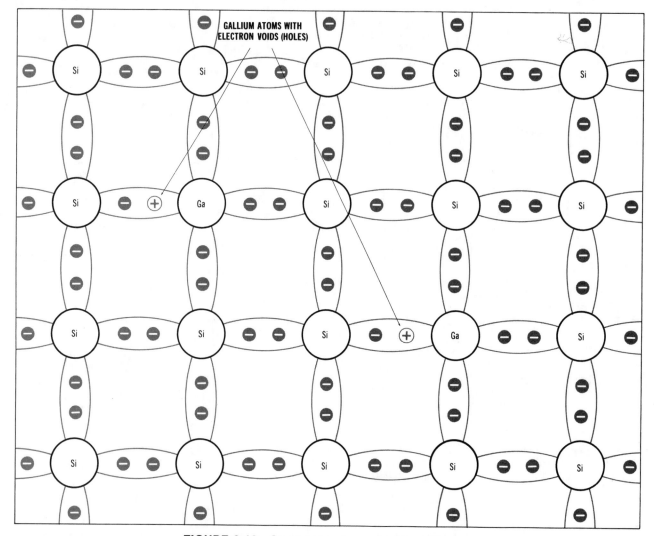

FIGURE 2-12   Crystal structure of p-type silicon.

the center or *junction* area. Holes, or electron voids, in the p-type material are driven to the junction area because the p-type material is connected to the positive side of the dc source.

When holes and electrons meet at the junction of a silicon diode, they combine and permit conduction of current. The process is continuous. For each hole filled with an electron, a new hole is formed by removing an electron from the p-type material. Electrons are also supplied to the n-type material as a result of its connection to the negative side of the dc source. The resistor in the circuit (R_L) is used to limit the current to a reasonable value. Excessive current has a tendency to damage diodes.

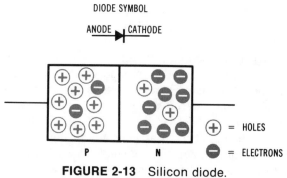

FIGURE 2-13   Silicon diode.

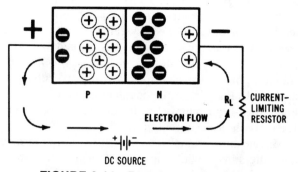

FIGURE 2-14   Forward-biased diode.

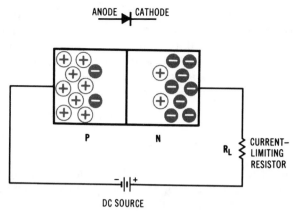

FIGURE 2-15 Reversed-biased diode.

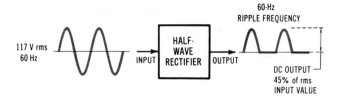

(A) Half-wave rectification.

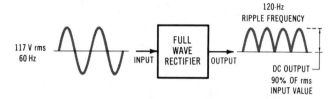

(B) Full-wave rectification.

FIGURE 2-16 Comparison of half-wave and full-wave rectification.

When a silicon diode is connected so that the anode, or p-type material, is positive and the cathode, or n-type material, is negative, it is called *forward biasing*. Forward biasing causes a pn junction to become rather low in resistance, which in turn permits a substantial amount of current to be conducted through the junction.

When the voltage of a dc source is connected as indicated in Fig. 2-15, the diode is *reverse biased*. In this condition, electrons of the n-type material are attracted to the positive side of the dc source, while holes are attracted to the negative side of the source. As a result of this, the electrons and holes are drawn away from the junction. They cannot effectively combine and produce any significant amount of current when connected in this manner. Reverse biasing therefore forms a very high-resistance junction that permits no significant conduction of current.

When alternating current is applied to a silicon diode, it forward biases the pn junction during one alternation, and reverse biases it during the other alternation. As a result, the diode conducts heavily in one direction and is nonconductive in the other direction. Alternating current is therefore changed into pulsating direct current by the action of the diode. The terms *diode* and *rectifier* are frequently used interchangeably in electronic discussions.

HALF-WAVE AND FULL-WAVE RECTIFICATION

Rectified power supplies today are either of the half-wave or full-wave type, depending on the needs of the circuit to which they are connected. Figure 2-16 graphically shows the output differences in half-wave and full-wave rectification. Note particularly the ripple frequency of the outputs and the potential output voltage values that can be developed.

Half-Wave Rectification

Figure 2-17 shows a simplified schematic diagram of a half-wave rectifier. In this circuit, the primary input voltage is 117 V rms. Through normal transformer action the primary voltage is stepped down, in this case to 25 V rms. The positive and negative alternations of the ac voltage across the secondary winding follow in step with those of the voltage applied across the primary winding.

Assume now that the top of the transformer secondary is positive and the bottom is negative during the positive alternation of the sine-wave input. When this occurs, the diode is forward biased and permits current conduction through the load resistor, R_L, as indicated. The positive alternation will appear across

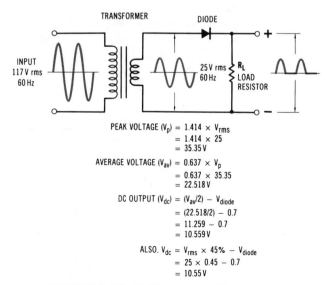

PEAK VOLTAGE (V_p) = 1.414 × V_{rms}
                                       = 1.414 × 25
                                       = 35.35 V

AVERAGE VOLTAGE (V_{av}) = 0.637 × V_p
                                               = 0.637 × 35.35
                                               = 22.518 V

DC OUTPUT (V_{dc}) = $(V_{av}/2)$ − V_{diode}
                                      = (22.518/2) − 0.7
                                      = 11.259 − 0.7
                                      = 10.559 V

ALSO, V_{dc} = V_{rms} × 45% − V_{diode}
                       = 25 × 0.45 − 0.7
                       = 10.55 V

FIGURE 2-17 Half-wave rectifier circuit.

the load resistor at nearly the same voltage value as that applied. The slight different in output is due to the resistance of the diode. A silicon diode will produce a 0.7-V drop. As a result, the output is reduced by this amount.

During the negative alternation of the sine-wave input, the top of the transformer will be negative and the bottom will be positive. By tracing the circuit back to the diode, you will notice that this voltage polarity reverse biases the diode. A reversed-biased diode represents an extremely high-resistance junction and serves as an open circuit. With no resulting current through R_L, the output voltage is zero for this alternation.

Through normal forward and reverse bias conditions, a half-wave rectifier will only produce one alternation in its output. The polarity of the output depends on the diode connections. In Fig. 2-17, the top of R_L will be positive and the bottom negative. By reversing the anode and cathode connections of the diode, the polarity of the output can be reversed. Depending on the needs of the circuit to which a power supply is connected, its output can be either positive or negative with respect to a common point or ground.

To determine the potential dc output of a half-wave rectifier, one needs to first determine the peak value of one alternation. This is achieved by multiplying the rms input voltage by 1.414. The average value of this alternation is then determined by multiplying the peak value by 0.637. Since only one alternation of the two in a cycle appears in the output, this value must be divided by 2. Combining these values, we find that 1.414 times 0.637 divided by 2 equals 0.450 359, or 45% of the rms input voltage. This means that the potential dc output of a half-wave rectifier is 45% of the rms ac value minus the diode voltage drop of 0.7 V. A dc voltmeter would read this value across the load resistor, R_L, in our simple half-wave rectifier.

The output pulses of a half-wave rectifier occur at the same rate as they do in the applied ac input. With 60-Hz input, the ripple frequency of a half-wave rectifier is 60 Hz. Only one pulse is produced, how-

ever, for a complete sine-wave input that has two alternations per cycle. Potentially, only 45% of the ac input is transformed into a usable output in half-wave rectification. Due to the low efficiency factor, half-wave rectification is not commonly used in electronic instruments that are used in industry.

Full-Wave Rectification

Full-wave rectification, as the name implies, changes both alternations of a sine-wave input into a pulsating dc output. In practice, full-wave rectification can be achieved with two diodes and a center-tapped transformer, or four diodes in a bridge circuit. Applications of both types are quite numerous today.

A basic full-wave rectifier using two diodes is shown in Fig. 2-18. The cathodes of both diodes, as you will note, are connected commonly together to form the positive output. The anodes of each diode are connected to opposite ends of the transformer secondary winding. The load resistor (R_L) completes the circuit path between the cathodes and the transformer center-tap connection.

In our example circuit of Fig. 2-18, when ac is applied to the primary winding of the transformer, it steps down the voltage in the secondary winding. If a greater output is desired, a transformer with a larger secondary voltage could be selected. Since the center tap, or point C in the diagram, is the electrical center of the transformer secondary winding, half of the induced voltage will appear between points C and A and the other half between points C and B. The resulting voltages will always be 180° out of phase with respect to point C.

As an example, assume that the positive alternation of the input causes point A to be positive and point B to be negative with respect to point C. During the next alternation, the polarity will reverse with point A being negative and point B positive with respect to point C. Point C will always be negative with respect to the positive end of the winding regardless of whether

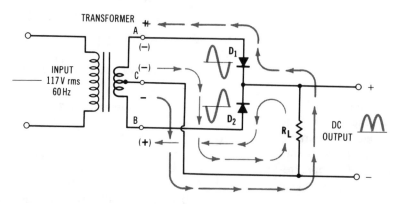

FIGURE 2-18 Full-wave rectifier.

it is point A or point B. Conduction of the two diodes changes back and forth with one conducting during the positive alternation and the other conducting during the negative alternation. All of this is done with respect to point C. In this case, point C is considered the negative output of the rectifier.

Referring again to Fig. 2-18, assume that the ac input is such that it causes point A to be positive and point B to be negative for one alternation. Starting at point C, electron flow is indicated by the solid arrows. With this polarity, note that D_1 is forward biased and D_2 is reverse biased. Conduction, therefore, occurs from point C, through R_L, through D_1, to point A. This produces a corresponding alternation of voltage across R_L and a resulting current through the indicated circuit path.

Using the procedure just mentioned, assume now that the next input alternation is applied to the transformer input. In this case, point A becomes negative and point B becomes positive with respect to point C. Starting at point C, electron flow is indicated by the dashed arrows. With this polarity, D_2 is forward biased and D_1 is reverse biased. Conduction now occurs from point C, through R_L, through D_2, to point B. This produces a corresponding alternation of voltage across R_L and a resulting current through the indicated circuit path.

An interesting point should be noted about the full-wave rectifier circuit. The current through R_L is in the same direction for each alternation of the input. This means that we have obtained dc output for both halves of the sine-wave input, or full-wave rectification.

The resulting output voltage of a full-wave rectifier is 90% of the ac rms voltage appearing between the center tap and the outer ends of the transformer. This value is twice the possible output of the half-wave rectifier discussed previously. This value is determined by calculating the peak value (V_p) of the rms voltage, then multiplying it by the average value (V_{av}).

The potential dc output will therefore be the rms value times 1.414 times 0.637. Since 1.414 times 0.637 equals 0.900 718, this is a very close approximation of the 90% value. Full-wave rectification is considered to be 50% more efficient than half-wave rectification.

The dc output voltage appearing across R_L of the circuit will be slightly less than the 90% times the rms value just described. Each diode, for example, has a voltage drop of 0.7 V. This means that a dc voltmeter would read the rms value times 90% minus 0.7 across R_L.

The ripple frequency of our full-wave rectifier is also different compared with the half-wave circuits. With each alternation producing an output across R_L, the ripple frequency will be twice the alternation frequency. In this case, the ripple frequency is 120 Hz or twice the input frequency. A higher ripple frequency and characteristic output of a full-wave rectifier is easier to filter than a similar half-wave output.

Full-Wave Bridge Rectifiers

A bridge structure of four diodes is commonly used in power supplies today to achieve full-wave rectification. In this rectifier configuration, two diodes will conduct during the positive alternation and two will conduct during the negative alternation. A bridge rectifier does not necessitate a center-tapped transformer as was used in the two-diode rectifier.

Figure 2-19 shows a schematic diagram of a full-wave bridge rectifier. In this circuit, note that the ac input is applied to the junction of diodes D_1 and D_2 at the top of the bridge, and to the junction of D_3 and D_4 at the bottom. The connections of the diodes at these two junctions are reversed with respect to the other. The junction of D_1 and D_4 has both cathodes connected together, while the junction of D_2 and D_3 has the anodes connected together. The common-anode connection serves as the negative output of the bridge, with

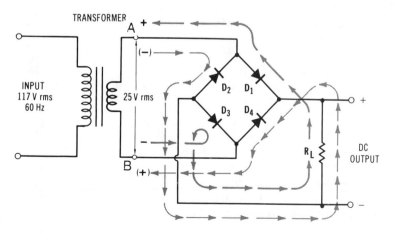

FIGURE 2-19 Full-wave bridge.

the common-cathode connection serving as the positive output.

When ac is applied to the primary winding of the power transformer, it can either be stepped down or up depending on the value of the dc needed. In our example, 25 V ac appears across the secondary winding. In normal operation, one alternation of the input voltage will cause the top of the transformer to be positive and the bottom negative. The next alternation will cause the polarity to reverse. Opposite ends of the transformer will therefore always be 180° out of phase with each other.

As an example, assume that the positive alternation of the input causes point A to be positive and point B to be negative in Fig. 2-19. With this polarity, diode D_1 will be forward biased and D_2 will be reverse biased at the top junction. At the bottom junction, diode D_3 will be forward biased and D_4 will be reverse biased. When this occurs, electrons flow from point B, through D_3, through R_L, through D_1, to point A. This path of electron flow, which is indicated by solid arrows, produces an alternation across R_L.

With the next alternation of the ac input, point A becomes negative and point B becomes positive. When this occurs, the bottom diode junction is of positive polarity and the top junction becomes negative. In this condition, D_2 and D_4 are forward biased, while D_1 and D_3 are reverse biased. The resulting electron flow, indicated by the dashed arrows, starts at point A, goes through D_2, through R_L, through D_4, to point B. This path of electron flow causes the second alternation to appear across R_L.

It is interesting to note that the current through R_L is in the same direction for each alternation of the applied ac input. This, of course, means that ac is changed or rectified into dc. The dc output, in this case, has a ripple frequency of 120 Hz. Since each alternation produces a resulting output pulse, the ripple frequency is twice the value of the alternation frequency, or 2×60 Hz = 120 Hz.

The dc output voltage appearing across R_L of the bridge circuit will be somewhat less than 90% of the applied rms value. Each diode, for example, produces a 0.7-V drop when conducting. This means that two diodes conducting during each alternation will reduce the output by 2 times 0.7, or 1.4 V. The resulting dc output will therefore be 90% of the rms value, less 1.4 V. In our circuit, this will be $0.90 \times 25 - 1.4 = 21.1$ V dc.

Bridge rectifiers are commonly used in electronic power supplies for instruments today. A very high percentage of all solid-state industrial instruments employ this type of rectifier because of its simplified operation and desirable output. Generally, the diode bridge is housed in a single device that has two input and two output connections.

Divided or Split Power Supplies

A rather new variation of the bridge power supply is presently being used in many industrial instruments. This supply has both negative and positive outputs with respect to a common point. Divided or split power supplies have been developed as a source for integrated circuits. The secondary winding of the input transformer is divided into two parts. The center tap or neutral serves as a common ground connection for the two outside windings. Each half of the winding has a complete full-wave rectifier. A schematic diagram of a dual power supply is shown in Fig. 2-20. This particular supply has outputs of +10.64 V and −10.64 V.

Operation of the power supply is similar to that of the two diode full-wave rectifier. Diodes D_1 and D_4 conduct during one alternation, while D_2 and D_3 conduct during the next alternation. Current flow through the two load resistors develops the output voltage. Point G serves as the ground or common point when positive and negative voltages are both developed by the power supply. This point also serves as a return to the transformer center tap.

The arrows of Fig. 2-20 are used to indicate the direction of current flow through the power supply for each alternation. For one alternation of the ac input, assume that the top of the transformer (A) is positive and the bottom (B) is negative. This causes current to flow from point B through D_4, R_{L2}, R_{L1}, D_1, and to point A. This makes point E negative and point D positive. With respect to point G, E is the most negative and D is the most positive. The solid arrows show the path of current for this alternation.

The next alternation makes the top of the transformer (A) negative and the bottom (B) positive. Current flows from point A through D_3, R_{L2}, R_{L1}, D_2 and returns to point B. This current flow again makes point

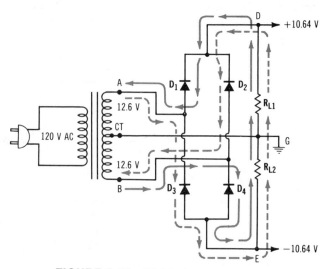

FIGURE 2-20   Divided power supply.

E negative and point D positive. The dashed arrows show the direction of current for this alternation. Since each alternation produces current flow through the two load resistors in the same direction, the output is considered to be full wave.

The dc output voltage of a divided power supply is somewhat less than that of a two diode full-wave rectifier power supply. Each diode, for example, produces a 0.7-V drop when conducting. This means that two diodes conducting during each alternation will cause the output voltage to be reduced by 1.4 V. The resulting output will therefore be 90% of the rms input, less 1.4 V. In our representative circuit, this will be $0.90 \times 25.2 - 1.4 = 21.28$ V of dc. Dividing this across the two load resistors will cause the output to be $+10.64$ V and -10.64 V. This output corresponds to that of a bridge rectifier.

FILTERING

After pulsating dc has been produced by a rectifier, it must be filtered for it to be usable in a power supply. Filtering involves changing the ripple frequency of a rectifier output into a constant dc voltage value. In practice, this is often called *smoothing out* the ripples of the pulsating dc voltage.

The filter circuit of Fig. 2-21 is composed of resistors R_1 and R_2, and capacitors C_1, C_2, and C_3. Resistor R_L represents the load of the power supply. This, in general, is representative of the combined electronic components receiving energy from the power supply. The filter circuit shown is called a *double pi-section filter*. Resistor R_1, capacitor C_1, and capacitor C_2 are connected together like the Greek letter pi (π), as are R_2, C_2, and C_3. We first discuss a single pi-section filter, but before that we need to learn something about capacitors and their effects in dc circuits.

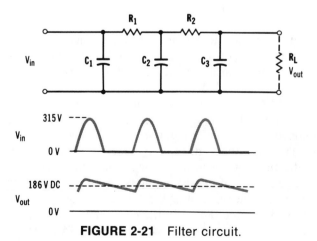

FIGURE 2-21 Filter circuit.

CAPACITANCE

Capacitance is a basic principle that must be understood in order to understand the operation of electronic circuits. The device that contributes capacitance to the circuit is a *capacitor*. A capacitor can be physically described as two conductors separated by an insulating material called a *dielectric*. Dielectric materials commonly used in capacitors are air, paper, plastic film, glass, mica, and oil. The unit for measuring capacitance is the farad. Since this is an extremely large and impractical unit, capacitors are normally measured in microfarads (millionths of farads) or picofarads (millionths of microfarads).

The capacitance of a capacitor is stated by the formula $C = KA/d$. The K in the formula stands for the dielectric constant. The dielectric constant of a vacuum is 1; for air it is slightly larger than 1. The dielectric constant for mica is about 7 to 9. The letter A represents the effective area of one plate of a capacitor. From the formula it can be seen that the larger this area is, the greater the capacitance will be. The letter d represents the distance between plates. It can be seen from the formula that the closer together the plates are, the greater is the capacitance. A paper capacitor can be easily examined in order to get an idea of its construction. It is formed by using two strips of metal foil as the plates, with leads (pigtails) connected to each of these "plates." Between the plates are inserted strips of paper that act as the dielectric. To conserve space, this laminated structure is rolled into a small cylinder and placed inside a cylindrical case made of paper, metal, or plastic.

Let us first examine the effect of such a capacitor in a dc circuit. The circuit of Fig. 2-22 demonstrates the ability of a capacitor to store energy in the form of an electrostatic field. When a battery is connected to a capacitor as in Fig. 2-22, the negative terminal of the battery provides electrons to the plate of the capacitor connected to it. When an excess of electrons appears at one plate of the capacitor, the plate takes on a negative charge that repels electrons from the other plate of the capacitor. These electrons are attracted to the positive terminal of the battery. The deficiency of electrons at this plate of the capacitor leaves it with a positive charge. This difference of potential

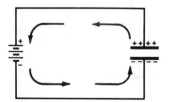

FIGURE 2-22 Capacitor charged by a battery.

across the capacitor will equal the voltage of the battery. We say the capacitor is *charged* when this occurs.

When a charged capacitor is disconnected from the battery, the difference of potential across the capacitor is unchanged. Energy has been stored in the capacitor and will remain until it is allowed to discharge through an external circuit. This discharge can be seen in a large capacitor if the leads are touched together. The presence of this stored energy can also be seen if the capacitor is allowed to discharge through an ammeter, as in Fig. 2-23. The meter provides a low-resistance path between the plates. The electrons on the negatively charged plate are attracted to the positive plate through this low-resistance path. This movement of electrons constitutes a current that will be indicated by the meter. After all excess electrons on the negative plate have moved to the positive plate, the capacitor is fully discharged. There is no longer a difference of potential across the capacitor and the electrostatic field across the capacitor no longer exists. It is the ability of the capacitor to store energy in an electrostatic field that enables it to oppose changes in voltage. A capacitor, therefore, has the ability to oppose a change in voltage.

In the preceding discussion, it was stated that the capacitor was charged instantaneously. However, in the circuit of Fig. 2-24, the current is limited by resistor R. With the switch in the off position there is no current in the circuit. There is also no charge on the capacitor. The instant the switch is closed, electrons flow in the circuit. The capacitor offers no opposition to this instantaneous change in current. When an electron appears at one plate of the capacitor, it immediately repels an electron from the opposite plate. The only opposition to electron flow at this instant is the resistor. The current in the circuit is determined by Ohm's law: $I = V/R$. The time it takes the capacitor to charge is thereby lengthened since the current is limited. The larger the resistor, the less the current. The less current in the circuit, the longer it takes the capacitor to charge. The time it takes capacitor C to charge is also determined by the size of the capacitor. This relationship is stated by the formula $C = QV$. C is measured in farads, Q indicates the quantity of charge measured in coulombs, and V represents the voltage (potential difference) across the capacitor. Notice that with a given voltage (V) across the capacitor, the larger the

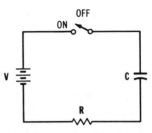

FIGURE 2-24   *RC* circuit.

capacitance (C) and the greater the charge (Q). The charge (Q) measured in coulombs indicates the quantity of electrons on one plate of the capacitor. Therefore, the larger the capacitance, the longer it takes the capacitor to charge. The length of time it takes the capacitor to charge to 63% of the applied voltage is represented by the formula $t = R \times C$. This value is called the time constant of the *RC* circuit. If R equals 1 megohm (1 million ohms; MΩ) and C equals 1 microfarad (one-millionth of a farad; μF), *RC* equals 10^{-6} F times 10^6 Ω equals 1 s. In 1 s the capacitor will charge to approximately 63% of the applied voltage. In about five time constants, or 5 s in this case, the capacitor in Fig. 2-24 will be fully charged.

Let us assume that in Fig. 2-24, $R = 1$ MΩ, $C = 1$ μF, and $V = 100$ V. The instant the switch is closed, the current in the circuit is 0.1 mA, since

$$I = \frac{V}{R} = \frac{100}{1\ 000\ 000} = 0.1 \text{ mA}$$

In a short period of time, the capacitor will have assumed a charge. Let us assume that this charge has risen to 10 V. The polarity of this voltage opposes the battery voltage. This means that the voltage drop across the resistor has decreased from 100 to 90 V. With 90 V across the 1-MΩ resistor, the current has now decreased to 0.09 mA:

$$I = \frac{V}{R} = \frac{90}{1\ 000\ 000} = 0.09 \text{ mA}$$

As the charge across the capacitor increases, the charging current decreases. It can be seen in Fig. 2-25 that the charge on the capacitor opposes further change.

As stated previously, in one time constant (in this case, 1 s), the capacitor has charged to 63% of the applied voltage. Since in this circuit the applied voltage is 100 V dc, the voltage across the capacitor is 63 V at the end of 1 s. During the next time constant, it will charge to 63% of the remaining voltage (37 V dc). This means that at the end of 2 s, the capacitor will have charged to about 86 V. In the next time constant, the capacitor charges to 63% of the remaining 14 V. At the

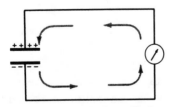

FIGURE 2-23   Capacitor discharging.

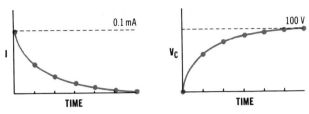

FIGURE 2-25 Charging current and accumulated charge on a capacitor.

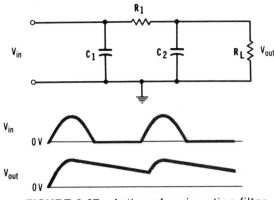

FIGURE 2-27 Action of a pi-section filter.

end of 3 s, the capacitor voltage is about 95 V; at the end of 4 s, it is about 98.1 V; after 5 s, it is 99.3 V, or almost the full battery voltage. Notice that the capacitor actually never fully charges, although the value is so close to the applied voltage as to be indistinguishable.

A charged capacitor will also take five time constants to discharge through a resistor. If the capacitor and resistor in Fig. 2-26 are the same size as those used in Fig. 2-24, it will take the same time for the capacitor to discharge. In the first time constant, it will discharge 63% of the voltage. During the next time constant, it will discharge 63% of the remainder, and so on. At the end of five time constants, the capacitor will be almost fully discharged. Notice that the amount of charge has no effect on the time it takes the capacitor to discharge.

Figure 2-27 shows the action of a pi-section filter in a power supply circuit. The input to the filter circuit is the pulsating dc voltage produced by the rectifier. When the first pulsation of voltage appears across C_1, it charges to the peak voltage of this pulsation. Since there is no resistor through which C_1 must charge, it will assume this charge instantaneously. Capacitor C_1, however, must discharge through R_1 and R_L. Since C_1, R_1, and R_L are all large values, this time constant is exceedingly long. The capacitor will not discharge instantly. During the time that the rectifier output is below the peak, the capacitor (C_1) discharges. Since the time constant is long, the discharge is not great. When the next positive pulse of voltage appears, C_1 again charges. The result is that the voltage across C_1 varies very little. These variations are further filtered by the charge and discharge of C_2. Capacitor C_2 will charge to the peak voltage dropped across R_L. C_2 also discharges through R_L. This time constant is still rather long, which provides a further filtering action. Those variations that do exist in the output are called the

ripple voltage. The filter should eliminate as much of this ripple as possible.

Before leaving the discussion involving the filter, a comment should be made concerning the effect of a capacitor on an ac circuit. The ratio between voltage and current (V_C/I_C) of a capacitor is called its *reactance*. Capacitive reactance (X_C) is measured in ohms. You will remember that a capacitor in a purely dc circuit acts as an open circuit as soon as it is fully charged; that is, the dc current becomes zero. You will notice, however, that the output of the rectifier is not a constant dc voltage; it has both an ac and a dc component. We have seen the effect of the capacitor charging to the peak voltage and maintaining this charge, due to its long time constant, during the interval between these peaks. This is also affected by its reactance. Capacitive reactance is determined by the formula

$$X_C = \frac{1}{2\pi f C}$$

You will notice that an increase in frequency or capacitance causes a corresponding decrease in capacitive reactance. If capacitor C_1 of Fig. 2-27 is 100 μF and the ac component is 60 Hz, the capacitive reactance of C_1 can be calculated by substituting these values in the formula:

$$X_C = \frac{1}{2\pi f C}$$

$$= \frac{1}{6.28 \times 60 \times 100 \times 10^{-6}}$$

$$= \frac{1}{3.768 \times 10^2}$$

$$= 26.54 \ \Omega$$

This means that the capacitor offers a very low reactance path to the ac component of the rectifier output, which results in a very small ac voltage drop across

FIGURE 2-26 Discharge in an *RC* circuit.

C_1. The same results would occur at C_2, which would explain the very small ac voltage component of the power supply output.

A COMPLETE POWER SUPPLY

The schematic diagram of a typical electronic instrument power supply is shown in Fig. 2-28. In this schematic, there are several differences from the simplified diagrams used in our discussion of power supplies. The most obvious difference is in the alternate connections provided in the primary winding of transformer T_1. Only one secondary winding is used in the actual circuit. In this case, the manufacturer probably uses this transformer in several different power supplies for similar equipment.

The power supply will develop approximately 57 V of dc. This is determined by taking 90% of the secondary voltage and subtracting the voltage drop across two diodes. The output is also adjustable to some extent by resistor R_{100}, which is connected in series with the positive output line. The circuit is also fused in both the primary and secondary lines by F_{101} and F_{102}, respectively.

REGULATION

According to our discussion of dc power supplies thus far, it is essential that they employ a rectifier and a filter circuit. In addition to this, some power supplies have an added circuit called a *regulator*. The primary purpose of a regulator is to aid the rectifier and filter in providing a more constant dc voltage to the load device. Power supplies without regulation have an inherent problem of changing dc voltage values due to variations in load resistance or fluctuations in the ac line voltage. With a regulator connected in the dc output, the voltage can be maintained within a rather close tolerance of its designed output.

In its simplest form, regulation of dc can be achieved to some extent by connecting a single resistor across the power supply output terminals (see Fig. 2-29). When variations in load resistance connected in parallel with this regulator resistor are small, some degree of regulation occurs. In this case, the regulator resistor serves as a constant load for the power supply. Under normal operating conditions, current from the power supply must pass through both the load resistor and the regulator resistor. This means that the regulator resistor serves as a fixed load regardless of the value of R_L. When the value of R_L or the input voltage changes to any real extent, this type of regulation is rather ineffective.

Zener Diode Regulators

A zener diode regulator employs a special type of semiconductor device that has an unusual conduction characteristic. When forward biased, a zener diode conducts as a conventional silicon diode. Normally, this type of diode is not used in the forward-bias direction.

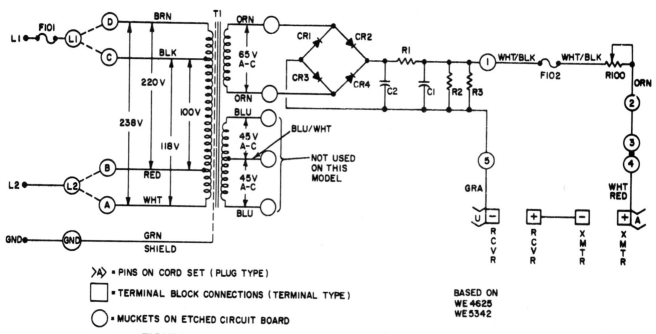

FIGURE 2-28 Schematic diagram of a typical electronic instrument power supply. (Courtesy of The Foxboro Co.)

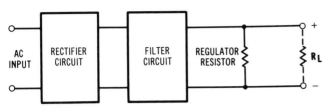

FIGURE 2-29 Regulator resistor across power supply output.

When reverse biased, however, it goes into conduction at a specific voltage value depending on the ratio of its doping material and silicon. Manufacturers can alter the reverse-bias conduction point, or zener voltage, when the device is being constructed. The percentage of zener voltage (V_Z) variation ranges from $\pm 10\%$ to $\pm 0.1\%$ according to the tolerance level of the selected device.

Figure 2-30 shows a zener diode regulator connected across the dc output of a power supply. In this circuit, all of the current from the power supply must pass through the series resistor (R_8). If, for example, a 10-V regulated output is desired, a 10-V zener diode would be selected. The diode would be connected in parallel with R_L as shown. As you will note, the diode is connected in a reverse-bias direction. When 10 V appear across the diode, it goes into conduction. This voltage will be maintained for a wide range of applied voltage values.

The unique current conduction characteristic of a zener diode permits it to achieve voltage regulation. Essentially, increasing or decreasing the applied source voltage is compensated for by increasing or decreasing the current through the zener diode. This, in turn, causes the voltage drop across the series resistor (R_8) to change accordingly. An increase in line voltage would immediately cause an increase in power supply output. This would, in turn, cause a corresponding increase in current and voltage drop across R_8. As a result, the voltage across the zener diode remains at 10 V, and the voltage across R_8 changes with circuit conduction.

Similarly, a decrease in line voltage would cause a reduced current through the zener diode. This would, in turn, cause a smaller voltage drop across R_8 which would maintain the voltage across the zener diode at 10 V. In effect, this means that variations in input voltage cause a corresponding change in current which is compensated for by changes in voltage drop across R_8. Through this action, the voltage across the zener diode is maintained at a constant level.

Changes in load resistance are also reduced with a zener diode regulator. Normally, an increase in load resistance would cause a rise in dc output voltage. With a zener diode installed across the load resistor, the voltage remains at a constant value. Essentially, the zener diode compensates for this change by increasing the current through it. This, in turn, maintains the load voltage at 10 V by the increasing voltage drop across R_8.

Without regulation, an increase in load current caused by a smaller R_L normally causes a decrease in load voltage. With the zener installed, however, the load voltage is maintained at the rated zener-voltage value. In this case, more current will be supplied to the load resistor, with less to the zener diode. As a result, a reduction in total current through R_8 will cause less voltage drop. This, in turn, increases the load voltage to compensate for the increase in load current.

Zener diodes are normally rated by wattage and zener voltage. Wattage, for example, indicates the ability of a diode to dissipate or give off heat. In a sense, the wattage rating is an indication of the maximum current-handling capability of the diode as its rated V_Z. A 1-W, 10-V zener diode would pass $I = W/V_Z$ current without being destroyed. In this case, 1 W divided by 10 V equals 0.1 A, or 100 mA, maximum current-handling capability. In practice, larger wattage ratings can be substituted in place of smaller wattage ratings. You should avoid using smaller wattage ratings when substituting any zener diode in a circuit.

Integrated Circuit Regulators

A rather recent trend in industrial instrument power supplies is the integrated circuit (IC) regulator. This chip accepts an unregulated dc input and develops a regulated output. IC regulators are available in a variety of fixed and variable values. The variable units seem to be more widely used today. A representative fixed value IC regulator is shown in Fig. 2-31. Notice that this regulator is housed in a three-terminal TO-202 package.

The internal structure of an IC regulator is shown in Fig. 2-32. In general, it contains a series-pass transistor, reference voltage source, feedback amplifier, and a thermal overload protection circuit. The circuitry is primarily the same for all fixed value units. This particular unit has three terminals. These are identified as V_{in}, V_{out}, and ground.

Operation of the three-terminal regulator is based

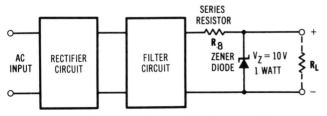

FIGURE 2-30 Zener diode regulator across power supply output.

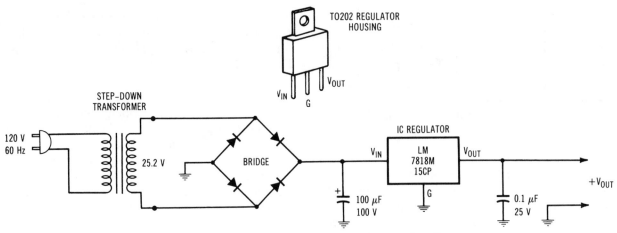

FIGURE 2-31 Fixed-value IC regulator circuit.

on the applied input voltage and the developed output voltage. Output voltage is compared with a reference voltage by a high-gain error amplifier. Any difference in voltage will be sensed by the error amplifier. The output of the error amplifier then controls conduction of the series-pass transistor. This transistor responds as a variable resistor. It changes inversely with the load current to maintain the output voltage at a constant value. An increase in output voltage would cause a corresponding increase in the conduction of the series-pass transistor. This, in turn, causes more voltage to

drop across the series resistor, which causes the output voltage to be lower. A decrease in output voltage, in turn, causes less conduction of the series pass transistor. Reduced current through the series resistor will cause less voltage drop. The output voltage, in this case, increases in value to compensate for the decrease in voltage drop.

Integrated circuit regulators are generally equipped with some type of current overload protection. This function is designed to prevent damage to the series-pass transistor. The three-terminal regulator

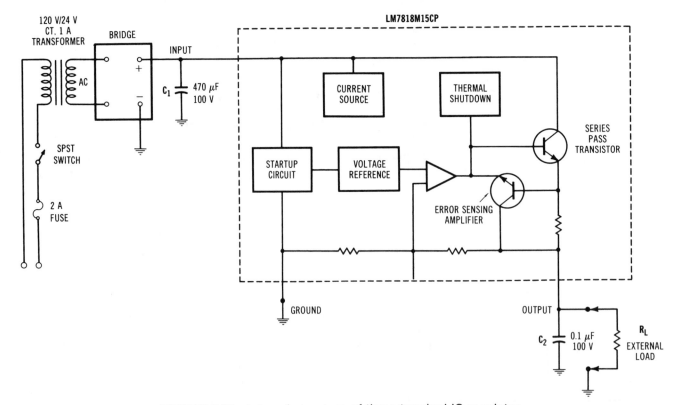

FIGURE 2-32 Internal structure of three-terminal IC regulator.

of Fig. 2-32 has three thermal shutdown capabilities. The temperature of the series-pass transistor is automatically sensed by the internal structure of the IC. If its operating temperature exceeds 175°C, the entire regulator turns off. Removing the load from the power source automatically reduces the output which causes the transistor to cool down. Operation is restored when the temperature of the series transistor returns to normal. Protection of this type makes the chip practically immune to overload damage.

SUMMARY

Electronic instruments have a variety of specific functions that must be performed in order to achieve a particular operation. The power supply is responsible for the function of rectification.

A common source of input voltage to a power supply is ac. Alternating current moves first in one direction and then in the other for a short period of time. This is described as a cycle of ac. Frequency, which is the number of cycles that occur in one second, is expressed in hertz (Hz). Alternating current is produced by rotating a coil of wire in a magnetic field.

Transformers are used in power supplies to step up or step down the applied voltage. The turns ratio of a transformer determines its ability to produce a change. The output voltage and current of a transformer are inversely related.

Rectifiers are used to change ac to dc in a power supply. A two-electrode device known as a diode is used to achieve rectification. A silicon diode is made of p-type and n-type material connected together.

When forward biased a diode will conduct current easily. Reverse biasing restricts current to a negligible value.

Half-wave rectification clips off an alternation of the applied sine wave, which results in a single-directional output current. Full-wave rectification transposes both alternations of an applied sine wave into a single-directional output. Full-wave rectification is achieved with two diodes and a center-tapped transformer, or with four diodes in a bridge arrangement.

The pulsating dc output of a rectifier must be filtered in order for it to be usable in a power supply. Filtering removes the ripple and changes the output into a constant dc voltage value.

A capacitor is two or more metal plates separated by a dielectric or insulating material. A capacitor has the ability to hold an electric charge. After it has been charged it has the ability to oppose a change in voltage. The fundamental unit of capacitance is the farad, with practical values measured in microfarads or picofarads.

When a capacitor, resistor, and a capacitor are connected together, they form a pi-section filter circuit. In operation, the first capacitor will charge, then discharge through the resistor, and the other capacitor discharges a short time later. This sequential action produces good filtering.

Regulators are frequently used in power supplies to aid the rectifier and filter in providing a more constant dc voltage to the load device. Zener diode regulators are used to produce a constant output voltage across the load resistor of a power supply. These diodes maintain a constant voltage through variations in current through a series resistor. Zener diodes are rated according to their wattage and V_Z values.

ACTIVITIES

HALF-WAVE RECTIFIERS

1. Construct the half-wave rectifier of Fig. 2-17 with a 1-A silicon diode (1N4004). The load resistor should be 1 kΩ, 1 W.
2. Connect ac to the circuit. Measure and record the primary ac voltage applied to the transformer. Measure and record the secondary ac voltage of the transformer.
3. Measure and record the dc voltage across the load resistor.
4. Measure and record the ac ripple voltage appearing across the load resistor.
5. If an oscilloscope is available connect it across the transformer secondary winding and observe the ac voltage. Make a sketch of the observed waveform.

6. Connect the oscilloscope across the load resistor and observe the output of the rectifier. Make a sketch of the observed dc waveform.

TWO-DIODE FULL-WAVE RECTIFIERS

1. Construct the two-diode full-wave rectifier of Fig. 2-18. The transformer should be 25.2 V center-tapped and the diodes should be 1N4004 or equivalent. The load resistor should be 1 kΩ, 1 W.
2. Measure and record the secondary voltage at points A–C, C–B, and A–B.
3. Measure and record the dc voltage appearing across the load resistor.
4. If an oscilloscope is available, connect the vertical

input probes across reference points A–C, C–B, and A–B. Use line triggering and note the phase of the ac wave at the reference points.

5. Make a sketch of the observed ac waveforms at the reference points showing the phase of the waves and the peak-to-peak ac voltage values.

6. Connect the vertical probes of the oscilloscope across the load resistor and observe the dc output of the full-wave rectifier circuit. Make a sketch of the observed display, noting the peak voltage value of the waveform.

FULL-WAVE BRIDGE RECTIFIERS

1. Construct the bridge rectifier circuit of Fig. 2-19. The bridge should be an SK3647 or any 200-V, 2-A equivalent. The load resistor should be 1 kΩ, 1 W.

2. Connect the primary of the transformer to a 120-V ac source.

CAUTION: Do not connect the oscilloscope probes to the 120-V ac transformer primary winding. This may cause severe damage to the oscilloscope and possible injury to the operator.

3. Measure and record the ac voltage of the power-line input and the secondary voltage of the transformer.

4. Measure and record the dc output voltage appearing across the load resistor.

5. Measure and record the ac ripple voltage across the load resistor.

6. If an oscilloscope is available, connect the vertical input probes across transformer secondary winding.

Make a sketch of the observed secondary winding voltage. Indicate the peak-to-peak ac voltage value of the observed wave.

CAUTION: Do not connect the oscilloscope probes to the 120-V ac transformer primary winding. This may cause severe damage to the oscilloscope and possible injury to the operator.

7. Connect the vertical probes of the oscilloscope across the load resistor. Make a sketch of the observed dc output wave. Indicate the peak voltage value of the wave.

DIVIDED POWER SUPPLY

1. Construct the divided power supply of Fig. 2-20. The transformer should be 25.6 V center-tapped. The bridge should be an SK 3647 or any 200-V, 2-A equivalent. The load resistor should be 1 kΩ, 1 W.

2. Apply ac input to the primary winding. Measure and record the secondary voltage at points A–CT, B–CT, and A–B.

3. Measure and record the dc voltage at points D–G, G–E, and D–E. Note the polarity of the dc voltage.

4. If an oscilloscope is available, observe the ac voltage at points A–CT, B–CT, and A–B. Note the phase relationship of the observed waves. Determine the peak-to-peak voltage value at these test points.

5. Connect the common or ground probe of the oscilloscope to test point G. Then connect the vertical probe to point D and alternately to point E. Note the polarity difference indicated by the horizontal-line position.

QUESTIONS

1. What is the function of a power supply in an electronic instrument?

2. What is meant by the term *rectification*?

3. Explain the difference between ac and dc electricity.

4. How is ac generated?

5. Referring to Fig. 2-4, identify the period of time (by letters) needed to complete the positive alternation, the negative alternation, and one complete cycle of operation.

6. Explain how a transformer steps down voltage between the primary and secondary windings.

7. Explain how the half-wave rectifier of Fig. 2-17 produces dc across the load resistor.

8. Explain how the full-wave rectifier of Fig. 2-18 produces dc across the load resistor.

9. Explain why the voltage output of the divided power supply of Fig. 2-20 has a positive and a negative voltage with respect to the common or ground (G) point.

10. What is the function of a filter in a rectifier power supply?

11. Describe the type of electronic instrument power supply shown in Fig. 2-28.

12. How does a zener diode achieve regulation?

CHAPTER 3

AMPLIFIERS

OBJECTIVES

Upon completion of this chapter, you will be able to:

1. Define amplification.
2. Explain the operation of a linear operational amplifier.
3. Identify the elements, schematic symbols, and circuitry of a linear operational amplifier.
4. Describe the physical construction of a bipolar transistor.
5. Explain how the elements of a bipolar transistor are biased.
6. Identify the common circuit configurations of a bipolar transistor.
7. Explain the construction differences between bipolar and unipolar transistors.
8. Describe the physical construction of a unipolar transistor.
9. Identify the elements, schematic symbols, and circuitry of different field-effect transistors.
10. Analyze the ID/VDS characteristics of a JFET and a MOSFET.

IMPORTANT TERMS

In the study of electronic amplifiers one frequently encounters a number of new and somewhat unusual terms. These terms play an important role in the presentation of this material. As a result, it is very helpful to review these terms before proceeding with the chapter.

Active component: An electronic device that has gain, such as a transistor or operational amplifier.

Amplification: The process of increasing the current, voltage, or power component of an electronic signal.

Amplifier: A device that increases the current, voltage, or power of an electronic signal.

Base: A thin layer of semiconductor material between the emitter and collector of a bipolar transistor.

Bipolar: A type of transistor with two p-n junctions. Current carriers have two polarities, including holes and electrons.

Channel: The controlled conduction path of a field-effect transistor.

Closed loop: A completed path for signal feedback between the output and input of an operational amplifier.

Collector: A section of a bipolar transistor that collects majority current carriers.

Depletion mode: A field-effect transistor conduction function that reduces the number of current carriers passing through its channel.

Drain: The output terminal of a field-effect transistor.

Emitter: A semiconductor section that is responsible for the release of majority current carriers.

Enhancement mode: A conduction function where current carriers are pulled from the substrate into the channel of a MOSFET.

Gain: A ratio of the output voltage, current, or power to the input voltage, current, or power of a circuit or electronic device.

Gate: The control element of a field-effect transistor.

JFET: The abbreviation for junction field-effect transistor.

Majority current carrier: Electrons of an n-material and holes of a p-material.

Minority current carrier: A conduction vehicle opposite to the majority current carrier, which has holes in an n-material and electrons in the p-material.

MOSFET: The abbreviation for metal-oxide-semiconductor field-effect transistor.

Noninverting amplifier: An amplifier circuit that keeps the input and output signals in phase.

Op amp: The abbreviation for operational amplifier.

Operational amplifier: An integrated circuit amplifier with high input impedance and low output impedance that will achieve many mathematical operations.

Pinch-off voltage: The voltage of a field-effect transistor that stops the flow of current carriers passing through the channel.

Source: A field-effect transistor terminal that is responsible for current carrier injection into the channel.

Stage of amplification: A transistor or IC and all the components needed to achieve amplification.

Transconductance: A field-effect transistor ratio showing the change in drain current caused by a change in gate–source voltage. The unit of transconductance is the siemens (S).

Unipolar: A semiconductor device that has only one p-n junction and one type of current carrier. Field-effect transistors are unipolar devices.

V-MOS: The abbreviation for vertical-grooved metal-oxide-semiconductor construction. This type of construction is used in one group of field-effect transistors.

INTRODUCTION

Electronic instruments must perform a variety of basic functions in order to accomplish a particular operation. An understanding of these functions is essential to understanding the operational theory of the instrument. In this chapter we investigate the amplification function. Amplifiers are devices that produce a change in signal amplitude.

Amplification refers to the process of accepting a weak signal and increasing its amplitude to a higher level. In general, amplifiers employ an active device such as a transistor, vacuum tube, or integrated circuit. The amount of amplification achieved by the active device is called *gain*. Gain is the ratio of the output signal amplitude to the input signal amplitude. Ideally, the output signal should only reflect a change in signal amplitude. Improper component selection, voltage

values, or circuit design, however, may cause the output to be somewhat distorted. The amount of distortion permitted is largely determined by the application of the amplifier.

For an active device to achieve gain it must have a source of operating energy and a signal to process. Direct current is used as the operating energy source for this type of device. Amplifiers in portable instruments usually receive their operating energy from chemical cells or batteries. Rectifier power supplies provide operational energy for most of the electronic instruments used in industry today.

When a signal is applied to an amplifier it causes the operating energy source to change somewhat. A small change in signal strength will, for example, cause a rather significant change in the output operating energy. The signal source may be either ac or dc and still be processed through the amplifier.

INTEGRATED CIRCUIT AMPLIFIERS

Within the last few years integrated circuits, or ICs, have practically taken over all small-signal linear amplifier applications in electronic instruments. This type of active component contains a large number of transistors, diodes, and resistors built into a single unit. A device of this type simply requires input connections, output terminals, and an energy source. When an electrical signal is applied to the input, an amplified version of the signal appears at the output. The circuitry of this device is practically all self-contained. Figure 3-1 shows the manufacturer's data sheet for a typical IC, with a schematic diagram of its internal components, and pin connection diagrams. This particular IC is housed in four distinct kinds of packages.

The triangle-shaped symbol is commonly used to indicate the amplification function. The point of the triangle is the output and the flat side with two leads is the input. The plus sign indicates the noninverting input and the negative sign indicates the inverting input.

The internal structure of an actual IC is quite small and very complex. It cannot be repaired when it fails to operate properly. The defective IC is simply removed from the circuit and replaced. Testing is achieved by observing the amount of gain achieved at the output of the device with respect to its input.

A large number of ICs being used today are called *operational amplifiers* or simply *op amps*. The LM741 of Fig. 3-1 is an amplifier of this type. The gain of this op amp can be controlled externally by connecting feedback resistors between the output and the input. A number of different amplifier applications can also be achieved by selecting different feedback components and combinations.

Operational Amplifiers/Buffers

LM741/LM741A/LM741C/LM741E operational amplifier

general description

The LM741 series are general purpose operational amplifiers which feature improved performance over industry standards like the LM709. They are direct, plug-in replacements for the 709C, LM201, MC1439 and 748 in most applications.

The amplifiers offer many features which make their application nearly foolproof: overload pro-

tection on the input and output, no latch-up when the common mode range is exceeded, as well as freedom from oscillations.

The LM741C/LM741E are identical to the LM741/LM741A except that the LM741C/LM741E have their performance guaranteed over a 0°C to +70°C temperature range, instead of −55°C to +125°C.

schematic and connection diagrams (Top Views)

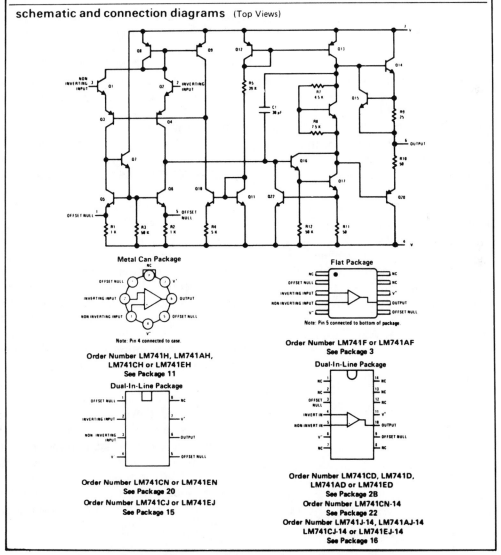

FIGURE 3-1 Schematic and connection diagrams for a typical IC. (Courtesy of National Semiconductor Corp.)

Figure 3-2 shows an LM741 IC connected without feedback resistors. When connected in this manner the IC may achieve a gain of 100 000. This type of circuit construction represents the open-loop characteristic of an op amp. When the input voltage of this amplifier is zero the output voltage will also be zero. If an input signal is applied, the output signal will rise to a value not exceeding the source voltage. If the output rises to the source voltage the op amp is said to be *saturated*. An input signal of only a few millivolts is needed to cause an op amp to reach saturation.

A closed-loop circuit configuration of an op amp connected as a noninverting amplifier is shown in Fig. 3-3. In this configuration, the gain of the amplifier is determined by the resistance ratio of R_1 and R_f. Gain or voltage amplification of a noninverting IC op amp is expressed by the formula

$$A_V = \frac{R_1 + R_f}{R_1}$$

where A_V is the voltage amplification, R_1 is the input resistance, and R_f is the feedback resistance.

If the input resistance of this op amp is 1000 Ω and the feedback resistor is 99 000 Ω, a voltage gain of 100 can be achieved. With a device of this type the circuit designer can simply select a combination of resistors that will permit a desired amount of amplification to be achieved.

A single IC amplifier and its associated components is commonly called a *stage* of amplification. Each amplifier stage is therefore designed to achieve a certain amount of gain. When two amplifier stages are connected together so that the output of the first feeds the input of the second, more gain is achieved. The total gain of a two-stage amplifier is then the product of the individual stage gains. In amplification applications that necessitate a great deal of gain, two or

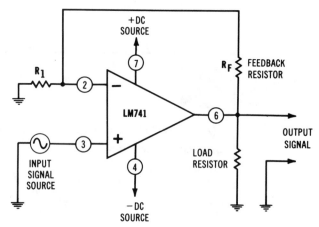

FIGURE 3-3 Closed-type IC op-amp circuit.

more IC stages may be connected together. Amplifiers of this type are often described as being *cascaded*. A number of IC designs are available today that contain four or more op amps similar to the LM741 in a single package. Extremely high gain capabilities and multifunction applications can be easily achieved with a single IC chip of this type. In the future, practically all small-signal amplifiers will be built on single-IC chips.

BIPOLAR TRANSISTOR AMPLIFIERS

A single-bipolar-transistor amplifier is somewhat more complex to use and understand than is an IC op amp. The op amp simply uses an input, an output, and feedback loop resistance to produce amplification. Such things as forward and reverse biasing, coupling, element current, and component selection are built into the IC structure and cannot be altered. In a discrete bipolar transistor amplifier, all of these things are directly influenced by external component selection. Component selection is therefore very critical in the operation of this type of amplifier. The internal structure of a bipolar transistor is, however, rather simple compared with that of an IC op amp (see Fig. 3-4).

In the following discussion of bipolar transistor amplifiers, a great deal of the crystal theory of the diode is directly applicable to the transistor. To avoid repetition of diode theory, we approach the subject from a slightly different point of view using the same general ideas. Hopefully, this will reinforce the diode theory and present some new transistor theory in a slightly different manner.

Crystal Structure

A large majority of bipolar transistors in use today are made by starting with a pure form of silicon. Impurity elements added to silicon when it is being processed

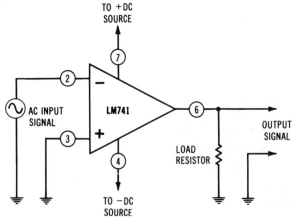

FIGURE 3-2 Op amp connected for open-loop operation.

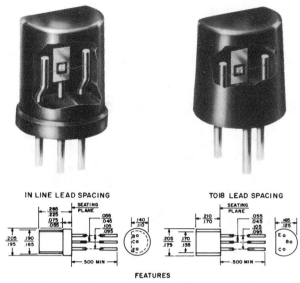

FEATURES

- MONOPLASTIC SOLID CONSTRUCTION FOR NPN AND PNP SILICON PLANAR PASSIVATED TRANSISTORS.
- PATENTED ASSEMBLY AND MOLDED ENCAPSULATION.
- CAPABILITY TO WITHSTAND EXTREME LEVELS OF MECHANICAL ACCELERATION, SHOCK AND VIBRATION.
- DEMONSTRATED MOISTURE RESISTANCE UNDER LONG TERM HUMIDITY AND TEMPERATURE.
- TEMPERATURE CYCLING CAPABILITY DEMONSTRATED OVER 300 CYCLES.
- PROVED LONG TERM OPERATING LIFE RELIABILITY.
- LEAD SOLDERABILITY.

FIGURE 3-4 Internal workings of a plastic encapsulated bipolar transistor. (Courtesy of General Electric Co.)

cause it to be transposed into n-type or p-type materials. An n-type material has an abundance of extra electrons in its structure that are not covalently bonded with other atoms. These electrons constitute the majority current carriers in an n-type crystal. In addition to this, there are also a few holes, or electron voids, in the covalent bonding of the structure. In an n-type material, holes constitute the minority current carriers. Figure 3-5A illustrates the electron/hole content of a piece of n-type material.

The p-type material of a transistor is made from silicon with added element impurities such as boron (B), gallium (Ga), or indium (In). These elements have only three electrons in their valence band or outer shell. Mixing a trivalent element with silicon upsets the normal covalent bonding structure. As a result of this added impurity, a p-type of silicon material has an abundance of electron holes that do not take place in covalent bonding. These holes constitute the majority current carriers of a p-type crystal. Any extra or free electrons that exist in the p-type material constitute the minority current carriers. Figure 3-5B illustrates the electron–hole content of a piece of p-type material.

The general classification *bipolar transistor* refers to a device that employs two diode junctions formed into a sandwich-like structure. Transistors of this type are made of npn or pnp crystal formations. In this type of structure, the crystals are laminated

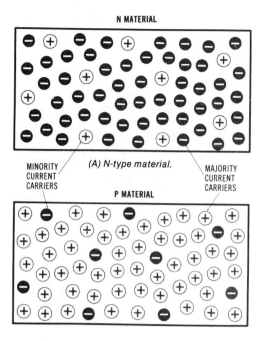

FIGURE 3-5 Electron–hole content of n- and p-type materials.

together into a permanent unit. Figure 3-6 shows these two structures with the element names normally associated with the respective n- and p-type materials.

Current and Biasing

To simplify our understanding of the electron/hole structure of transistors, let us think of the bipolar transistor as a three-section "egg crate." Figure 3-7 illustrates the comparison. Consider the eggs as electrons (n-type charges) and the empty spaces as holes (p-type

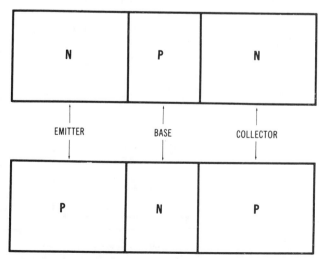

FIGURE 3-6 Element names and structures of npn and pnp transistors.

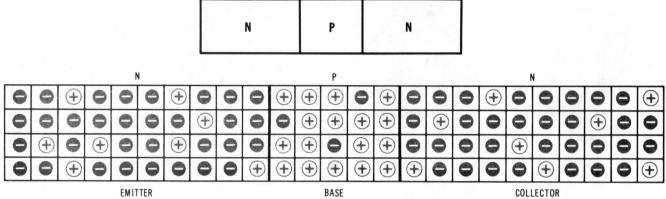

FIGURE 3-7 "Egg-crate" concept of the electron–hole structure of transistors.

charges). Notice that the two n-type sections have many eggs (electrons) but very few holes. The p-type section in the middle is the opposite; it has many holes but very few electrons. If someone took several electrons from the n-type section (marked emitter) and placed them in the p-type section (marked base), what would happen? Moving the electrons from the emitter to the base would increase the number of holes in the emitter and reduce the number of holes in the base. So, in effect, what would happen is this:

1. Electrons would move from emitter to base.
2. Holes would move from base to emitter.

The purpose of this comparison is to make sure we realize that holes and electrons move in opposite directions in a transistor, as shown in Fig. 3-8. In the explanations that follow, the egg-crate idea will make it easier to visualize the movements of holes and electrons.

Figure 3-9 shows that the electrons (negative charges) are attracted to the positive terminal and on into the battery. For every electron that leaves the crystal, another electron enters from the negative terminal of the battery, as shown in Fig. 3-10.

Each electron that goes into the battery leaves a hole, as shown in Fig. 3-11. These and all other holes

in the crystal are attracted to the negative terminal. The holes do not flow into the battery—they move only inside the crystal. As the holes arrive at the negative terminal of the crystal (Fig. 3-12) they are refilled by the electrons coming into the crystal from the negative terminal of the battery.

Reversing the battery, as in Fig. 3-13, reverses the direction of movement of holes and electrons in the crystal. The electrons will, of course, still move toward the positive terminal, and the holes will move toward the negative terminal. Reversing the battery does not, however, have any effect on the amount of current in the circuit. This statement is valid only if we are considering one crystal element. This is a very important point in our study of transistor operation. Remember, reversing a battery connected to a single crystal element does not change the amount of current—only its direction. A current-controlling effect is realized only when two or more crystal elements are used together.

As we have learned, a bipolar transistor has three semiconductor (crystal) elements: emitter, base, and collector. All bipolar transistors have these three elements, whether they are npn or pnp types. To get an idea of how these semiconductor elements control current, let us cut the transistor in half and consider each

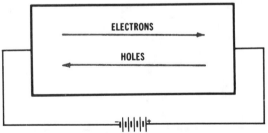

FIGURE 3-8 Movement of electrons and holes in a semiconductor.

FIGURE 3-9 Electrons attracted to a positive battery terminal.

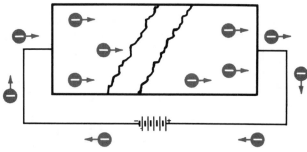

FIGURE 3-10 Electrons entering crystal balance electrons leaving crystal.

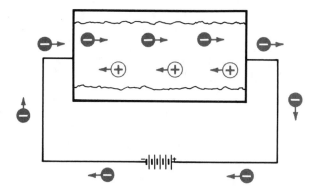

FIGURE 3-12 Movement of electrons and holes in a single crystal.

half separately. When this is completed, we will put the halves together and apply what we have learned to the transistor as a whole.

Figure 3-14 shows an npn transistor cut in half. Each half forms a crystal diode and is capable of controlling current. Notice that each diode section consists of an n-type and a p-type crystal. As we learned previously, an n-type crystal has a majority of negative charges (more free electrons), and a p-type crystal has a majority of positive charges (more holes). It is this difference that makes the crystal diode act as a unidirectional (one-way) current device.

There are obviously two ways in which a battery can be connected to a diode. The first method results in maximum current. It is called *forward bias*. The second method of connecting the battery causes the diode to act as an open circuit. This is called *reverse bias*. In Fig. 3-14A, forward bias is accomplished by connecting the negative battery terminal to the n-type crystal and the positive terminal to the p-type crystal. This produces maximum current in the diode. Reverse bias is accomplished by connecting the negative battery terminal to the p-type crystal and the positive terminal to the n-type crystal, as shown in Fig. 3-14B. This produces only negligible current in the diode.

In all cases in the following discussion, the laws of attraction and repulsion hold true:

1. Electrons inside the crystal are repelled from the

negative voltage terminal and are attracted to the positive voltage terminal.

2. Holes inside the crystal are repelled from the positive voltage terminal and are attracted to the negative voltage terminal.

Figure 3-15 shows the majority current carriers in both crystals. Let us see what happens when the diode is forward biased.

1. The negative battery voltage pushes electrons in the n-type area toward the p-type area.

2. The positive battery voltage pushes holes in the p-type area toward the n-type area.

3. Because of the force applied by the voltage source, the electrons and holes penetrate the junction between the two crystals.

4. Electrons getting through to the p-type area are quickly attracted to the positive terminal and move through the conductor to the battery.

5. Every electron that moves out of the crystal leaves a hole.

6. Positive voltage pushes these holes into the n-type area.

7. The holes getting into the n-type area are attracted toward the negative terminal.

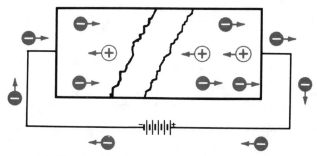

FIGURE 3-11 Holes attracted to negative battery terminal.

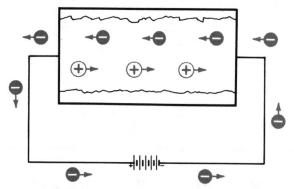

FIGURE 3-13 Effect of battery reversal.

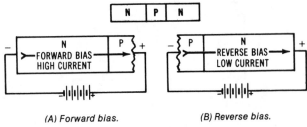

(A) Forward bias. (B) Reverse bias.

FIGURE 3-14 Forward and reverse bias.

8. Electrons from the battery fill the holes arriving at the negative voltage terminal.
9. Note the fact that for every electron leaving the crystal, another one moves into the crystal.
10. It should also be noted that current consists of majority current carriers.

The majority current carriers offer a good path for current. Since there is a large number of current carriers completing the path through the diode junction, a large current occurs.

Now let us examine the other half of the transistor—the pn diode of Fig. 3-16. Notice that the negative terminal of the battery connects to the p-type crystal and the positive terminal connects to the n-type crystal. This is a reverse-bias connection. With reverse bias, the minority current carriers in each crystal are forced to the center of the diode. Therefore, only the majority current carriers can penetrate the junction to get to the opposite side.

1. The minority current carriers, or electrons, in the p-type area penetrate the junction to get into the n-type area.
2. The minority current carriers, or holes, in the n-

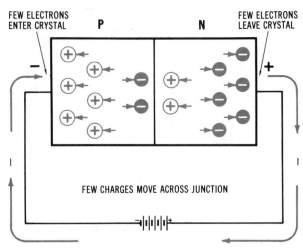

FIGURE 3-16 Electron and hole movement in a reverse-biased diode.

type area penetrate the junction to get into the p-type area.
3. Only a small number of current carriers complete the path between both crystals.
4. The few electrons that go from the p-type area into the n-type area move toward the positive terminal.
5. These few electrons go to the battery, leaving a few holes behind.
6. Since only a few electrons have left the crystal, only a few electrons can enter the crystal from the negative voltage side to fill the holes.

The important thing to remember is that with reverse bias only a few current carriers exchange places between the two crystals. These are the minority current carriers. A very minute current occurs due to this action. This means that the reverse-biased junction offers a very high resistance to current.

BIPOLAR TRANSISTOR ACTION

We have discussed how current carriers flow in the individual crystal elements of a bipolar transistor. We have also considered the operation of crystal diodes biased in the forward direction and in the reverse di-

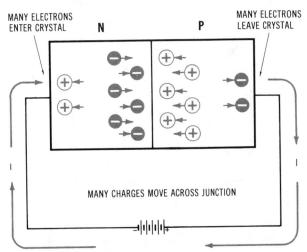

FIGURE 3-15 Electron and hole movement in a forward-biased diode.

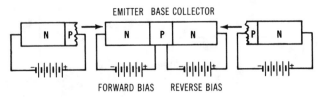

FIGURE 3-17 Forward and reverse bias in a transistor.

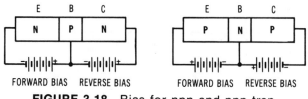

FIGURE 3-18 Bias for npn and pnp transistors.

rection. We are now going to consider what occurs when crystals are combined to form a transistor.

Putting the forward- and reverse-biased diodes back into the transistor, as shown in Fig. 3-17, we notice that the emitter-to-base (np) diode section is forward biased, and the base-to-collector (pn) diode is reverse biased. Transistors are always biased in this manner. This bias method is used for both npn and pnp transistors, as shown in Fig. 3-18. Remember:

1. Emitter-to-base is always forward biased.
2. Base-to-collector is always reverse biased.

Let us now consider how the charges move in an npn transistor. Consider the action of the npn transistor shown in Fig. 3-19. The transistor is correctly biased with forward bias on the emitter-to-base junction and reverse bias on the base-to-collector junction. The charge movement is as follows:

1. The bias voltage on the emitter-to-base combination causes electrons to move toward the base crystal. The base is a much thinner [about 0.001 in. (0.0254 mm)] crystal than either the emitter or collector

crystals. Therefore, since the electrons are moving at a tremendous rate of speed, most of them (actually 95 to 99%) pass through the thin base crystal and go to the collector. The few electrons (from 1 to 5%) that do not penetrate the base are attracted to the positive voltage on the base. These few electrons result in a very small base current.

2. The great number of electrons that go to the collector are attracted to the positive terminal of the battery. These electrons leaving the collector to enter the battery make up the collector current. Every electron that moves out of the collector leaves a hole that is forced in the opposite direction by the positive collector voltage. The holes penetrate the thin base crystal and go to the emitter. Every hole that reaches the emitter is filled by another electron, which proceeds from emitter to collector.

The idea of the current-controlling action of a bipolar transistor should now become apparent. In other words, if the base is made more positive, the collector current increases. If the base is made less positive, the collector current decreases. The most important thing to remember about the npn transistor is that the forward-biased emitter–base junction controls the amount of collector current. A little later we will see how a weak signal voltage can be superimposed on the forward-bias voltage. Transistor action then produces an amplified signal in the collector circuit.

Figure 3-20 shows the charge movement in a pnp transistor. The principles of operation for the npn and pnp transistors are basically the same; the major difference is in the crystal arrangements. The charge movement in a pnp transistor is as follows:

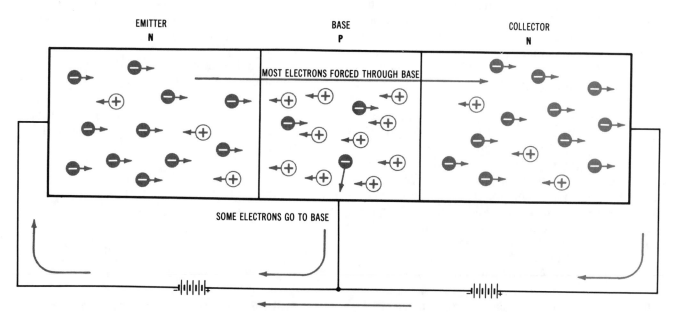

FIGURE 3-19 Charge movement in an npn transistor.

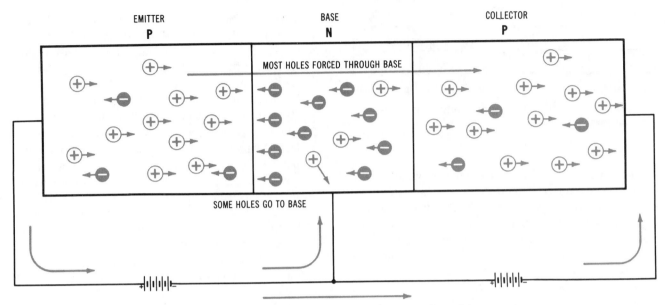

EMITTER
P

BASE
N

COLLECTOR
P

MOST HOLES FORCED THROUGH BASE

SOME HOLES GO TO BASE

FIGURE 3-20   Charge movement in a pnp transistor.

1. Forward bias on the emitter–base junction causes holes in the p-type material to move toward the base. Most of the holes penetrate the thin base crystal and enter the collector area.

2. As the holes arrive in the collector area, they are filled by electrons coming from the negative collector terminals. These electrons, because of the positive potential on the emitter, move through the thin base crystal and proceed to the emitter. There are also a few electrons coming from the negative base potential to fill the few holes that didn't get from emitter to collector.

3. All the electrons that go to the emitter are attracted to the positive emitter terminal and flow into the battery. Every electron that moves out of the emitter leaves a hole behind. All the holes left behind move from the emitter and most of them pass through the base crystal to the collector. A very few move to the base terminal.

Considering the foregoing explanation, it should be apparent that the major difference between npn and pnp transistor operation is in the current carriers. In the npn transistor the majority current carriers are electrons, holes and electrons in their respective crystal material. In the pnp transistor, the majority current carriers are holes, electrons, and holes in their respective crystal material. Again, the important point to remember is that the forward bias on the emitter–base junction controls the collector current. Thus a small-signal voltage connected in series with the emitter–base junction produces an amplified signal at the collector. We go into this action in more detail when we analyze the operation of a basic amplifier circuit.

BIPOLAR TRANSISTOR CIRCUITS

Figure 3-21 shows the two schematic symbols used for bipolar transistors—one is for the npn and the other is for the pnp transistor. Both symbols show the emitter, base, and collector. The collector and emitter are drawn at an angle to the base in both symbols. Notice, however, that the emitter is designated by the use of an arrow, but the collector does not have an arrow. The direction of the emitter arrow points away from the base in the npn transistor, but toward the base in the pnp transistor. Remember that the emitter arrow always points away from the direction of electron flow.

Now that we have learned how current moves in a transistor, we shall consider the three basic bipolar transistor circuit arrangements.

1. The common-emitter circuit of Fig. 3-22A has a common connection between the base input and the collector output.

2. The common-base circuit of Fig. 3-22B has the base commonly connected to the emitter input and the collector output.

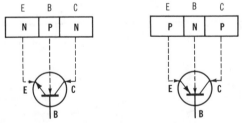

FIGURE 3-21   Schematic symbols for bipolar transistors.

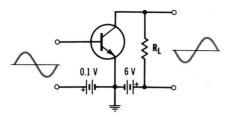

(A) Common-emitter circuit.

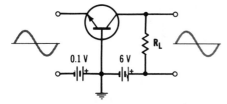

(B) Common-base circuit.

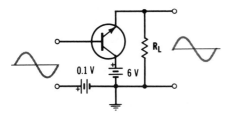

(C) Common-collector circuit.

FIGURE 3-22 Signals in common-emitter transistor circuit.

3. The common-collector circuit of Fig. 3-22C has the collector commonly connected to the base input and the emitter output.

To illustrate how each of the three types of transistor circuits works, we will discuss each one separately. The circuits of Fig. 3-22 will be used. The input to each circuit will be achieved by a transformer.

Figure 3-23 might cause some confusion if you recall that the base-to-collector junction must be reverse biased. Actually, if the schematic is studied, it will be noticed that the base-to-collector junction really is reverse biased at a value of −5.9 V. This can be understood by visualizing the +6 V of the collector battery as appearing at the collector, while the +0.1

V of the bias battery appears at the base. The difference between the two voltages is 5.9 V. The base is negative with respect to the collector, as required for the reverse-bias condition.

In the steady-state condition, the positive 0.1 V on the base will cause a certain amount of current to move from the emitter to the collector. The actual amount of this current will depend on the characteristics of the transistor involved. We know from our previous discussion of transistor action that most of the current passes to the collector and very little leaves through the base. The current in the collector circuit passes through the load resistor on its way back to the battery. This current through R_L produces a voltage drop across the resistor. The output for this circuit in Fig. 3-23 is taken between the collector and the emitter.

Consider now how a signal applied to the base affects the circuit. As the input signal rises in the positive direction, the signal voltage causes the transistor to conduct more. When the input signal drops in the negative direction, the signal subtracts from the +0.1-V bias voltage. This causes the base to become less positive with respect to the emitter; therefore, fewer electrons (less current) move from emitter to collector. This changing collector current will develop a signal across R_L. This amplified output is coupled through C_C to the output.

We recall that in the common-emitter circuit, it was difficult to detect the presence of reverse bias on the base-to-collector junction. In Fig. 3-24, the common-base circuit, the reverse-bias condition is quite obvious. In this circuit, during the steady-state condition shown in the diagram, the +0.1-V forward bias on the base–emitter junction would produce a small amount of current. Here, as before, the magnitude of this current would be dependent on the characteristics of the transistor.

When an ac signal voltage is applied to the circuit, the positive alternation subtracts from the 0.1-V battery potential. The emitter, therefore, becomes less negative, or more positive. This reduces the forward-bias voltage, and the transistor current decreases, resulting in an increase in the output voltage. The emitter becomes more negative during the negative alternation

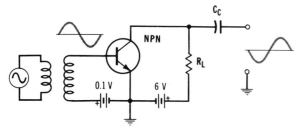

FIGURE 3-23 Signals in common-emitter transistor circuit.

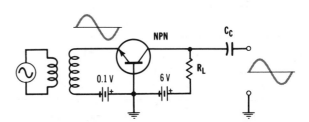

FIGURE 3-24 Signals in common-base transistor circuit.

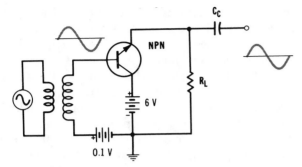

FIGURE 3-25 Signals in common-collector (emitter-follower) transistor circuit.

of the input signal. This increases the collector current. A corresponding decrease in the output signal results.

In the common-collector circuit, we have a bias condition similar to that used in the common-emitter circuit. Notice in Fig. 3-25 that the reverse bias on the base-to-collector junction is accomplished by the +0.1-V battery making the base 5.9 V less positive (negative) with respect to the collector. Thus the reverse-bias condition is accomplished. In this circuit the input is applied to the base and the output is taken from the emitter. The collector is the common element. In other words, it is associated with both the input and output circuits.

During the positive portion of the input signal, the forward bias between the emitter and base increases, producing more current in the transistor. This current moves through the load resistor (in the emitter circuit), causing an output to be developed. When the input signal swings into the negative region, it subtracts from the forward bias on the base-to-emitter junction. This decreases the amount of current in the transistor; consequently, less voltage is developed across the load resistor.

In the common-collector configuration, the output signal is less than the input signal. The output also follows the input in all respects. In this regard, a common-collector circuit is frequently called an emitter-follower amplifier. Circuits of this type are normally used to achieve impedance matching.

FIELD-EFFECT (UNIPOLAR) TRANSISTORS

Field-effect transistors (FETs) represent the second major type of discrete solid-state amplifying devices. These devices, unlike their bipolar counterparts, conduct current in a single piece of semiconductor material known as a channel; thus they are also known as *unipolar transistors*. Junction FETs and metal-oxide-semiconductor FETs are the two general classifications of FETs. Element names and theory of op-

eration are significantly different than the bipolar type of transistor.

The junction field-effect transistor (JFET) is a modification of the bipolar transistor. Structurally, JFETs conduct current through a single piece of semiconductor material called a *channel*. An additional piece of semiconductor material attached to the side of the channel is called a *gate*. Current conduction through the channel is controlled by reverse biasing on the gate. Figure 3-26 shows representative crystal types, element names, and schematic symbols of the junction field-effect transistor.

A simple JFET amplifier circuit is shown in Fig. 3-27. In this circuit, the positive side of the battery, V_{DD}, is connected to the drain through R_D, and the negative side is connected through R_S. With these two connections made, current easily passes through the channel. A voltage drop across R_S makes the source electrode slightly positive with respect to the negative battery terminal.

When resistor R_G is attached to the negative side of V_{DD}, it reverse biases the gate–source junction. Action of this type is used to control conduction of the channel current. A positive signal applied to the input would reduce this bias voltage and cause increased channel current conduction. A negative signal would add to the reverse biasing of the gate and cause less channel conduction. Ac signals applied to the gate will

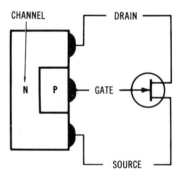

(A) N-channel JFET.

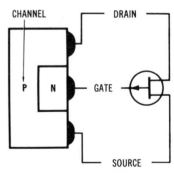

(B) P-channel JFET.

FIGURE 3-26 Junction field-effect transistors.

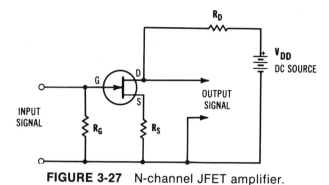

FIGURE 3-27 N-channel JFET amplifier.

cause a constant variation of bias voltage above and below a certain operating point. The output would then reflect this as an I_D that changes in step with the ac input signal.

When the gate bias voltage of a FET becomes high enough it will completely stop the drain current, I_D. The term *pinch-off voltage* is commonly used to describe this operating condition. This condition corresponds to the reverse-biasing action of the base–emitter junction of a bipolar transistor.

Since the gate of a FET responds very quickly to changes in voltage, it is classified as a voltage-sensitive device. This represents a unique feature of the FET compared with bipolar devices, which are con-

sidered current sensitive. Because of this feature the n-channel FET has operating characteristics that are very similar to those of a vacuum tube.

JFET Characteristic Curves

A family of JFET characteristic curves is shown in Fig. 3-28. The *x*-axis of this graph shows the voltage applied between drain and source as V_{DS}. The *y*-axis of the graph is used to display the drain current, I_D, in milliamperes. The pinch-off voltage of this device is -7 V dc. Operation of this particular JFET at gate control voltages between $V_{GS} = 0$ V and $V_{GS} = -5$ V is permissible.

The gain of a FET is commonly described by the term *transconductance*, which has the letter symbol g_m. This term refers to the input/output relationship of the device. A change in V_{GS}, for example, causes a corresponding change in output current, I_D. The formula $g_m = I_D/V_{GS}$ permits easy calculation of transconductance.

Assume now that the transconductance of a JFET amplifier having the characteristic curves of Fig. 3-28 is to be determined. If the V_{DS} appearing across the device is held at 8 V, this will serve as a point of reference. A change of V_{GS} from -1 to -2 V along the reference line would cause a corresponding change

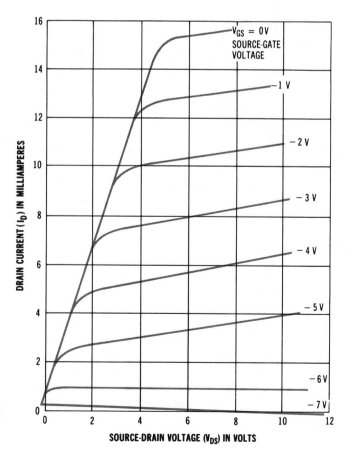

FIGURE 3-28 Family of I_D/V_{DS} characteristic curves for an n-channel JFET.

in I_D of 13 to 10.5 mA. Entering these values into the transconductance formula will show that

$$g_m = \frac{I_D}{V_{GS}}$$

$$= \frac{13 \text{ to } 10.5 \text{ mA}}{-1 \text{ to } -2V}$$

$$= \frac{0.0025}{1}$$

$$= 2500 \ \mu S$$

The transconductance of a JFET is a measure of the ease with which current carriers pass through the channel. Conductance, in this case, is the reciprocal of resistance, or $G = 1/R$. The unit of conductance is the *siemens*. The ease with which current carriers ease through a FET is, therefore, measured in siemens or, more commonly, microsiemens. This same term is used to show the gain of a vacuum tube.

MOS Field-Effect Transistors

Metal-oxide-semiconductor FETs (MOSFETs) are a variation of the unipolar transistor family. MOSFETs are designed so that the gate electrode is insulated from the channel by a layer of metal oxide. Reverse biasing of the gate–source electrodes causes a depletion of drain current carriers in the channel, like that of the JFET. The insulated-gate feature of this device can also permit an increase in drain current to take place when appropriate bias voltages are applied. This type of operation is described as the *enhancement mode* of operation. This condition of operation occurs when the gate is forward biased. JFETs, by comparison, cannot be operated when the gate is forward biased because it would cause a large gate current. With the insulated-gate characteristic of the MOSFET, the polarity of the gate bias voltage is not a problem. Figure 3-29 shows representative crystal types, element names, and schematic symbols for the MOSFET.

Figure 3-30 shows the circuit diagram of an n-channel MOSFET amplifier. This circuit is very similar to that of the JFET amplifier of Fig. 3-27. Resistors R_{G1} and R_{G2} form a voltage divider across the power source to bias the gate. The ratio of these two resistors can be altered to provide different values of V_G voltage according to the demands of the circuit application. This resistance ratio can be altered to make the gate positive for forward biasing in the enhancement mode or negative for reverse biasing in the depletion region.

Figure 3-31 shows a drain family of characteristic curves for an enhancement–depletion type of MOSFET. Note that this device can be used in either a positive or negative gate-biasing application. Trans-

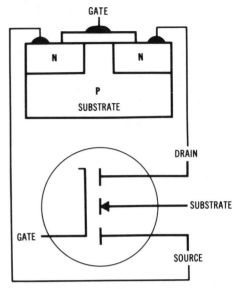

(A) N-channel enhancement MOSFET.

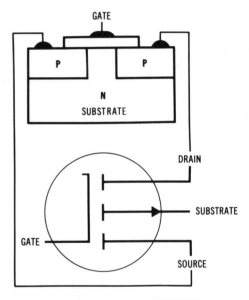

(B) P-channel enhancement MOSFET.

FIGURE 3-29   Metal-oxide semiconductor field-effect transistors.

conductance calculations and circuit operations of the MOSFET and JFET are practically identical in all respects.

V-MOS Field-Effect Transistors

A new generation of the field-effect transistor is finding its way into solid-state active device technology. This type of device is largely used to replace the older bipolar transistor. Power FETs are now being used in power supplies and solid-state switching applications. Collectively, these devices are called V-MOSFETs or V-MOSs. The "V" designation finds its origin in ver-

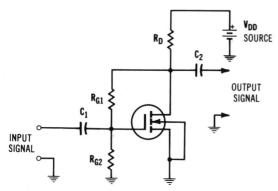

FIGURE 3-30 N-channel enhancement-mode MOSFET amplifier.

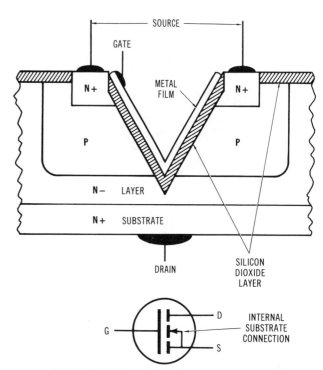

FIGURE 3-32 Cross-sectional view of an n-channel V-MOSFET.

tical-groove MOS technology. The device is constructed with a V-shaped groove etched into the substrate. Construction of this type requires less than a horizontally assembled device. The construction of a V-MOS device permits better heat dissipation and a high density of material in the channel area.

Figure 3-32 shows a cross-sectional view of the crystal structure, element names, and schematic symbol of an n-channel V-MOS transistor. Notice that a V-groove is etched in the surface of the structure. From the top, the V-cut penetrates through the n^+, p, and n^- layers and stops near the n^+ substrate. The two n^+ layers are heavily doped, and the n^- layer is lightly doped. A thin layer of silicon dioxide covers both the horizontal surface and the V-groove. A metal film deposited on top of the groove serves as the gate. The gate, therefore, is insulated from the groove. The source leads on each side of the groove are connected internally. The bottom layer of n^+ material serves as

a combined substrate and drain. Current carriers move vertically between the source and drain.

A V-MOS transistor is classified as an enhancement type of MOSFET. No current carriers exist in the source and drain regions until the gate is energized. An n-channel device, such as the one in Fig. 3-32, does not conduct until the gate is made positive with respect to the source. When this occurs, an n channel is induced between the two n^+ areas near the groove. Current carriers can then flow through the vertical channel from source to drain. When the gate of an n-channel device is made negative, no channel exists and the current carriers cease to flow.

SMALL-SIGNAL AMPLIFIERS VERSUS LARGE-SIGNAL AMPLIFIERS

Practically all of the active devices that will achieve amplitude control can be classified as either small-signal or large-signal amplifiers. This particular classification indicates the location of the amplifier in the system. Initially, the input signal of a system is quite small and requires a great deal of amplification. Small-signal amplifiers are therefore purposely designed to accept a weak signal and amplify it accordingly. After the signal has been amplified rather significantly, it can then be used to control larger amounts of current that are fed into the load device. Large-signal amplifiers are, therefore, designed to achieve this function. These de-

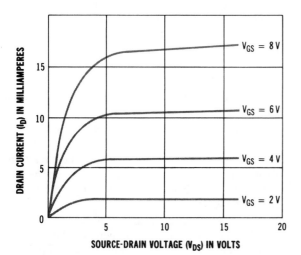

FIGURE 3-31 Family of I_D/V_{DS} characteristic curves for an enhancement–depletion type of MOSFET.

FIGURE 3-33 Internal structure of a power transistor. (Courtesy of General Electric Co.)

Integrated circuits are commonly used to achieve amplification in electronic instruments today. ICs such as the operational amplifier are built on a single IC chip. Without a feedback resistor, gains of 100 000 are common. With a feedback resistor, gain is determined by a resistance ratio of the input and feedback resistors.

Discrete transistor amplifiers must take into account such things as forward and reverse biasing, component selection, and element current. Bipolar transistors have two junctions in an npn or pnp structure. Current is primarily based on the flow of majority current carriers. A conventional connection method has the emitter–base forward biased and the base–collector reverse biased. Transistors can be connected into a circuit that can have any one of the three elements common to the input and output circuitry. Common-emitter, common-base, or common-collector amplifier configurations are in use today.

JFETs and MOSFETs are unipolar transistors. These devices respond to current carriers passing through a crystal channel that has the source and drain electrodes attached to it. Control is achieved by biasing the gate element.

V-MOSFETs are a rather new version of MOS technology that is being used to replace bipolar power transistors. A V-groove is etched in the substrate of this device. Characteristically, its operation is similar to that of an enhancement-mode MOSFET. V-MOS transistors have good heat dissipation because of the greater concentration of material in the V-groove area.

vices are substantially larger in physical size and often include special heat-dissipating connections. Figure 3-33 shows the internal structure of a power transistor. This device is capable of controlling several amperes of collector current as compared with a few milliamperes by the small-signal transistor.

SUMMARY

Amplification is a process in which a small signal is applied to the input of an active device such as a transistor or integrated circuit, and its output amplitude is increased to a higher level.

ACTIVITIES

BIPOLAR TRANSISTOR TESTING

1. Determine the lead polarity of the ohmmeter being used in this activity. Use the $R \times 100$ range if the ohmmeter is an analog instrument. If a digital meter is used, set the ohmmeter range to the 2-kΩ position.

2. In the first part of this procedure the ohmmeter is used to identify a transistor as pnp or npn. The ohmmeter's voltage source is used to bias the transistor junctions. A junction will indicate low resistance when forward biased and high resistance when reverse biased. The base lead of a transistor will show low resistance between the emitter and the collector when the ohmmeter is alternately switched between these two leads. For a pnp transistor the negative ohmmeter lead goes to the base, and the positive lead will show low resistance when it is switched between the emitter and collector. For an npn transistor the positive ohmmeter lead goes to the base and the negative lead will show low resistance when switched between the emitter and collector.

3. Select a transistor and identify its polarity as npn or pnp. Note the location of the center material of each transistor. This is the base of a bipolar transistor. The two outside material leads represent the emitter and the collector.

4. Use this test procedure to identify the polarity of a transistor's material.

5. This identification procedure applies only to good transistors.

6. Apply the identification procedure to several transistors.

7. Another test procedure that can be applied to a transistor determines if there is leakage between the outside leads. These leads are assumed to be the emitter and the collector.

8. Connect the ohmmeter leads to the two outside material leads of the transistor. In general, there should be no leakage current or resistance in either polarity direction that the ohmmeter is connected. This is an indication of a good transistor.

9. Apply the leakage test to several transistors.

JUNCTION FIELD-EFFECT TRANSISTOR TESTING

1. Determine the lead polarity of the ohmmeter being used in this activity. Use the $R \times 100$ range if the ohmmeter is an analog instrument. If a digital meter is used, set the ohmmeter range to the 2-kΩ position.

2. In the first part of this procedure the ohmmeter is used to identify the leads of a JFET. The ohmmeter's voltage source is used to bias the channel, the gate–source junction, or the gate–drain junction. The leads can be identified by the response produced by the materials. The source and drain will show the same resistance in either polarity direction of the ohmmeter. The gate–source and gate–drain both respond as a diode. They will show low resistance in the forward direction and high or infinite resistance in the reverse direction. The polarity of the ohmmeter is used to identify the device as p- or n-channel.

3. Select a JFET to be tested. Make a sketch of the bottom view of the transistor lead layout.

4. Determine the source (S) and drain (D) leads of the transistor by connecting the ohmmeter between two randomly selected leads. If the two selected leads indicate a resistance, reverse the ohmmeter probes. Two leads that show the same resistance in either direction of polarity are the source and the drain. The remaining lead is the gate. Measure and record the source–drain resistance of the device being evaluated.

5. The gate–channel of a JFET responds as a diode. This can be observed by connecting the ohmmeter between two leads and then reversing the ohmmeter leads. The gate–source or gate–drain will respond as a diode by showing low forward resistance and high or infinite reverse resistance.

6. If the gate shows low resistance when the positive probe is connected to the gate and the negative to either the source or drain, the JFET is an n-channel device. A p-channel device will produce low resistance when the gate is negative and the source or drain is positive.

7. An approximate test of the JFET's ability to amplify is displayed by connecting the ohmmeter to the source and drain. The polarity of the ohmmeter is not significant. Touch the gate and either the source or drain with your finger. This should cause a decrease in resistance of the channel. A more significant resistance change occurs when a 10-kΩ resistor is connected between the gate and either the source or drain. Very little or no resistance change during the test indicates an open condition or improper lead identification.

8. Apply this test procedure to several JFETs.

NONINVERTING OPERATIONAL AMPLIFIER

1. Construct the noninverting operational amplifier of Fig. 3-3. The circuit needs to be energized by a split or divided power supply. Use two 9-V radio batteries. The common or interconnected battery terminal of the power supply is connected to the ground symbol of the circuit. Resistor R_1 is 10 kΩ, R_F is 100 kΩ, and the load resistor is 1 kΩ.

2. Apply an ac signal to pin 3. Adjust the ac input source to 1.0 $V_{p\text{-}p}$. With an ac voltmeter, measure and record the ac output voltage developed across the load resistor.

3. Calculate the voltage gain (A_v) for this circuit. How closely does the calculated voltage gain compare with the measured voltage gain?

4. Adjust the ac input to produce 0.1 $V_{p\text{-}p}$. Measure and record the ac output voltage developed across the load resistor. How closely do the calculated and measured voltage values compare?

5. If an oscilloscope is available, turn it on and prepare it for operation. Connect the vertical probe to pin 3 and the common lead to the ground. Make a sketch of the observed input signal, noting its phase and peak-to-peak voltage value. Move the vertical probe to the output (pin 6) of the op amp. Make a sketch of the observed output signal while noting its phase and peak-to-peak voltage value. Describe your findings.

6. Increase the value of the input signal while observing the output. Watch for the input value that causes the output to be distorted. What does this represent?

QUESTIONS

1. What are the two inputs of an op amp called?

2. What is meant by the term *amplification*?

3. What is a single IC amplifier and all of its associated components commonly called?

4. What is meant by the term *bipolar transistor*?

5. How is a bipolar transistor biased for operation?

6. What is meant by the term *common* in common-emitter, common-base, and common-collector amplifiers?

7. Why is a junction field-effect transistor classified as a unipolar device?

8. What is transconductance?

9. What insulating material separates the gate from the channel of a MOSFET?

10. Which field-effect transistor employs the enhancement mode in its operation?

11. What is a V-MOS field-effect transistor?

12. Explain the difference between majority and minority current carriers.

CHAPTER **4**

OSCILLOSCOPES

OBJECTIVES

Upon completion of this chapter, you will be able to:

1. Draw a block diagram of an oscilloscope.
2. Explain the fundamental operation of an oscilloscope block diagram.
3. Describe the operation of a cathode-ray tube (CRT).
4. Make a list of the vertical deflection controls of an oscilloscope.
5. Explain the functional operation of basic horizontal sweep controls.
6. Define triggering and identify the controls that are used to achieve this operation in an oscilloscope.

IMPORTANT TERMS

In the study of oscilloscopes one frequently encounters a number of new and somewhat unusual terms in common usage. These terms play an important role in the presentation of this material. As a rule, it is very helpful to review these terms before proceeding with the chapter.

Alternate: A vertical operating mode that displays one vertical channel then the other interchangeably. The alternate mode of operation is best suited for sweep rates of 1 ms/division or faster.

Aperture: A small opening or hole in construction of a cathode-ray tube electrode.

Cathode-ray tube: An evacuated glass tube in which electrons emitted from one end cause a visual display to be produced on the viewing area.

Chop: A vertical operating mode that breaks the two channel traces into small segments and switches quickly between the two. The chop operation should be used to display slow sweep rates.

Control: The part of an electronic system that alters an operation. Full control is achieved with a switch that turns a circuit on or off. Variable control is achieved with a device that changes its value gradually.

Deflection: A process that bends or changes the course of an electron beam after it is emitted from the cathode of a cathode-ray tube.

Electron beam: A stream of electrons emitted from the cathode or gun of a cathode-ray tube.

Electron gun: The electron emitting assembly of a cathode-ray tube located near the rear of the tube.

Electrostatic field: A space in which static electricity causes invisible lines of force to exist.

Focus control: An oscilloscope control that causes the electron beam to converge into a fine line or trace on the viewing area.

Graticule: A grid-line scale that appears on the viewing area of a cathode-ray tube and is used for measuring displayed signals.

Grid: A cathode-ray tube element having one or more openings for the passage of electrons or ions.

Holdoff time: The time or space of a horizontal time base waveform that follows the retrace period. Triggering generally occurs during this part of the waveform.

Horizontal deflection: The sweeping action of an electron beam that causes it to move from left to right to left on the viewing area.

Horizontal operating mode: An optional oscilloscope function that permits intensified, delayed, and normal horizontal sweep.

Intensity control: An oscilloscope control that alters the brightness level of the display.

Oscillator: An electronic circuit that changes dc operating power into ac or some type of time-varying signal.

Probe: The input device of an oscilloscope that receives voltage and current signals that will produce or control the visual display.

Ramp time: The linear rise time of a horizontal sweep signal.

Retrace time: The falling or dropping time of a horizontal sweep signal.

Sweep: A process that changes the course of action of the electron beam of an oscilloscope when it produces a visual display.

Synchronization: An oscilloscope operation that permits vertical and horizontal sweep signals to be in step or occur at the same point in time.

Time-base generator: An oscilloscope sawtooth-wave-producing circuit that deflects the electron beam horizontally.

Time/division switch: A control that changes the sweep rate of the time-base generator.

Trace time: A period of time during horizontal sweep of the electron beam where a display appears on the viewing area of an oscilloscope.

Trigger: An oscilloscope signal that initiates the sweep function and determines the beginning point of a trace.

Trigger operating mode: A switch used to select different triggering operations for the time-base generator of an oscilloscope.

Vertical deflection: The process of moving the electron beam of an oscilloscope up and down.

Vertical operating mode: An optional oscilloscope control function that permits a dual-channel instrument to have one, both, add, alternate, or chop channel selection.

X-axis: A graphic representation of the horizontal display operation of an oscilloscope.

Y-axis: A graphic representation of the vertical display operation of an oscilloscope.

INTRODUCTION

The oscilloscope is a unique instrument designed to graphically measure time-varying voltage and current values. This method of measurement permits the operator to actually see voltage and current signal traces instead of viewing the results on a deflection meter or a digital display instrument. In making measurements of this type, the oscilloscope takes very little energy away from the circuit being analyzed. It also responds well to irregularly shaped signal voltages, high frequency, and phase relationships. The oscilloscope is an indispensable tool for instrument calibration, equipment performance evaluation, and troubleshooting procedures.

The primary function of an oscilloscope is to measure time-varying signals and to display these signals so that they can be analyzed. As a general rule, a user must initially prepare this instrument for operation before it can be used to make a suitable display. For the experienced operator, the setup procedure is usually rather easy to achieve. This type of operator generally has a good understanding of essential control functions. To the inexperienced operator, the setup procedure may seem to be confusing and somewhat time consuming, and is generally made by trial-and-error methods. It would be helpful if the inexperienced operator knew something about basic oscilloscope control functions.

The primary purpose of this presentation is to discuss the fundamental operation of an oscilloscope. In this regard, the oscilloscope is first viewed as a series of functional blocks connected together to form an operating system. Presentation of the block structure is an organization procedure that permits the system to be viewed as a functional instrument. We will then take a look at the specific role played by each block in the operating system. Next, we will look at some of the controls and the setup procedure needed to make the oscilloscope operational.

The fundamental blocks or parts of an oscilloscope connect together and cause it to respond as an operational system. In general terms, the system is composed of a display device, the horizontal sweep section, vertical amplification and sweep, triggering, power supply, and probes. As a rule, these functional parts are the same for nearly all oscilloscopes. Figure 4-1 shows these parts in a block diagram.

The blocks of a basic oscilloscope are named according to the specific function that they achieve in the operation of the instrument. The vertical section, for example, controls the Y-axis or vertical part of the display. Any up-or-down motion of the electron beam is controlled by the vertical section. The horizontal section of the basic instrument controls the left to right movement of the electron beam. This part of the instrument produces the X-axis of the display. The trigger section determines the specific point in time where horizontal sweep begins. Triggering is achieved by some type of switching action. The cathode-ray tube (CRT) is ultimately responsible for a graphic display of the signal, the end result of all parts of the system working together. The power supply develops all of the operational voltages needed to energize the circuit

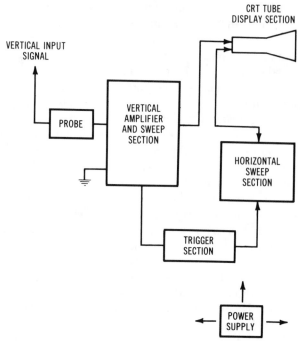

VERTICAL INPUT SIGNAL

PROBE

VERTICAL AMPLIFIER AND SWEEP SECTION

CRT TUBE DISPLAY SECTION

HORIZONTAL SWEEP SECTION

TRIGGER SECTION

POWER SUPPLY

FIGURE 4-1   Functional parts of an oscilloscope.

components. The probe serves as an external input receptacle for the instrument. The combined actions of these functions are needed to make the oscilloscope operational.

CATHODE-RAY TUBE

The cathode-ray tube (CRT) is responsible for the display function of an oscilloscope. Structurally, the CRT is a long evacuated glass tube in which electrons produced at the neck end of the device cause an image to appear on a glass surface in the display area. Electrons are produced, accelerated, and focused in the rear assembly of the tube. This part of the tube is called the *electron gun* or simply the *gun*. Horizontal and vertical deflection plates are located near the center of the neck

area. The electron beam passes through these plates as it moves toward the display area. The large-diameter end of the tube is the display or viewing area. The inside glass face of the viewing area is coated with a phosphorescent material. When the high-velocity electron beam strikes this material it produces a characteristic glow. The CRT, in a strict sense, is an electronic transducer. It changes an invisible electron beam into light energy that is displayed on a glass surface.

Figure 4-2 shows the construction details of a CRT, while Fig. 4-3 illustrates the operation of the electron-gun assembly in the neck of the CRT. A beam of electrons is produced by the cathode of the tube when it is heated by the application of a filament voltage. Electrons emitted from the cathode are initially attracted by the positive potential of anode 1. The quantity of electrons passing toward anode 1 is determined by the amount of negative bias voltage applied to the control grid. Anode 2 is operated at a higher positive potential than anode 1 to further accelerate the electron beam toward the screen of the CRT. High voltage in the display area of the tube serves as the final attracting force for the electron beam. When the electron beam strikes the phosphorescent screen, light is produced.

Control of the electron beam is achieved by a number of different processes. The quantity or number of electrons reaching the display area of the CRT determines its brightness or intensity level. The intensity control, usually connected to the grid, determines the negative voltage value of the grid. The grid is a small cylinder-like structure with the end nearest the cathode open and the other end having a small aperture. The number of electrons forced to pass through the aperture is dependent on the amount of negative voltage on the grid. A high negative voltage will repel large numbers of electrons, thus reducing the level of intensity. Reduced negative voltage increases the quantity of electrons reaching the display area, thus increasing the intensity. The intensity control is generally located on the front panel of the oscilloscope.

The sharpness of the display image or trace is

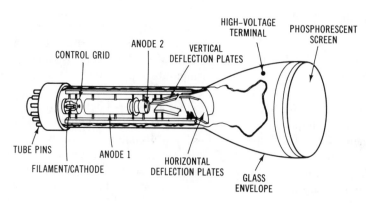

CONTROL GRID
ANODE 2
VERTICAL DEFLECTION PLATES
HIGH-VOLTAGE TERMINAL
PHOSPHORESCENT SCREEN
TUBE PINS
FILAMENT/CATHODE
ANODE 1
HORIZONTAL DEFLECTION PLATES
GLASS ENVELOPE

FIGURE 4-2   Construction details of a cathode-ray tube.

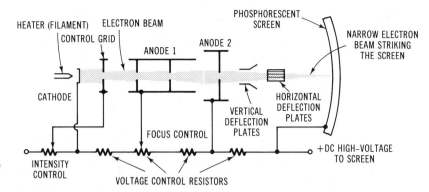

FIGURE 4-3 Cathode-ray-tube gun operation.

determined by the focus control. In a CRT, focus is controlled electrostatically. Electrons emitted from the cathode have a natural tendency to separate or spread apart as they move toward the face of the CRT. Each electron, being of a negative charge will be repelled by its neighboring electrons. This action would normally cause the trace to appear as a large fuzzy ball on the face of the CRT. To alter this condition, focus is achieved by altering the difference in the positive voltage between anodes 1 and 2. As a rule, anode 1 is varied while anode 2 remains at a fixed value. Variations in the positive voltage cause the trajectory angle of the electron beam to change by altering the shape of the electrostatic field. As a result of this adjustment, the point of electron-beam convergence can be altered by changing a voltage value within the tube. The focus control is generally located on the front panel of the oscilloscope. In most oscilloscopes the focus and intensity control adjustments are usually interrelated. Adjustment of one control usually necessitates some adjustment of the other control.

DEFLECTION

The position of an electron beam in the display area of a CRT is controlled by two sets of deflection plates. These plates are housed in the neck of the CRT and are located between the electron gun and the display area. Figure 4-2 shows the position of these deflection plates near the center of the CRT. The set of plates closest to the gun assembly controls the vertical deflection. The second set of plates controls horizontal deflection. The combined effect of these two sets of plates causes the electron beam to have both vertical and horizontal deflection at the same time.

Electron-beam deflection of a CRT is accomplished by electrostatic charge energy. This is achieved by applying voltages of the correct polarity to the deflection plates. Normally, vertical deflection voltages are derived from an external source of energy. This generally represents the signal being viewed on

the CRT face. Horizontal sweep voltages are developed internally by a time-base generator. Most oscilloscopes have optional circuitry that will permit the horizontal sweep signal to be developed externally and applied to the deflection plates. External sweep or internal sweep is selected by a switch.

An end view of a CRT being observed from the display area is shown in Fig. 4-4. This representation shows the location of the vertical and horizontal deflection plates with respect to the electron beam. If no voltage is applied to two deflection plates, the electron beam will position itself in the center of the display area. To demonstrate how deflection is achieved, let us see how the electron beam responds when voltage is applied to the deflection plates. Initially, we describe the response when voltage is applied to only one set of plates. We then discuss the response of the electron beam when sweep voltage is applied to both sets of plates at the same time.

Figure 4-5 shows electron-beam response when voltage is applied only to the vertical deflection plates. If the top vertical plate is made positive and the bottom plate negative, the electron beam is attracted by the top plate and repelled by the bottom plate, as shown in Fig. 4-5A. Reversing this polarity causes the beam to move toward the bottom of the CRT, as shown in Fig. 4-5B. Ac voltage applied to the two plates, as

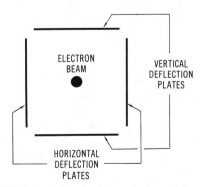

FIGURE 4-4 End view of the deflection plates.

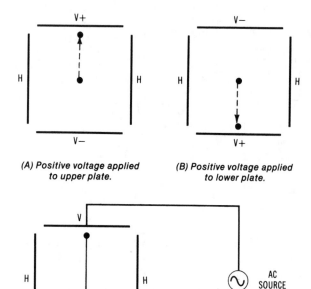

(A) Positive voltage applied
to upper plate.

(B) Positive voltage applied
to lower plate.

(C) An ac voltage applied between both plates.

FIGURE 4-5 Vertical deflection response.

shown in Fig. 4-5C, causes the electron beam to sweep from top to bottom according to the applied frequency. High-frequency ac will cause the electron beam to appear as a solid vertical line. Low-frequency ac will cause a dot produced by the electron beam to move slowly between the top and bottom of the CRT. The amount of applied voltage determines the length of the vertical sweep pattern.

Figure 4-6 shows electron-beam response when voltage is applied only to the horizontal deflection plates. In Fig. 4-6A the left horizontal plate is positive and the right plate is negative. This causes the electron beam to be deflected to the left. Figure 4-6B shows the response of the electron beam when the voltage polarity is reversed. Note that the beam is deflected to the right by this polarity. Figure 4-6C shows the response when ac is applied. In this case, the electron beam sweeps back and forth producing a horizontal line. The frequency of the applied ac determines the sweep rate.

Internally generated horizontal sweep signals generally have a sawtooth shape. This signal, shown in Fig. 4-7, has a linear rise time and a rapid fall time. The rising portion of the wave is called the ramp and the falling part of the wave is the retrace. The time or space after retrace is called the *holdoff area*. The ramp portion of the wave causes sweep from left to right and retrace causes return of the electron beam to the left side of the display. The holdoff part of the wave de-

termines how long the trace must wait between trigger pulses. The frequency of the sawtooth wave is determined by the sweep rate of the internal horizontal time-base generator.

When sweep signals are applied simultaneously to the horizontal and vertical deflection plates, they cause a specific pattern to appear on the face of the CRT. Assume now that an unknown sine wave is to be viewed on the oscilloscope. This wave is supplied to the vertical input for application to the deflection plates. Generally, the oscilloscope is set up for internal horizontal sweep signal generation. A sawtooth wave is generated by the time-base generator and applied to the horizontal deflection plates. Figure 4-8 shows the display pattern that appears on the CRT as a result of the applied signals. The beam sweeps from left to right according to the horizontal signals. In the same time frame, the vertical signal causes the electron beam to be deflected up and down according to its voltage value. At any specific point in time, the position of the electron beam is determined by the combined forces of the vertical and horizontal deflection voltage values. These forces cause the sine wave of the vertical input signal to be produced on the CRT. The resulting display is four complete sine waves.

If the time period of the horizontal sweep voltage

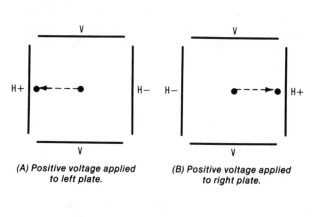

(A) Positive voltage applied
to left plate.

(B) Positive voltage applied
to right plate.

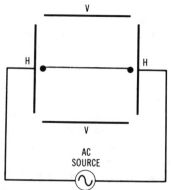

(C) An ac voltage applied to the horizontal plates.

FIGURE 4-6 Horizontal deflection response.

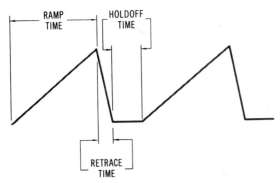

FIGURE 4-7 Sawtooth waveform of the horizontal time-based generator.

is changed it will alter the display appearing on the CRT. In Fig. 4-9, the horizontal sweep frequency has been doubled without changing the vertical input signal. In this case, doubling the horizontal frequency causes the time base to be halved. Time is a function of frequency and is expressed as $T = 1/F$. When the horizontal sweep is reduced to half of its previous value, fewer sine waves are displayed on the CRT. This means that the frequency of the horizontal time-base generator determines the scale of the time axis. The scale or operational range of the time base generator is controlled by a switch on the front panel of the oscilloscope. This switch is calibrated in time/division units. Typical ranges are seconds/centimeter, milliseconds/centimeter, or microseconds/centimeter. Specific values in an operating range might be 0.5, 0.2, 0.1 s, 50, 20, 10, 5, 2, 1, 0.5, 0.2, 0.1 ms/division, or 50, 20, 10, 5, 2, 1, 0.5, 0.2, 0.1, 0.05 µs/division.

Triggering and Synchronization

To produce a steady waveform on the face of the CRT, an oscilloscope must repeat the same trace path. For this to be achieved, the displayed signal must be periodic and the electron beam must always begin its trace at the same point on the wave. Another way of describing this function is to say that the start of the sawtooth voltage must be synchronized with a specific point on the displayed signal. Synchronization of the vertical and horizontal signals is a function of the trigger system.

Figure 4-10 shows signals that are applied to the vertical and horizontal deflection systems and the resulting display produced by an oscilloscope. The top display is the waveform supplied to the vertical deflection system. The middle waveform is produced by the time-base generator and applied to the horizontal deflection system. The bottom trace is the resulting waveform that will be displayed on the CRT. Note that the time-base signal starts its sweep at point (a) of the sine wave. When the time-base signal completes one trace period, it drops back to its initial voltage value and waits for the next sine wave to return to its beginning point before starting the next trace. This is indicated as point (a') on the display. Since the resulting image is of a periodic nature and because the time base is synchronized to produce exactly the same portion of the signal on each trace, the display image appears to be a stationary sine wave. If the applied signal voltage were to change during the sweep period or if the sweep began at a random point on the input signal voltage, the resulting image would be different for each trace period. This would make the display

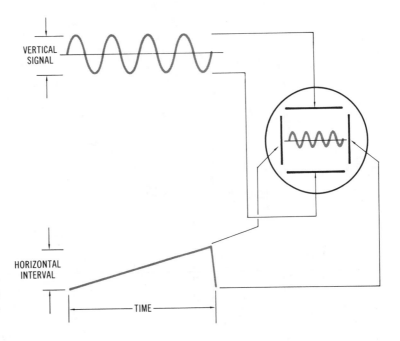

FIGURE 4-8 Time-varying signals simultaneously applied to the horizontal and vertical deflection plates.

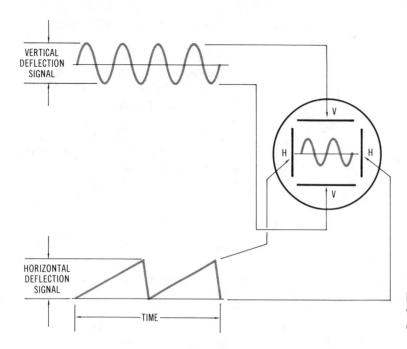

FIGURE 4-9 Display of Fig. 4-8 with horizontal deflection frequency doubled.

unsteady or appear to be in a continuous state of motion.

There are two controls of an oscilloscope that permit the operator to select the point on the display signal where sweep begins. The trigger level control sets the voltage value which causes the sweep to start. This control alters the starting point or beginning of the display. Generally, this function is achieved by a variable control adjustment. The slope control adjusts the polarity of the voltage where sweep begins. On the display of Fig. 4-10, triggering occurs on the positive slope. If the negative slope were used as the trigger point, sweep would occur at points *b* and *b'*. This would cause the vertical part of the trace to start with a downward motion instead of an upward motion. The slope function is achieved by changing a switch. The slope is either positive or negative according to the desires of the operator.

An oscilloscope generally has some additional trigger controls depending on its design. One of these is described as the triggering mode of operation. Trigger-mode selection is achieved by a switch. If the EXT

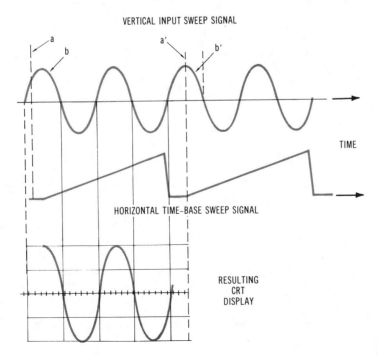

FIGURE 4-10 Synchronized vertical and horizontal sweep signals and resulting CRT display.

trigger mode is selected, the oscilloscope is triggered by an external signal. This signal is derived from the circuit under test or from an external generating source that is used as a reference. With external triggering, it would be possible to observe the phase relationship of amplifier input and output signals. If the LINE trigger mode is selected, it permits the time-base generator to be synchronized by the ac line voltage. This would permit the oscilloscope to detect the source of unwanted signals. If noise appearing on a waveform can be made stationary when the signal is synchronized with the line voltage, the noise source is in the line voltage. If the AUTO mode switch is selected it causes the oscilloscope to be placed in the automatic triggering mode of operation. Triggering in this manner is considered to be in the automatic signal seeking mode of operation. Assume that a trigger pulse starts the sweep signal. This action causes the electron beam to sweep during the ramp time and the retrace period and until the holdoff period ends. At this point a timer begins to run. If another trigger does not occur before the timer runs out, a pulse is generated automatically. This permits most signals to be displayed automatically because triggering is duplicated by the internal circuitry of the oscilloscope. This mode of operation also permits the operator to trigger on signals with changing voltage amplitudes or shapes without making level control adjustments. Automatic triggering is probably the most widely used mode of operation for general oscilloscope work.

POWER SUPPLY

The power supply of an oscilloscope is primarily responsible for supplying dc and in some cases ac voltage to the active and passive components of the instrument. These voltage values are derived from the ac power line. Typically, the primary line voltage is 120 V at 60 Hz. This voltage is applied to a transformer which steps up or down the voltage according to the needs of the circuit. The voltage values developed by the supply depend on the circuitry of the oscilloscope and the components being utilized. Modern oscilloscopes are predominately solid-state device actuated. Bipolar transistors, junction field-effect transistors, and integrated circuit devices are used in the circuits. As a rule, these devices respond to low voltage dc energy. The CRT of an oscilloscope, however, requires some rather sizable dc voltage values. Dc voltages of 100 to 2000 V are needed to energize the electrodes. In addition to this, the anode of the CRT may necessitate a dc voltage in the range of 5 to 10 kV. This part of the power supply requires some rather unusual circuitry such as voltage multipliers or a spe-

cial high-voltage transformer to develop the necessary voltage values.

Figure 4-11 shows the low-voltage power supply of a typical oscilloscope. Ac power is applied to the primary winding of T_{700}. This particular oscilloscope has both 120- and 240-V input capabilities. Selection is made through a switch located on the back side of the chassis near the power-line input. The supply transformer has two secondary windings that are used to develop the appropriate ac for the rectifier circuits. The top winding has an ac value of 70 V rms. A bridge rectifier connected to this secondary winding develops an unregulated dc output of $+100$ V, a $+100$-V regulated output, and a $+33$-V output. The lower secondary winding is center tapped and used in a split or divider type of supply circuit. The resulting dc output of this supply is regulated at $+8$ V and -8 V. Operational amplifiers U_{742A} and U_{742B} have differential inputs that monitor the changes in output voltage. Correction voltage signals applied to series transistors Q_{754} and Q_{756} and to the negative series transistors Q_{774} and Q_{776} provide the necessary level of regulation for the $+8$- and -8-V supplies.

The high-voltage supply of an oscilloscope is somewhat unusual when compared with the low voltage supply. Figure 4-12 shows the high-voltage supply of a representative oscilloscope. Transistor Q_{458} and its associated circuitry responds as an oscillator that drives the high voltage transformer T_{460}. When the instrument is initially turned on, current through transistor Q_{454} provides forward bias for the base of the oscillator transistor Q_{458}. Conduction of Q_{458} causes an increase in collector current which develops voltage across the primary winding of T_{460}. This action causes a corresponding increase in induced voltage to the feedback winding of the transformer. The polarity of the voltage supplied to the base of Q_{458} causes the transistor to be more conductive. Collector current through Q_{458} is cumulative and continues to increase. Operation continues until the transistor reaches saturation. The primary current of the transformer then levels off. This condition causes the magnetic field of the transformer to reverse polarity. A polarity change in feedback voltage causes Q_{458} to be reverse biased. This action immediately causes the transistor to be nonconductive. The process is then repeated on a continuous basis. The resulting current flow through the primary winding of T_{460} has a sine-wave characteristic with a frequency of approximately 50 kHz. The developed secondary voltage is increased due to the turns ratio of the secondary coils. The resulting ac is eventually rectified by diodes CR_{465} and CR_{463}. Dc voltages of 2 kV or more are supplied to the control electrodes of the CRT from this source.

One secondary winding of transformer T_{460} is also used as a source for the voltage multiplier U_{460}. This

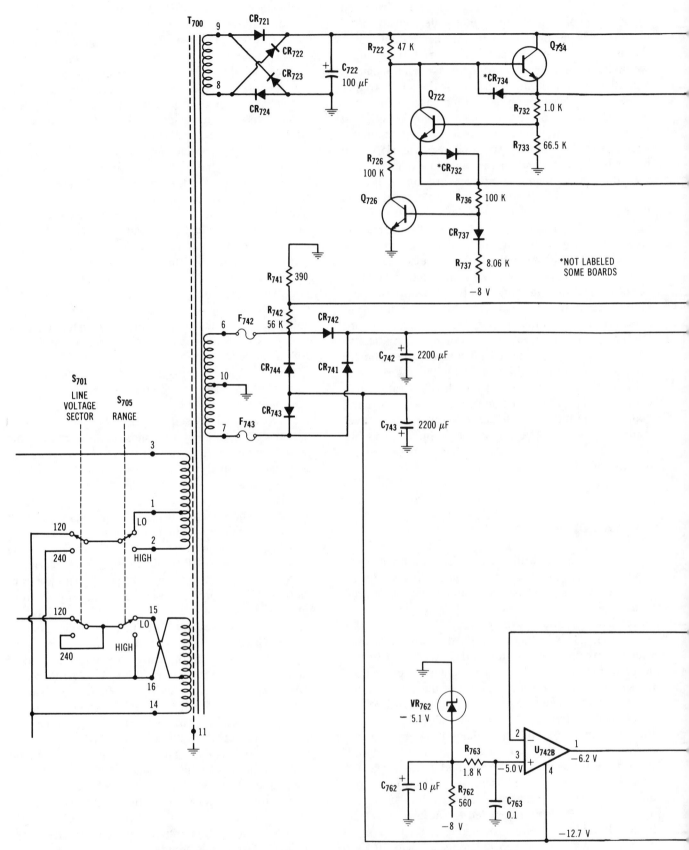

FIGURE 4-11 Low-voltage power supply of a typical oscilloscope. (Courtesy of Tektronix.)

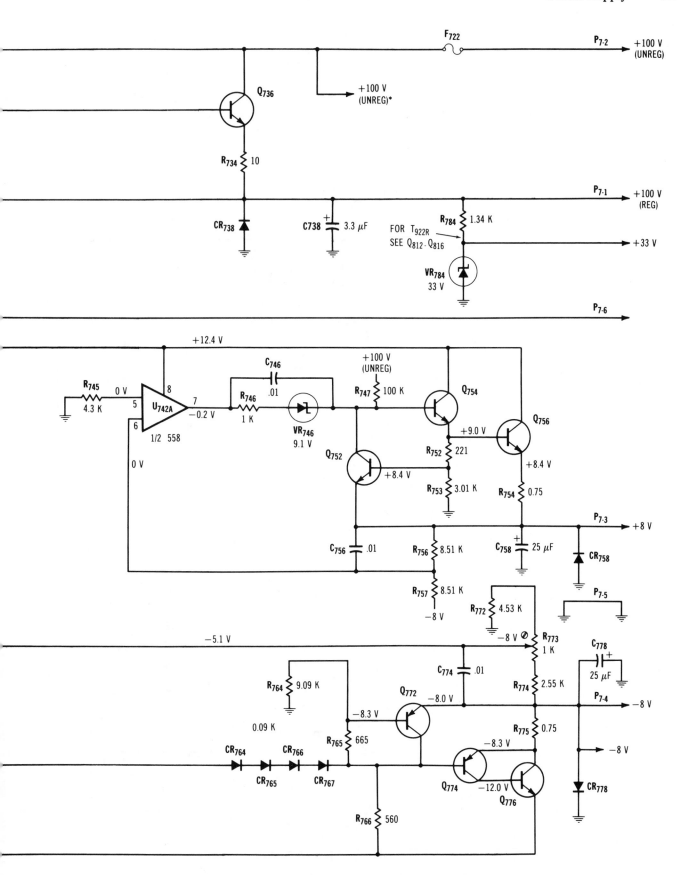

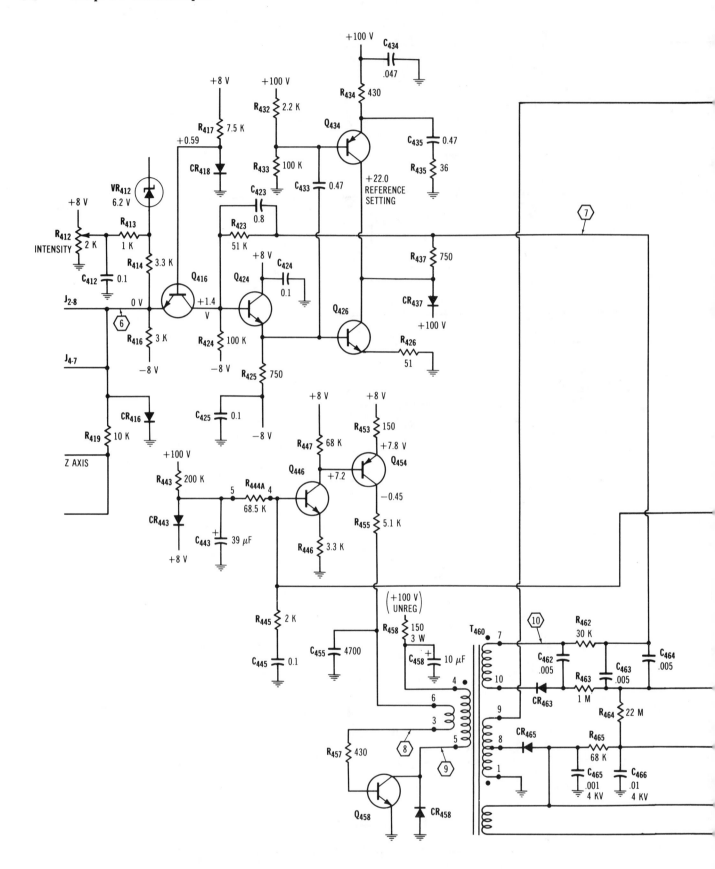

FIGURE 4-12 High-voltage power supply for an oscilloscope.
(Courtesy of Tektronix.)

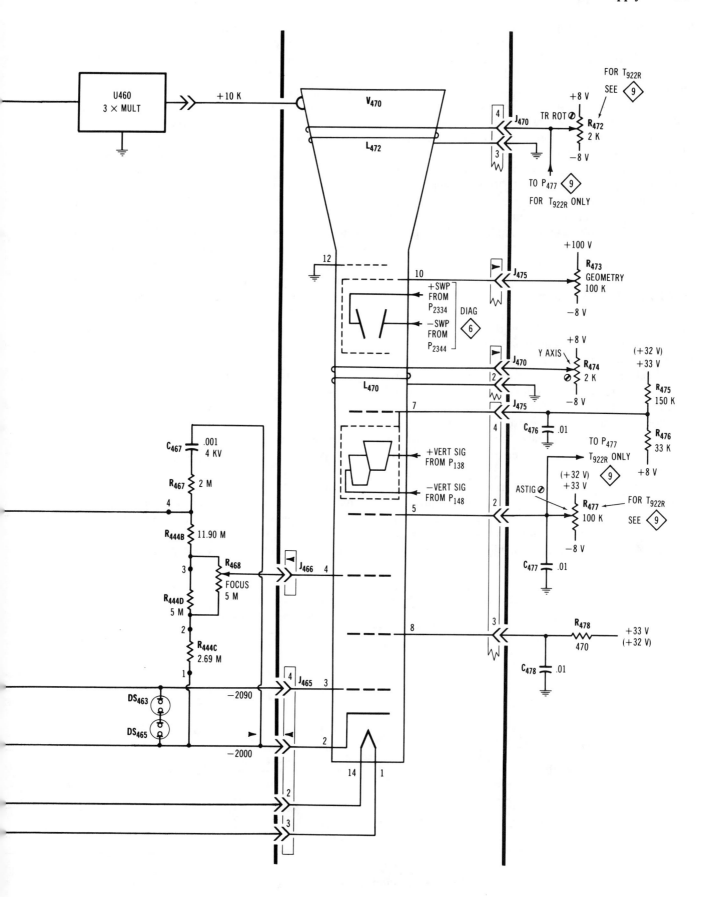

particular device responds as a voltage tripler. The re-sulting dc output is increased by a factor of 3. This part of the circuit is used as a supply for the CRT anode. Specifically, this circuit develops +10 kV for the anode.

The high-voltage supply of an oscilloscope is gen-erally described as a generated voltage source. It de-velops high-voltage ac from a low-voltage dc source. The dc energizes a transistor that responds as an os-cillator. High-frequency ac is supplied to the primary winding of the transformer. The transformer of this circuit can be made smaller in size because of the high-frequency ac. The circuit, in general, is rather efficient and very reliable for long periods of operational time.

OSCILLOSCOPE PROBES

The probe of an oscilloscope is responsible for con-necting voltage and current signals to the vertical input terminals without loading or disturbing the circuit under test. To meet these requirements, a variety of probes is available, from simple passive units to so-phisticated active probes for special measurement ap-plications. In each case, the probe must not degrade the performance of the oscilloscope, and it should be properly matched and calibrated to ensure measure-ment accuracy.

Figure 4-13A shows a general block diagram of an oscilloscope probe. The probe head contains the signal sensing device. In a passive probe this is achieved by a 10-MΩ resistor shunted by a 7-pF ca-pacitor. Active probes house a bipolar transistor or

JFET in the probe head. A coaxial cable is used to couple the probe head to the termination circuitry. The termination provides the oscilloscope with the source impedance needed to connect the coaxial cable to the vertical input circuitry.

Passive probes are widely used in most oscillo-scope applications. The simplest passive probe is a nonattenuating or $\times 1$ unit. This type of assembly sim-ply consists of a length of coaxial cable with a probe tip at one end and a BNC connector at the other end. Connection is achieved directly through the coaxial cable. Electrically, the coaxial cable has some shunt-ing capacitance that must be taken into account. A 50-Ω coaxial cable has a shunting capacitance of 30 pF per foot. A 5-ft length of coax would therefore offer approximately 150 pF of shunt capacitance to the os-cilloscope input. This type of probe offers a rather large shunt capacitance to high-frequency ac. As a rule, $\times 1$ or direct probes are used exclusively for low-frequency and power-line circuit measurement appli-cations.

One of the most widely used passive probes is the compensating $\times 10$ attenuating unit of Fig. 4-13B. This particular probe will provide signal attenuation by a ratio of 10:1 over a wide range of frequencies. The probe head has a 9-MΩ resistor shunted by a small variable capacitor. Adjustment of the variable capac-itor permits the probe to compensate for impedance changes over a wide range of frequency. An applied signal should be attenuated by a factor of 10 but still maintain its original shape and phase without distor-tion. Compensation adjustments permit the probe to couple sample signals to the vertical input without pro-ducing distortion and adversely loading down the cir-

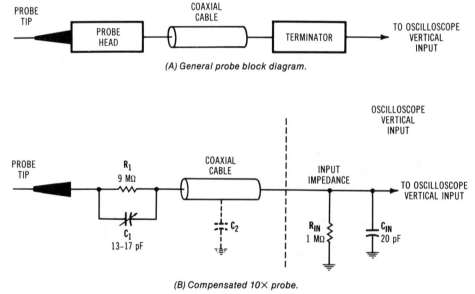

(A) General probe block diagram.

(B) Compensated 10× probe.

FIGURE 4-13 Oscilloscope probe diagrams.

cuit being evaluated. This adjustment is achieved by applying a representative square wave signal of the frequency being analyzed by the probe. The compensating capacitor is then adjusted to produce the best reproduction of the square wave on the CRT. Compensation can be made by a screwdriver adjustment or by twisting a barrel capacitor over the probe tip. The structure of the variable capacitor of the probe usually dictates the adjustment procedure.

OSCILLOSCOPE CONTROLS

The operating controls of an oscilloscope are usually divided into convenient groups that alter or control a specific instrument function. This includes CRT control, vertical deflection, horizontal sweep, triggering, modes of operation, and probes. Each group has a number of unique controls in its makeup. In general, these controls are designed to perform some type of operational procedure. Operational controls are usually placed in a convenient location where they can be easily adjusted.

CRT Display Controls

The CRT display group of oscilloscope controls consists of intensity, focus, trace rotation, and beam finder. These controls are generally located in a position relatively close to the viewing area of the CRT. On the reference oscilloscope of Fig. 4-14, these controls are grouped together and located in an area to the immediate right of the CRT.

Intensity Control

The intensity control of an oscilloscope is responsible for adjusting the brightness level of the display. This is achieved by altering the amount of negative voltage going to the control grid of the CRT. Intensity of the electron beam is determined by the quantity or number of electrons reaching the viewing area. When the negative voltage of the grid is reduced in value a larger number of electrons are permitted to reach the display area. Increased negative voltage will reduce the brightness or intensity of the electron beam. In practice, the intensity control should be adjusted to produce the lowest level of brightness that will permit the display to be effectively viewed. The operational life of the CRT can be prolonged when the intensity of the electron beam is kept at a minimum level. Functionally, this control adjusts the brightness level of the trace so that it can be viewed in different ambient light conditions.

The intensity control of the reference oscilloscope of Fig. 4-14 is located in a group of controls to the immediate right of the CRT. The intensity control is located near the top of the control group. If a functioning oscilloscope is available, locate the intensity control and adjust it through its operating range. This should demonstrate the effectiveness of the control by altering the intensity of the electron beam.

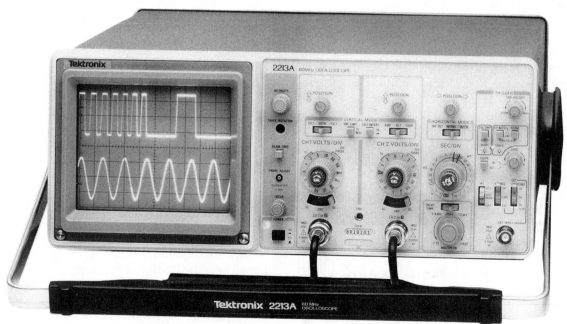

FIGURE 4-14 Oscilloscope. (Courtesy of Tektronix.)

Focus

The focus control of an oscilloscope is primarily responsible for altering the sharpness of the display image. Focus is achieved by altering the voltage level of the CRT's first anode. The potential difference in voltage between first and second anodes of the CRT determines the strength of an electrostatic field. This condition alters the trajectory of the electron beam. A rather large difference in voltage will cause the beam to converge into a fine trace near the surface of the viewing area. Focus is simply a voltage adjustment that permits the trace to have fine detail.

The focus control of the reference oscilloscope is located in the CRT group of controls to the right of the display. The focus control is at the bottom of the group. If a functioning oscilloscope is available, locate the focus control and adjust it through its operating range. In most instruments, focus and intensity adjustments are interrelated. Adjust these controls to see the influence that each control has on the other.

Beam Finder

The beam finder of an oscilloscope is a convenience control that permits the user to locate the electron beam when it is offscreen. The beam finder is a pushbutton switch located in the CRT group of controls. When this button is depressed, it causes the vertical and horizontal deflection voltages to be reduced. This permits the trace to be displayed on the limited space of the CRT face. When the operator sees the location of the beam, the vertical and horizontal position controls can be adjusted to center the trace.

The beam finder is usually included in the CRT group of controls of an oscilloscope. For the reference oscilloscope, the beam finder is in the middle of the CRT control group. If an operating oscilloscope is available, locate the beam finder button. Depress the button and hold it down while adjusting the vertical and horizontal position controls. When the trace is centered, release the pushbutton. The trace should then appear in the approximate center of the display. Further adjustment of the vertical and horizontal position controls will permit the trace to be centered properly in the display area. Some oscilloscopes do not have a beam finding switch.

Trace Rotation

The trace rotation control of an oscilloscope is another one of the CRT group of controls included on the front panel of the instrument. Trace rotation permits the user to electrically align the horizontal trace of the display with fixed lines of the graticule. This control is generally less accessible than the other controls.

This is purposely done to prevent accidental misalignment of the control. In most oscilloscopes, this adjustment is made with a small screwdriver. Once the adjustment has been accomplished, it generally does not need to be made again unless the instrument is subjected to a stray magnetic field or moved to a different location. In portable instruments this control is very handy because of the different locations where the instrument is used.

The trace rotation control of the reference oscilloscope is located in the first column of controls to the right of the CRT. A small circle or dot indicates the location of the control. It is recessed from the surface and must be adjusted with a screwdriver. To adjust this control, turn on the instrument and make the necessary adjustments that will permit a single horizontal line to be displayed across the center of the display. Position the line so that it is aligned with one of the horizontal lines of the graticule. Adjust the control so that the trace line is parallel with the graticule line. It may be necessary to adjust the vertical position of the trace again to assure that the two are parallel.

Vertical Deflection Controls

The vertical section of an oscilloscope supplies the display part of the instrument with vertical information that will ultimately appear on the CRT. To do this, the vertical section takes the input signal, amplifies it, and develops a suitable voltage value that will deflect the electron beam. The input signal is usually the signal or voltage being analyzed by the oscilloscope. Figure 4-15 shows a block diagram of the vertical section of a typical oscilloscope. Controls are attached to different parts of the system. Typical controls are vertical position, input coupling, vertical operating mode, input sensitivity or volts/division selector, and variable volts/division control.

Vertical Position Control

The vertical position control permits the operator to adjust the trace to a desired viewing location. This control adjusts the distribution of voltage between the two deflection plates. If the voltage is equally distributed between the deflection plates, the trace will be centered vertically. Adjusting the control so that the top plate is more positive than the bottom plate will move the display toward the top of the CRT. Reversing the voltage distribution will cause the trace to be positioned near the bottom of the display area. The vertical position control of our reference instrument is located near the top of the vertical group of controls. If an operating oscilloscope is available, adjust the vertical position control to see the influence that this control has on the position of the trace.

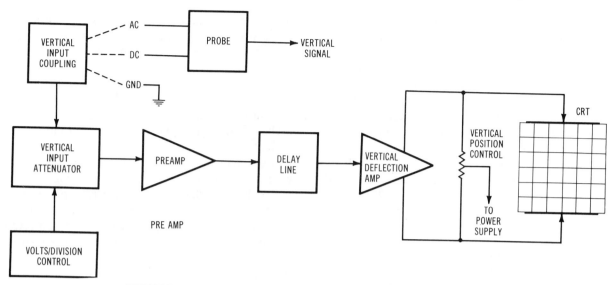

FIGURE 4-15 Block diagram of the horizontal sweep section.

Vertical Input Coupling

The input coupling function of an oscilloscope is achieved by a switch. This lets the operator control how the input signal is coupled to the vertical input section of the instrument. In the AC position, the input signal voltage must pass through a capacitor. In the DC position, the input signal is directly coupled to the input of the vertical amplifier. The middle position or GND refers to the ground. Placing the switch in this position disconnects the external input signal from the vertical section and grounds the vertical input.

The input coupling switch of the reference oscilloscope is located just above the probe connector near the bottom of the vertical group of controls. The influence that this control has on the operation of the oscilloscope can be observed when an appropriate signal is applied to the vertical input. An ac signal will adjust its zero reference point to a value determined by the vertical position control. A dc signal will adjust its reference to some level determined by the voltage value of the applied dc. The ground switch position selects the chassis ground as a reference operating point.

Volts/Division

The volts/division control or vertical sensitivity of an oscilloscope is controlled by a rotary switch. This control permits the instrument to extend its range of operation so that signals of a few millivolts to several volts can be displayed on the CRT. Using the volts/division switch also changes the scale factor of the display screen. Each position setting of the switch has a number value that represents the scale factor. The 10 position means that each major vertical division

represents 10 V. Other position settings cause a corresponding volts per division function to be established.

For our reference oscilloscope, the volts/division switch is located in the center of the vertical deflection group of controls. The smallest volts/division value is 2 mV and the largest value is 100 V. These values are all dependent on the variable control being set to the calibrate position. Locate the volts/division control and study its range settings.

If an operating oscilloscope is available, locate the volts/division switch and rotate it through the operating range. Apply an ac signal to the vertical input and adjust the instrument to produce a suitable sine-wave display. Adjust the range of the volts/division switch while observing the influence that each switch position has on the size of the displayed signal. Determine the voltage value of the display according to an appropriate volts/division switch setting. The total amount of vertical sweep or deflection is based on the peak-to-peak voltage of the ac signal being displayed. The probe of an oscilloscope may also have some influence on the amount of vertical sweep produced by the instrument. A ×1 probe will provide a direct reading of the volts/division ranges, while a ×10 probe will scale down the input by a factor of 10. The probe and volts/division switch setting of the instrument determine the range of vertical sweep.

Variable Volts/Division Control

Most oscilloscopes have a variable volts/division control that can be used to change the volts/division range setting by a factor of ×2 or more. This control is used to make quick amplitude comparisons on a series of signals. It obviously changes the range of the volts/division setting. In most instruments this control has a notch or position setting where calibration is actuated.

In the reference oscilloscope, the variable volts/division control is located in the center of the volts/division switch. To see the influence that this control has on the displayed waveform, apply a signal to the vertical input. Select a volts/division range that will produce a display of four or more divisions. Then adjust the variable volts/division control through its range while observing the display. Notice that the amplitude of the display is reduced by this adjustment. For most oscilloscope applications, the variable control should remain in the calibrate position.

Vertical Operating Mode

The vertical operating mode of an oscilloscope is an optional control for instruments that have two vertical channels. This type of oscilloscope has a duplicate set of vertical controls for each channel. Two independent traces can be displayed on the CRT at the same time. The vertical operating mode is a switch function that permits the operator to select a desired channel or a combination of channel options.

The reference oscilloscope has its vertical mode switches located just below the vertical position controls of each channel. Channel 1 is located on the left side of the control panel and channel 2 is on the right. The vertical mode switch of channel one selects channel 1, both, or channel 2. The mode switch of channel 2 has add, alternate, and chop modes of operation.

To demonstrate the operation of the vertical mode switch, turn on an oscilloscope and prepare it to display a trace. Adjust the two channels so that the vertical controls are at the same setting. Place the channel 1 mode switch in the BOTH position and the channel 2 switch in the alternate or ALT position. Adjust the position controls so that the trace of channel 1 is at the top and channel 2 is near the bottom. Apply an ac signal to each channel. If necessary, adjust the volts/division switch of each channel to produce a display of suitable size. Switch the mode switch to CH1, BOTH, and CH2 while observing the display. Return the switch to the BOTH position. Then switch the channel two-mode switch to the ADD position. This will add the two traces. If the inversion button of channel 2 is depressed, the display will show channel 1 minus channel 2. The CHOP and ALT mode switch is used to observe two signals at any sweep speed. The alternate mode displays one channel and then the other in an alternate sequence. At high sweep rates this type of display is very desirable. At slow sweep rates there is a noticeable alternating effect in the two displays. The CHOP mode breaks the two traces into small segments and switches between the two traces very quickly. This is generally noticeable when 60-Hz signals are displayed on the instrument.

Horizontal Sweep Controls

For an oscilloscope to make a display on the face of a CRT it needs both vertical and horizontal sweep to deflect the electron beam. Horizontal sweep is normally provided by an internal generator that produces a sawtooth-shaped waveform. The rising part of this waveform is called the ramp or trace period and the falling part is called the *retrace interval*. The trace period causes beam deflection from left to right. Retrace causes the beam to return to the left in preparation for the next trace period. The horizontal sweep rate of an oscilloscope is operator controlled, which permits it to display different frequencies.

Figure 4-16 shows a block diagram of the horizontal sweep circuitry of an oscilloscope. Note that the circuitry is divided into two distinct sections. The generator is responsible for sweep signal development. The amplifier increases the amplitude of the signal so that it will drive the deflection plates. Controls of this section are attached to the part of the circuit that has the greatest influence on its operation.

The horizontal sweep of an oscilloscope has a variety of different controls that regulate its operation. These include horizontal position, operational mode, seconds/division, variable sweep, and magnification. These controls are generally grouped together in a special area of the control panel. On the reference oscilloscope of Fig. 4-14, the horizontal group of controls is located in the fourth column to the right of the CRT.

Horizontal Position Control

The horizontal position control is designed to change or alter the location of the horizontal trace on the face of the CRT. This adjustment is made by shifting the distribution of voltage to the horizontal deflection plates. The trace shifts in the direction of the deflection plate with the highest positive voltage value. As a rule, the horizontal trace should be positioned in the center of the viewing area.

The horizontal position control of the reference oscilloscope is located at the top of the horizontal sweep controls. To demonstrate the operation of this control, turn on an oscilloscope and prepare it for operation. A horizontal line should appear on the CRT. When this occurs, alter the position control through its range. This should cause the trace to shift from right to left according to the setting of the control. If a dual-channel oscilloscope is used, the position control alters both channels in the same manner.

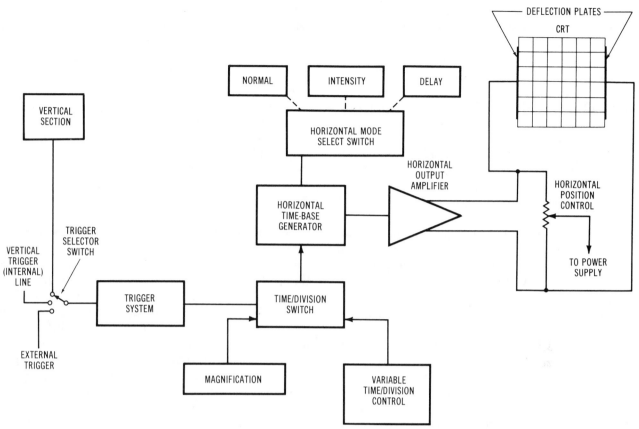

FIGURE 4-16 Horizontal sweep section.

Horizontal Operating Mode

The horizontal operating mode is an optional function that depends on the degree of sophistication involved in an oscilloscope's structure. Single time-base oscilloscopes usually have only one mode of operation. Our reference oscilloscope however, is equipped with three modes of operation, called normal, intensified, and delayed sweep. As a rule, the instrument is used in the normal mode of operation for most applications. In this mode, the horizontal time base responds as an energy source for the horizontal sweep system.

The intensified mode of operation permits the operator to alter the electron beam with a signal voltage that will cause its intensity to vary according to an external signal. The external modulating signal is applied to the back of the instrument panel through a special BNC connector. This mode of operation permits some rather unusual tests to be performed with the oscilloscope. The electron beam of the CRT of a television receiver is intensity modulated to produce light variations in the display.

Delayed sweep permits the instrument to add a precise amount of time between the trigger point and the beginning of the sweep period. With delayed

sweep, the operator may choose to trigger the trace anywhere along the displayed waveform. The delay time is used to control the start or beginning of the waveform being displayed. When the oscilloscope is placed in the delay-time mode of operation, the delay-time switch and multiplier are actuated. These controls are located on the bottom of the horizontal group of controls.

The horizontal sweep mode function of the reference oscilloscope of Fig. 4-14 is achieved by a switch. This switch is located just below the position control. For normal measurement applications the mode switch is placed in the no delay or NO DLY mode of operation. If an oscilloscope is available, examine its operational mode capabilities and place it in the normal or no-delay sweep position. Apply a signal to see how it responds.

Horizontal Time-Base Control

The horizontal time base of an oscilloscope is responsible for generating a sawtooth wave that deflects the trace horizontally. This action is achieved by a switch that controls the sweep rate of the time-base generator. The switch positions are identified in time/division values. Three ranges of sweep are generally included in an oscilloscope: seconds/division, milliseconds/division, and microseconds/division. A number of discrete values are included in each of these ranges.

Locate the time/division switch of the reference oscilloscope of Fig. 4-14. Note the three operational ranges of the switch and the divisions in each range. Keep in mind that the function of this switch is to select the sweep frequency of the horizontal time base. When the frequency of the time base coincides with the frequency of the signal being applied to the vertical input, a suitable display will be produced. If the time-base rate is greater or less than the frequency of the observed signal, an unintelligible signal will be displayed. The frequency of the time-base generator must be adjusted to a reasonable approximation of the observed signal frequency to produce a meaningful display. Operation of this control is simply a matter of selecting an appropriate sweep range that will produce a usable display.

Variable Time/Division Control

The switch positions of the time/division control are designed to provide calibrated time-base values. On occasions there is a need for variable control of the time base. The variable time/division control permits the operator to adjust the time base with a ratio of something in the range of 2.5 : 1. This control is located at the center of the time/division switch. A detent in the extreme clockwise position places the variable control in its calibrate position. When the control is out of the detent position, its variable condition is in effect. Locate the variable time/division control on the reference oscilloscope.

If an operating oscilloscope is available you may want to evaluate the operation of the variable time/division control. To do this, a suitable signal must be applied. With a display of the signal on the instrument, turn the variable time/division control through its range. This should cause a noticeable change in the display of the wave. Then return the control to its detent position. Essentially, this control is left in the calibrate position for most measurement applications.

Horizontal Magnification

Most of the oscilloscopes in use today offer some means of horizontally magnifying the waveforms that appear on the screen. Magnification is achieved by multiplying the time base by some fixed factor. A factor of 10 is typical. This is achieved on the reference oscilloscope by pulling out the variable time/division control when it is in the calibrate position. This action changes the time base components by a factor of 10.

A 0.05-μs signal could be extended to a 5-ns signal by engaging the magnification switch. Magnification is very useful when analyzing specific parts of a signal.

Refer to the reference oscilloscope of Fig. 4-14. Locate the magnification function switch on the variable time/division control. It is actuated by pulling out the control knob. This is typical of the magnification function of many oscilloscopes. If an oscilloscope is available, locate the magnification control. See how the switch function of the instrument is actuated. Test its response when a signal is displayed on the CRT.

Triggering Control

The time-base generator of the horizontal sweep system is considered to be free running. The frequency of the generated sawtooth waveform is based on the *RC* time constant of circuit components. For this signal to be displayed, there must be some form of synchronization. The horizontal sweep signal must be in step with the vertical signal. The triggering function of an oscilloscope is responsible for selecting a synchronizing signal and applying it to the horizontal sweep generator. Most oscilloscopes are equipped with internal, line, and external triggering capabilities. The trigger system simply tells the oscilloscope which trigger source to use according to its switch selection. The trigger is then adjusted with the slope and level controls to recognize a particular voltage level and polarity.

The controls of a trigger system are level, slope, variable holdoff, source selection switching, trigger modes, and coupling. On our reference oscilloscope of Fig. 4-14, these controls are located at the right edge of the control panel. Notice the terminology used to describe the functions.

Variable Holdoff Control

Not every triggering event can be accepted as a trigger pulse for the time-base generator. The trigger system will not recognize triggering during the trace time, the retrace period, or for a short time that follows retrace, called the *holdoff time*. The holdoff period does, however, provide additional time after the retrace period that it used to produce stability. In some applications, the holdoff time may not be long enough to provide good stability of the display. The variable holdoff control permits adjustment of the holdoff time. By changing the holdoff time, it becomes possible for the instrument to accommodate a trigger point that will appear at the same position on the wave for each repetition of the signal.

On the reference oscilloscope, the variable holdoff control is located at the top of the trigger group of controls. To demonstrate the operation of this control, a signal must be applied to the vertical input. Adjust it to produce a stable display. The trigger mode should be placed in the normal position and the trigger source switched to internal. Then adjust the variable holdoff control while observing the display. As a rule, the display will become unstable and lose its synchronization at some point in the adjustment span. If an operating oscilloscope is available, it would be helpful to make this adjustment to see the response of the holdoff control.

Trigger Operating Modes

The trigger operating mode of an oscilloscope is used to select different triggering methods for the time-base generator. In the reference oscilloscope, the MODE switch has three positions: automatic, normal, and TV Field. The normal trigger mode is the most useful. It accommodates the widest range of signals. This mode of operation will not permit a trace to be displayed on the CRT unless the time-base generator is triggered.

In the automatic or AUTO mode, a trigger pulse starts the trace, retrace, and holdoff sequence. At the completion of the sequence, a timer begins its operation. If another trigger pulse does not occur before the timer runs out, a pulse is generated to trigger the next sweep sequence. The automatic mode of triggering is considered to be the signal seeking mode of operation. This means that for most measuring applications, the automatic mode will match the trigger level to the trigger signal value. Trigger levels in the automatic mode do not require a value setting outside of the signal range. This mode also lets the time base generator trigger on signals that have changing amplitudes or waveshapes without making level adjustments.

When the trigger operating mode switch is placed in the TV Field position the oscilloscope becomes useful in television signal analysis. In this operating mode, the time-base generator will trigger on TV fields at 100 μs/division and TV lines at 50 μs/division. This permits the instrument to display video signals, horizontal frequency, vertical, and color burst signals with good synchronization.

Trigger Signal Sources

The trigger signal source of an oscilloscope is divided into three groups: internal, line, and external. The trigger source used by an oscilloscope does not effectively alter the operation of the trigger circuit. An internal trigger signal however, means that the signal being displayed by the CRT is also used to trigger the time-base generator.

The triggering source and its switching procedure varies a great deal among different oscilloscope makes and models. For the reference instrument, two switches on the front panel are used to select the trigger source. These switches are labeled SOURCE and INT. The internal triggering source is enabled when the switch is placed in the INT position. In this position, triggering can be from either vertical channel, or the vertical mode of operation. Triggering in either case is determined by the vertical signal. In the VERT MODE position, the trigger source is selected for any of the vertical combinations such as channel A + B, A − B, chopped, A only, or B only. In a sense, this trigger mode selection procedure is considered to be an automatic internal source selection procedure.

The LINE source of triggering permits an alternative to the internal trigger source derived from the vertical input signal. Line triggering is very useful when analyzing signals that derive their energy from the ac power source. Line triggering is enabled by placing the source switch in the LINE position. This selects its trigger signal from a sample of the ac power line.

An alternative to internal triggering is external triggering. This triggering source comes from an externally supplied signal that is applied to a BNC connector on the front panel of the instrument. External triggering usually gives the operator greater control over the display. To use this triggering source, the selector switch is placed in the EXT position. The trigger signal must then be supplied to the instrument from some outside source. External triggering is very useful when analyzing digital signals. An operator might want to look at a long train of very similar pulses while triggering with a signal derived from a clock or another part of the circuit. The external trigger signal can also be used as a reference source in phase analysis of amplifier circuits.

Trigger Coupling

When an external trigger source is applied to an oscilloscope, it generally has signal coupling. The external trigger coupling circuit can be dc, dc with attenuation, or ac. The dc coupling circuit permits both ac and dc signals to be applied to the external trigger source. The dc with attenuation input is used to accommodate signals with voltage values greater than those needed for normal signal input. This external trigger coupling circuit divides the input by a factor of 10. A 100-V input signal divided by 10 equals 10 V. The ac coupling circuit blocks the dc components of the signal and couples only the ac component.

In our reference oscilloscope, trigger coupling is switch selected. This switch is located on the lower right corner of the control panel just above the external input connector. Locate this switch and notice its coupling switch possibilities.

SUMMARY

The oscilloscope is an instrument designed to graphically measure time-varying voltage and current values. It is composed of a number of functional blocks that are used to make the instrument operational. The vertical section of the instrument is used to control the Y-axis of the display. The horizontal section controls the X-axis of the display. The trigger section determines the specific point in time where horizontal sweep begins. The cathode-ray tube is ultimately responsible for the display. These sections of the instrument are all energized by a power supply. Probes are used to connect an external input signal to the instrument.

The CRT is responsible for the display function of the oscilloscope. This part of the instrument operates on the principle of thermionic emission. Electrons emitted from the cathode pass through a number of electrodes before reaching the CRT face or display area. The control grid alters the quantity of electrons, which determines the intensity of the display. Focusing is achieved by passing the electron beam through two anodes. An electrostatic field varies the convergence point of the electron beam on the display area.

Electron-beam deflection of a CRT is accomplished by electrostatic charge energy applied to the vertical and horizontal deflection plates. The vertical plates cause the trace to move in an up-down motion.

Horizontal deflection plates move the trace from right to left. In normal usage, a signal to be displayed on the CRT is applied to the vertical deflection plates. The horizontal signal is generated internally by a time-base generator. To produce a usable display, the frequency of the time-base generator must closely approximate the vertical input frequency.

Triggering or synchronization of the vertical and horizontal display signals is extremely important when producing a stable display on the CRT. Triggering is used to start the operational sequence of the time-base generator. This generator produces a sawtooth wave that has trace, retrace, and holdoff periods of time. Triggering is required to start the operational sequence produced by the sweep system. The triggering source can be derived internally or be selected from an external source. This function of the oscilloscope is controlled by a switch.

The operating controls of an oscilloscope are divided into convenient groups that control a specific instrument function. The CRT controls consist of intensity, focus, trace rotation, and beam finder. The vertical group of controls consist of position, vertical mode of operation, volts/division, and input coupling. The horizontal group of controls deals with position, horizontal mode of operation, time/division switch, variable time/division, magnification, and delay time. Triggering involves variable holdoff, trigger mode, level, slope, source selection, and external coupling.

To use an oscilloscope effectively, the operator must have some understanding of the instrument's operating controls. These controls set up the instrument for measuring applications.

ACTIVITIES

OSCILLOSCOPE CONTROL
FAMILIARIZATION

1. Locate the CRT display controls. Turn on the power and adjust the intensity and focus controls to produce a suitable display.

2. Locate the vertical control section of the oscilloscope. Does your instrument have single- or dual-channel capabilities?

3. Adjust the instrument to produce a single-channel display. Alter the position control while observing the influence that it has on the display. Rotate the VOLTS/DIV switch to its least sensitive position. The CAL control should be in the detent position. Note the value of the volts needed to make one vertical division of display.

4. Adjust the instrument to produce a display for each channel. Position the display for channel 1 two divisions from the top of the CRT. Position channel 2 up two divisions from the bottom of the CRT. The

VOLTS/DIV switch of each channel should be in the least sensitive position. The CAL control should be in the detent position.

5. Locate the horizontal controls of the oscilloscope. If the instrument has a horizontal mode switch, set it for no delay. Rotate the SEC/DIV switch to the 0.5 ms position. The CAL control should be in the detent position. Notice the sweep rate of the horizontal lines displayed by each channel. Change the SEC/DIV switch to a slower sweep rate. How does this alter the horizontal display lines? Change the SEC/DIV switch to a faster sweep rate. How does this alter the horizontal display lines?

6. Locate the trigger controls of the oscilloscope. Set the holdoff control to the normal or full counterclockwise position. Set the trigger mode switch to the automatic or AUTO position. Set the source switch to the internal or INT position.

7. This is a general setup procedure for initializing the operation of an oscilloscope.

MEASURING VOLTAGE WITH AN OSCILLOSCOPE

1. Initialize the oscilloscope for operation. Place the trigger source in the LINE position.
2. Connect three resistors such as 100, 470, and 1 kΩ in series. Apply some value of low voltage ac such as 6.3, 12.6, or 25.2 V to the circuit.
3. Prepare the oscilloscope to measure ac voltage on channel 1.
4. Measure and record the peak-to-peak ac voltage applied to the series circuit.
5. Measure and record the peak-to-peak ac voltage across each resistor.

DETERMINING FREQUENCY WITH AN OSCILLOSCOPE

1. Initialize the oscilloscope for operation. Place the trigger source in the INT position.
2. Using the series circuit of the previous activity, connect the vertical probe across the 1-kΩ resistor.
3. Adjust the SEC/DIV switch to produce two steady sine waves. Center the waveform with the vertical position control. The zero reference line should be the center of the display. The CAL control of the SEC/DIV switch should be in the CAL position without magnification.
4. Count the number of horizontal divisions needed to make a complete sine wave. $F = 1/T$. $T =$ the number of divisions × the SEC/DIV setting. What is the calculated frequency?
5. If a signal generator is available, prepare it to produce a sine wave of approximately 100 Hz. Connect the signal generator output to the oscilloscope vertical input. Determine the frequency of the wave with the oscilloscope.
6. Set the signal generator to produce a frequency of approximately 10 kHz. Determine the frequency of the wave with the oscilloscope.

QUESTIONS

1. Draw a block diagram of an oscilloscope.
2. Explain the functional operation of the cathode-ray tube of the oscilloscope.
3. Explain the functional operation of the vertical sweep.
4. Explain the functional operation of the horizontal sweep.
5. What is meant by the term *triggering*?
6. What is internal or INT triggering?
7. What switch of an oscilloscope is used to prepare the instrument for two-channel display?
8. When is CHOP used?
9. When is alternate or ALT used?
10. What value of ac voltage is displayed by an oscilloscope?
11. How is frequency measured with a calibrated time-base oscilloscope?
12. How is voltage measured with an oscilloscope?

CHAPTER 5

DIGITAL ELECTRONICS

OBJECTIVES

Upon completion of this chapter, you will be able to:
1. Explain the differences between analog and digital systems.
2. Evaluate different numbering systems.
3. Explain the response of AND, OR, NOT, NOR, and NAND gates.
4. Identify logic gates by schematic symbol and function.
5. Identify basic flip-flops by schematic symbol and show the input/output response.
6. Explain the function of a counter, decoder, and digital display.
7. Identify the basic parts of a digital voltmeter and show how they are connected in a block diagram.

IMPORTANT TERMS

In this chapter we are going to investigate some of the basic principles of digital electronics. You will have a chance to become familiar with a number of basic digital system functions. New words, such as *flip-flop, gates, counters,* and *decoding,* will begin to have some meaning for you. A few of these terms are singled out for study. As a rule, the chapter will be more meaningful if these terms are reviewed before proceeding with the text.

Alphanumeric: A numbering system containing numbers and letters.

Analog: Being continuous or having a continuous range of values.

AND gate: A digital device whose output is true or 1 when all of its inputs are true or logic 1.

Base: The number of symbols in a number system. A decimal system has a base of 10.

Binary: A two-state of two-digit numbering system.

Binary-coded decimal (BCD): A code of 10 binary numbers for decimal values of 0 to 9.

Binary-coded octal (BCO): A code of eight binary numbers for decimal values of 0 to 7.

Counter: A digital circuit capable of responding to state changes through a series of binary changes.

Decade counter: A counter that achieves 10 states or discrete values.

Decoding: A digital circuit that changes coded data from one form to a different form.

Digital: A value or quantity related to digits or discrete values.

Digital voltmeter (DVM): A voltmeter that displays measured values as numbers instead of the deflection of a hand on a graduated scale.

Dot matrix: A type of display that uses discrete LEDs in a configuration to form letters and numbers.

Flip-flop: An electronic circuit that has two operating states and the ability to change from one state to the other with the application of an appropriate input signal.

Flip-flop, *J-K*: A flip-flop having two inputs, designated *J* and *K*. With the application of a clock pulse, a 1 on the *J* input will set the flip-flop to the 1 state. A 1 on the *K* input will reset it to the 0 state. A 1 on both *J* and *K* at the same time will cause a state change regardless of the previous state.

Flip-flop, *R-S*: A flip-flop having two inputs designated *R* (reset) and *S* (set). A 1 applied to the *S* input will set the flip-flop to the 1 state. A 1 applied to the *R* input will reset it to the 0 state. 1's applied to *S* and *R* simultaneously should not occur.

Flip-flop, *R-S-T*: A flip-flop having an *R* (reset) input, an *S* (set) input, and *T* (trigger) input. The *R* and *S* inputs are used to set and reset the flip-flop. The *T* input causes a state change with each applied pulse.

Gate: An electronic circuit that performs special logic operations such as AND, OR, NOT, NAND, and NOR. This circuit is commonly built on an IC.

Hexadecimal: A base 16 numbering system represented by the numbers 0, 1, 2, 3, 4, 5, 6, 7, 8, 9, A, B, C, D, E, and F.

Light-emitting diode (LED): A semiconductor device in which the energy of minority current carriers when combining with holes produces visible light.

Liquid-crystal display (LCD): A digital display that has a layer of liquid crystal sandwiched between two pieces of polarized glass plates.

Logic: A decision-making capability of electronic circuitry.

NAND gate: A logic function that produces a 1 output for all input combinations except when the inputs are all 1.

NOR gate: A logic function that produces 0 output for all input combinations except when the inputs are all 0.

NOT gate: A logic function that produces a 1 output when the input is 0 and a 0 output when the input is 1. A NOT gate is also called an inverter.

Octal: A base 8 numbering system represented by the numbers 0, 1, 2, 3, 4, 5, 6, and 7.

OR gate: A logic function that produces a 1 output for all input combinations except when the inputs are all 0.

Radix: The base of a numbering system.

Ramp generator: An electronic circuit that produces a signal with a gradual inclined slope similar to the shape of a sawtooth.

Seven-segment display: A number display device that has seven bars or slits that are used to produce numbers.

Voltage-controlled oscillator (VCO): An electronic signal generator whose output frequency is a function of an applied input voltage.

INTRODUCTION

Many of the instruments that we use today respond to signals that contain some type of numerical information. Specific number elements, which are described as digits, become very important. A decimal system, for example, has 10 elements or digits that are used repeatedly in the counting process. Any device that employs a numerical signal in its operation is classified as a digital electronic system. Computers, calculators, digital display instruments, numerical control equipment, timers, controllers, and industrial process instruments are included in this classification.

Digital electronic instruments are unique in their response to signal processing. Unlimited quantities of digital information can be processed in nanoseconds of time. With the operational speed of industrial applications being of prime importance, digital electronic applications tend to be widely used over other instruments. We will therefore direct the attention of this chapter to digital electronics and some of its applications.

DIGITAL NUMBERS

The most common numbering system in use today is the decimal system. Ten digits are used in this numbering system to achieve counting: 0, 1, 2, 3, 4, 5, 6, 7, 8, and 9. The number of discrete digits of a system is commonly called its base or radix. The decimal system has a base or radix of 10.

Nearly all modern numbering systems are described as having *place value*. This term refers to the placement of a particular digit with respect to others in the counting process. The largest digit that can be used in a specific place or location is determined by the base of the system. In the decimal system, the first position to the left of the decimal point is called the *unit's place*. Any digit from 0 to 9 can be used in this place. When number values greater than 9 are to be used, they must be expressed in two or more places. The next position to the left of the unit's place is the *10's place* in a decimal system. Each place added to the left extends the capability of this system by a power of 10.

A specific number value of any base can be expressed by addition of weighted place values. The decimal number 1259, for example, would be expressed as $(1 \times 1000) + (2 \times 100) + (5 \times 10) + (9 \times 1)$. Note that these values increase progressively for each place extending to the left of the starting position or decimal point. These place or position factors can also be expressed as powers of the base number. In the decimal system this would be 10^0, 10^1, 10^2, and 10^3, with each succeeding place being expressed as the next power of the base 10. Mathematically, each place value is expressed as the digit number times a power of the radix of the numbering system base. The decimal number 3421 is expressed this way in Fig. 5-1.

The decimal numbering system is commonly

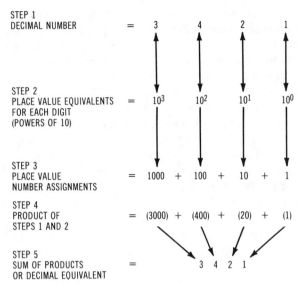

STEP 1
DECIMAL NUMBER   =   3   4   2   1

STEP 2
PLACE VALUE EQUIVALENTS   = 10^3 10^2 10^1 10^0
FOR EACH DIGIT
(POWERS OF 10)

STEP 3
PLACE VALUE   =   1000 + 100 + 10 + 1
NUMBER ASSIGNMENTS

STEP 4
PRODUCT OF   =   (3000) + (400) + (20) + (1)
STEPS 1 AND 2

STEP 5
SUM OF PRODUCTS   =   3   4   2   1
OR DECIMAL EQUIVALENT

FIGURE 5-1   Components of decimal number 3421.

used today and is very convenient in our daily lives. Electronically, however, it is rather difficult to employ. Each digit of a base 10 system, for example, would require a specific value associated with it. Electronically, a system using this numbering method would require a special detection process to distinguish between different number values. The problems associated with defining and maintaining these 10 levels are very difficult to solve.

Binary Numbering System

Practically all electronic digital systems in operation today are of the binary type. This type of system has 2 as its base or radix. The largest digital value that can be expressed in a specific place by this system is the number 1. Essentially, this means that only the numbers 0 or 1 are used in the binary system. Electronically, the value of zero can be expressed as a very low voltage value or no voltage. The number 1 can then be indicated by some voltage value assignment larger or more significant than zero. Binary systems that use this voltage value assignment are described as having positive logic.

The two operational states of a binary system, one and zero, can be considered as natural circuit conditions. When a circuit is turned off or has no voltage applied, it is considered to be in the off or 0 state. An electrical circuit that has voltage applied or is operational, is considered to be on or in the 1 state. A binary digit can therefore be either a 1 or a 0. The term *bit* (*bi*nary digi*t*) is commonly used to describe this condition.

The basic principles of numbering that are used by the decimal or base 10 numbers apply in general to binary numbers. The radix of the binary system, for example, is 2. This means that only the digits 0 and 1 can be used to express a specific place value. The first place to the left of the starting point, or, in this case, the binary point, represents the unit's or 1's location. Places that follow to the left of the binary point refer to the powers of 2. Some of the digital values of numbers to the left of the binary point are $2^0 = 1$, $2^1 = 2$, $2^2 = 4$, $2^3 = 8$, $2^4 = 16$, $2^5 = 32$, $2^6 = 64$, $2^7 = 128$, and so on.

When different numbering systems are used in a discussion, they usually incorporate a subscript number to identify the base of the numbering system being used. The number 101_2 is a typical expression of this type. This would be described as one–zero–one instead of the decimal equivalent of one hundred and one.

The number 101. is equivalent to five to the base 10, or 5_{10}. Starting at the first digit to the left of the binary point, this number would have a place value of $1 \times 2^0 + 0 \times 2^1 + 1 \times 2^2$ or $1_{10} + 0_{10} + 4_{10} = 5_{10}$. The conversion of a binary number to an equivalent decimal number is shown by steps in Fig. 5-2.

A simplified version of the binary-to-decimal conversion process is shown in Fig. 5-3. In this method of conversion, write down the binary number first. Starting at the binary point, indicate the decimal equivalent or powers-of-2 numbers for each binary place location where a 1 is indicated. For each zero in the binary number, leave a blank space or indicate a zero. Add the place-value assignments and record the decimal equivalent. Practice this method on several binary numbers until you are proficient in this conversion process.

Conversion of a decimal number to a binary equivalent is achieved by repetitive steps of division

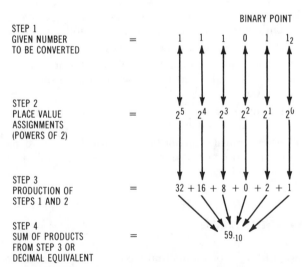

                                                      BINARY POINT
STEP 1
GIVEN NUMBER   =   1   1   1   0   1 1_2
TO BE CONVERTED

STEP 2
PLACE VALUE   = 2^5 2^4 2^3 2^2 2^1 2^0
ASSIGNMENTS
(POWERS OF 2)

STEP 3
PRODUCTION OF   =   32 + 16 + 8 + 0 + 2 + 1
STEPS 1 AND 2

STEP 4
SUM OF PRODUCTS   = $59._{10}$
FROM STEP 3 OR
DECIMAL EQUIVALENT

FIGURE 5-2   Conversion of a binary number to a decimal number.

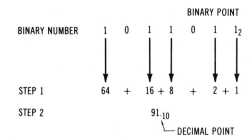

FIGURE 5-3 Simplified binary conversion process.

by the number 2. When the quotient is even with no remainder, a 0 is recorded. When the quotient has a remainder, a 1 is recorded. The steps needed to convert a decimal number to binary number are shown in Fig. 5-4.

The conversion process, in this case, is achieved by writing down the decimal number (35_{10}). Divide this number by the base of the system, or 2. Record the quotient and remainder as indicated. Move the quotient of step 1 to step 2 and repeat the process. The division process continues until the quotient becomes zero. The binary equivalent is simply the remainder values in their last-to-first placement order. You may want to practice this process on several numbers to gain some degree of proficiency.

Binary-Coded-Decimal Numbers

When large numbers are to be indicated by binary numbers, they become somewhat awkward and difficult to use. For this reason, the binary-coded-decimal method of counting was devised. In this type of system, four binary digits are used to represent each decimal digit. To illustrate this procedure, we have selected the number 329_{10} to be converted to a binary-coded-decimal (BCD) number. In straight binary numbers, $329_{10} = 101,001,001_2$.

To apply the BCD conversion process, the base 10 number is first divided into discrete digits according to the place values (see Fig. 5-4). The number 329_{10} therefore equals the digits 3-2-9. Converting each digit to binary would permit us to display this number as $0011\text{-}0010\text{-}1001_{BCD}$. Decimal numbers up to 999_{10} could be displayed and quickly interpreted by the process with only 12 binary numbers. The dashed line between each group of digits is extremely important when displaying BCD numbers.

The largest digit to be displayed by any group of BCD numbers is 9. This means that six digits of a number coding group are not being used at all in this system. Because of this the octal, or base 8, and the hexadecimal, or base 16, systems were devised. Digital systems still process numbers in binary form but usually display them in BCD, octal, or hexadecimal values.

Octal Numbering Systems

Octal, or base 8, numbering system is commonly used to process large numbers through digital systems. The octal system of numbers uses the same basic principles outlined with decimal and binary counting methods.

The octal numbering system has a radix or base of 8. The largest number displayed by the system before it changes the place value is seven. The digits 0, 1, 2, 3, 4, 5, 6, and 7 are used in the place positions. The place values of digits starting at the left of the octal point are the powers of 8: $8^0 =$ units or 1's, $8^1 = 8$'s, $8^2 = 64$'s, $8^3 = 512$'s, and $8^4 = 4096$'s, and so on. It is easy to indicate large number values with a minimum number of digits through this system.

Hexadecimal Numbering Systems

The hexadecimal, or base 16, numbering system is commonly used to process large number values. The hexadecimal system of numbers uses the same basic principles outlined with the decimal, binary, and octal

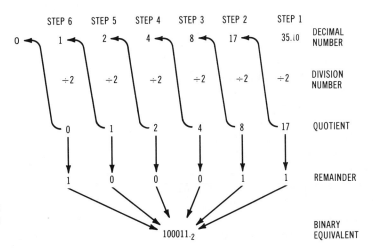

FIGURE 5-4 Conversion of a decimal number to a binary number.

counting systems. The largest number displayed by the system before it changes the place value is 15. The digits 0, 1, 2, 3, 4, 5, 6, 7, 8, 9 and letters A, B, C, D, E, and F are used in the place positions. The letter designations A through F are used to represent the number values of 10, 11, 12, 13, 14, and 15. This is purposely done to indicate two-digit numbers with a single-digit symbol. The place values of digits starting at the left of the hexadecimal point are the powers of 16: 16^0 = units or 1's, 16^1 = 16's, 16^2 = 256's, 16^3 = 4096's, and so on. Extremely large numbers can be indicated by this type of numbering with a minimum of digits. Hexadecimal numbers are widely used in microprocessors and computers.

DIGITAL ELECTRONIC OPERATIONAL STATES

Digital electronic circuits require a precise value definition or operational state in order for them to be useful. In digital circuits binary signals are considered to be far superior to those of the octal, decimal, or hexadecimal systems. In effect, binary signal data can be processed very easily through a circuit because they are represented by two stable states of operation. These states are defined as on or off, up or down, right or left, or any other two-condition designations. There is no in-between step or condition for this designation. The two states are decidedly different in value and can be easily distinguished.

The symbols or representations used to define the operational states of a binary system are extremely important in digital electronics. Voltage, the number 1, true, or a letter designation such as A are often used to denote the on or conducting operational state. The other alternative can be defined as no voltage, 0, false, or the letter designation A. Digital circuits must be capable of being set to either of these two operational states. The circuit must remain in this state until something causes a change in its operating condition.

Any electronic device that has two operational states is considered to be bistable. Switches, relays, transistors, diodes, and integrated circuits are widely used to achieve this type of operation. A bistable device, therefore, has the capability of storing one binary digit or bit of information. When a number of these devices are used, it is possible to build an electronic circuit that will make decisions based on the applied input signals. The resulting output is decision based on the status of the applied input signal. Since the device and its associated circuitry are capable of making decisions, the circuit is generally called a *logic circuit*.

The control capability of a bistable device to make decisions based on two-state input operations is called *binary logic*. A majority of these control func-

tions can be accomplished with different combinations of parallel and series switching operations. Three basic logic circuits have been devised to represent these operations. The circuits and resulting logic decisions are called AND, OR, and NOT. These decisions are unique and play a very important role in digital electronic circuitry.

Electronic circuits designed to achieve specific logic functions are commonly called *gates*. This term refers to the capability of the circuit to pass or block specific digital signals. In words, logic gate functions are often expressed by a simple if-then statement. If the inputs of an AND gate are all 1, the output will be a 1. If a 1 is applied to any input of the OR gate, the output will be 1. If any input is applied to a NOT gate, the output will be reversed.

AND Gates

An AND gate is designed to have two or more inputs and one resulting output. The logic decision of an AND gate is based on the status of its input. If both or all inputs are in the on state, the output will be on or 1. Mathematically, this expression achieves the multiplication function. One (1) times one (1) equals one (1). The other alternatives are zero (0) times zero (0), one (1) times zero (0), and zero (0) times one (1) all equal zero (0).

Figure 5-5 shows a simple switch-lamp analogy of the AND gate, its symbol, and an operational table. In this circuit when the switch is on it represents a one (1) condition and the off condition represents a zero (0). The lamp or output of the circuit responds in the same manner. The switches of this circuit are con-

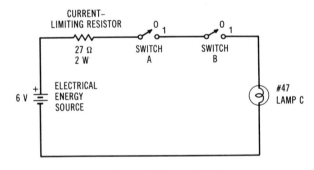

(A) Simple AND gate circuit.

SWITCH A	SWITCH B	LAMP
0	0	0
0	1	0
1	0	0
1	1	1

(B) AND gate symbol.   *(C) AND gate truth table.*

FIGURE 5-5   AND gate information.

nected in a series configuration. Since two switches are used in this circuit, the input has 2 × 2 or 4 possible input combinations. An output is generated for each input combination. The table of Fig. 5-5C shows input/output combinations of the gate. As a rule, tables of this type are called *truth tables*. This table tells how the logic circuit responds.

OR Gates

An OR gate is designed to have two or more inputs and one output. With two input gates, there will be 2 × 2, or four possible output combinations. The input states of an OR gate are generally described as 1 and 0 combinations like those of the AND gate. The output of an OR gate is significantly different from that of the AND gate. Figure 5-6 shows a simple switch-lamp analogy of the OR gate, its symbol, and a truth table describing its switching combinations.

Functionally, an OR gate will produce an on or 1 output when both inputs are 1, or when either input is one. Mathematically, this function is described as $A + B = C$. This expression shows OR addition. Circuit applications of this gate are used to make decisions as to whether a 1 or on condition appears at either input. The inputs of an OR gate are connected in parallel. As a result of this type of construction, the output can be energized or turned on by either input on an independent basis.

NOT Gates

The NOT gate is somewhat unusual when compared with other gate functions. A NOT gate has only one input and one output. The input and output both have two states or conditions of operation. The output of a NOT gate is, however, the reverse or inversion of the input. An on state at the input will produce an off state at the output. The reverse action of this state has an off input producing an on output. In effect, the NOT gate has an inversion capability.

Figure 5-7 shows a simple switch-lamp analogy of the NOT gate. When the switch is off, the lamp is on. Closing the switch shorts out the lamp and turns it off. The NOT gate in this case, achieves inversion of the switch action. A logic symbol and truth table of the NOT function are also shown in Fig. 5-7B and C.

Mathematically, a NOT gate inverts or complements a number. A word description of this gate has an A input, causing the output to NOT be A. This is expressed by the equation $A = \overline{A}$. Applications of the NOT function are used to invert circuit operational characteristics and in combinational logic circuits.

Combinational Logic Gates

When two or more basic logic gates are connected together, they form a combinational logic gate. In practice, the NOT function is generally combined with either the AND or the OR function. An AND gate followed by a NOT gate forms a NOT–AND combination called a NAND gate. The output of this gate is a reverse or inversion of the AND function. In the same manner, NOT–OR or NOR gates are formed by combining OR and NOT gates. A NOR gate is functionally an inversion of the OR gate.

A lamp-switch version of the NAND function, its symbol, and an operation state table are shown in Fig. 5-8. Control of this circuit is achieved by switches A

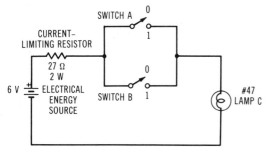

(A) Simple OR gate circuit.

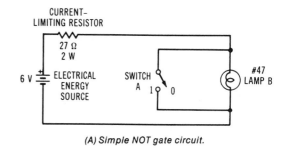

(A) Simple NOT gate circuit.

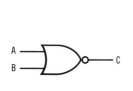

(B) OR gate symbol.

SWITCH A	SWITCH B	LAMP C
0	0	0
0	1	1
1	0	1
1	1	1

(C) OR gate truth table.

FIGURE 5-6 OR gate information.

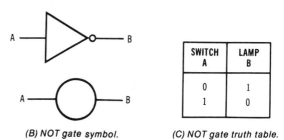

(B) NOT gate symbol.

SWITCH A	LAMP B
0	1
1	0

(C) NOT gate truth table.

FIGURE 5-7 NOT gate information.

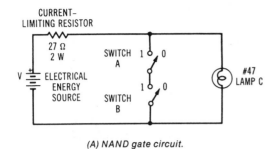

(A) NAND gate circuit.

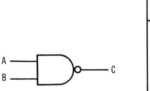

(B) NAND gate symbol.

SWITCH A	SWITCH B	LAMP C
0	0	1
0	1	1
1	0	1
1	1	0

(C) NAND gate truth table.

FIGURE 5-8 NAND gate information.

and B. The output is shown as the lamp and indicated by the letter C. Due to the circuit configuration, the lamp (C) will be on for three of the four switch combinations. When both switches are on at the same time, the lamp will be off. Mathematically, the NAND function is expressed as $A \times B = \overline{C}$. The bar symbol over a letter indicates the negation or inversion function. NAND gates are frequently combined with other logic functions to achieve specific counting and decoding operations.

A lamp-switch analogy of the NOR function, its symbol, and an operational state table are shown in Fig. 5-9. Control of this gate is achieved by switches

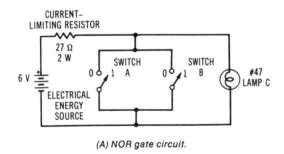

(A) NOR gate circuit.

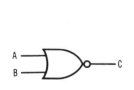

(B) NOR gate symbols.

SWITCH A	SWITCH B	LAMP C
0	0	1
0	1	0
1	0	0
1	1	0

(C) NOR gate truth table.

FIGURE 5-9 NOR gate information.

A and B with the output indicated by the lamp C. The lamp or output will only be on when switches A and B are both in the off position at the same time. The other three switch combinations will cause the lamp to be off. Mathematically, the operation of a NOR gate is expressed as $A + B = \overline{C}$. The output will only be active or on when switch A and B are off. NOR gates are frequently used to achieve a variety of decoding and counting functions in conjunction with other gates.

Logic Gate Circuits

Gate functions can be achieved today in a variety of different ways. The switch-lamp version of a gate circuit was only used to demonstrate the functional operation of the gate. Actual logic gates are significantly different in construction and circuit operation. A number of transistors are used in the construction of specific logic gates. Different families of ICs are now available to achieve practically any logic gate or function. The operation of this type of circuit is based on the components involved and the configuration in which they are connected. Through this type of construction it is now possible to build complex logic circuits by simply interconnecting different logic gates. An understanding of basic logic is therefore much more important today than it has been in the past. The truth table of a logic function and its symbol terminal connections tend to be the important items needed in selecting an IC to perform a particular function. This type of information can be readily found in a manufacturer's data manual.

DIGITAL ELECTRONIC DEVICES

Digital electronics employs a number of devices which are not classified specifically as logic gates. These devices usually play some unique role in the operation of a digital electronic circuit. Such things as flip-flops, counters, decoders, and memory devices are included in this classification. A presentation of truth tables, logic symbols, and the operational characteristics of these devices will be made so that they may be used more effectively when the need arises. In general, these devices are constructed on IC chips. Operation is based to a large extent on the internal circuit construction of the IC. Very little can be done to alter the operation of these devices other than modify the input or use its output to influence the operation of another device.

Flip-Flops

Flip-flops are commonly used to generate signals, shape waves, and achieve division. In addition to these operations, a flip-flop may also be used as a memory device. In this capacity, it can be made to hold an output state even when the input is completely removed. It can also be made to change its output when an appropriate input signal occurs.

The reset–set or *R-S* flip-flop is a typical digital electronic control device. The logic diagram, symbol, and truth table of this flip-flop are shown in Fig. 5-10. Note that the truth table of this device is somewhat more complicated than that of a simple logic gate. It, for example, must show the different states of the device before an input pulse occurs, and then show how it changes after the input has arrived. Note that two of the operating conditions produce an unpredictable output. In this state of operation the first arriving pulse will produce an output by coincidence.

In many digital electronic circuits flip-flops must be set and cleared at specific times with respect to other operating circuits. This type of operation can be achieved by manipulating flip-flops in step with a clock pulse. In this case, the appropriate *R-S* inputs and clock pulse must all be present to cause a state change. A device of this type is called an *R-S* triggered flip-flop or simply an *R-S-T* flip-flop.

The truth table of an *R-S-T* flip-flop is basically the same as that of the *R-S* flip-flop of Fig. 5-10. The

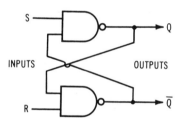

(A) Logic diagram with NAND gates.

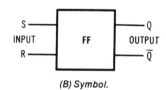

(B) Symbol.

APPLIED INPUTS		PREVIOUS OUTPUTS		RESULTING OUTPUTS		
S	R	Q	$\bar{Q}$	Q	$\bar{Q}$	
0	0	1	0	1/0	1/0	UNPREDICTABLE
0	1	1	0	1	0	
0	0	1	0	0	1	
1	1	1	0	1	0	
1	0	0	1	0/1	0/1	UNPREDICTABLE
0	1	0	1	1	0	
1	0	0	1	0	1	
1	1	0	1	0	1	

(C) Truth table.

FIGURE 5-10   *R-S* flip-flop information.

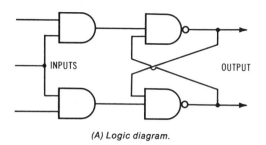

(A) Logic diagram.

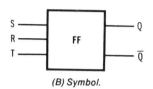

(B) Symbol.

APPLIED INPUTS			PREVIOUS OUTPUTS		RESULTING OUTPUTS		
S	R	T	Q	$\bar{Q}$	Q	$\bar{Q}$	
0	0	0	1	0	1/0	0/1	UNPREDICTABLE
0	1	1	1	0	1	0	
1	0	0	1	0	0	1	
1	1	1	1	0	1	0	
0	0	0	0	1	0/1	1/0	UNPREDICTABLE
0	1	1	0	1	1	0	
1	0	0	0	1	0	1	
1	1	1	0	1	0	1	

(C) Truth table.

FIGURE 5-11   *R-S-T* flip-flop information.

R-S-T flip-flop will only initiate a state change when a clock pulse arrives at the *T* or trigger input. A two-input AND gate is simply added to the set and reset inputs to accomplish this operation. Figure 5-11 shows this modification, the logic symbol, and truth table for an *R-S-T* flip-flop.

Another important flip-flop device that is commonly used in digital electronic circuits is the *J-K* flip-flop. This device is somewhat unusual because it has no unpredictable output states. It can be set by applying a 1 to the *J* input and can be cleared by feeding a 1 to the *K* input. A 1 signal applied to both *J* and *K* inputs simultaneously causes the output to change states or toggle. A 0 applied to both inputs at the same time does not initiate a state change. The inputs of a *J-K* flip-flop are controlled directly by the application of a clock pulse.

Figure 5-12 shows a logic diagram, symbol representation, and truth table for a *J-K* flip-flop. Notice that there are no unpredictable output states for this device. Several versions or modifications of the basic *J-K* flip-flop are now available. These include devices with preset and preclear inputs that can be used to establish sequential operations at a precise point in

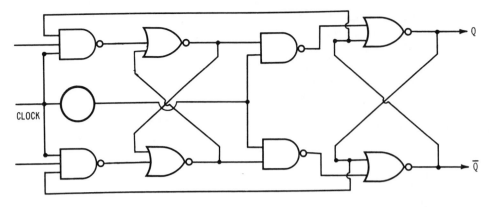

(A) Logic diagram.

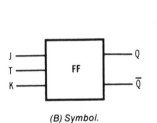

(B) Symbol.

APPLIED INPUTS			PREVIOUS OUTPUTS		RESULTING OUTPUT	
J	K	T	Q	$\bar{Q}$	Q	$\bar{Q}$
0	0	0	0	1	0	1
0	1	1	0	1	0	1
1	0	0	0	1	1	0
1	1	1	0	1	1	0 ← TOGGLE STATE
0	0	0	1	0	1	0
0	1	1	1	0	0	1
1	0	0	1	0	1	0
1	1	1	1	0	0	1

(C) Truth table.

FIGURE 5-12 *J-K* flip-flop information.

time. Flip-flops are commonly used as the basic logic element for counting operations, temporary memory, and sequential switching operations.

Digital Counters

One of the most versatile and important logic devices of a digital electronic circuit is the counter. This device can be employed to count a wide variety of objects in a number of different digital circuits. While this device may be called on to count an endless number of objects, it essentially counts only electronic pulses. These pulses may be produced electronically by a clock mechanism, electromechanically, photoelectrically, acoustically, or by a number of other processes. The basic operation of the counter, however, is completely independent of the pulse generator.

Binary Counters

A common application of the digital counter is used to count numerical information in binary form. This type of device simply employs a number of flip-flops con-

nected so that the Q output of the first device drives the trigger or clock input of the next device. Each flip-flop therefore has a divide by two function.

Figure 5-13A shows *J-K* flip-flops connected to achieve binary counting. This device is generally called a ripple counter. Each flip-flop in the circuit has its *J* and *K* inputs held at a logic 1 level. Each clock pulse applied to the input of FF$_A$ will then cause a change in state. The flip-flops only trigger on the negative going part of the clock pulse. The output of FF$_A$ will therefore alternate between 1 and 0 with each pulse. A 1 output will appear at the Q output of FF$_A$ for every two input pulses. This means that each flip-flop has a divide-by-2 function. Five flip-flops connected in this manner will produce a 2^5, or 32_{10}, count. The largest count is 11111_2 or 31_{10}. This occurs when a 1 appears at all the Q outputs.

By grouping three flip-flops together as in Fig. 5-13B, it is possible to develop the units part of a binary-coded-octal (BCO) counter. Therefore, 111_2 would be used to represent the seven count or seven units of an octal counter. Two groups of three flip-flops connected in this manner would produce a maximum count of 111-111_2, which represents 77_8 or 63_{10}.

By placing four flip-flops together in a group as in Fig. 5-13C,, it is possible to develop the units part of a binary-coded-hexadecimal (BCH) counter. Thus 1111_2 would be used to represent F_{16} or 15_{10}. Two groups of four flip-flops could be used to produce a maximum count of $1111\text{-}1111_2$, which would represent FF_{16} or 255_{10}. Each succeeding group of four flip-flops would be used to raise the counting possibility to the next power of 16.

Binary counters that contain four interconnected flip-flops are commonly built on one IC chip. Fig. 5-14 shows the logic connections of a four-bit binary counter. When used in this type of counter FF_A will show a maximum count of 1111_2 or 15_{10}. By disconnecting FF_A from FF_B and applying the clock to the input of FF_B, it is possible to have a three-bit or BCO

counter. The outputs of flip-flops FF_A through FF_D are labeled A, B, C, and D, respectively.

Decade Counters

Since most of the mathematics that we use today is based on the decimal or base 10 system, it is important to be able to count electronically by this method. Digital electronic circuits are, however, designed to process information in binary form because of the ease by which a two-state signal can be manipulated. The output of a binary counter must therefore be changed into decimal form before it can be used by a person not familiar with binary numbers. The first step in this process is to change binary signals into a binary-coded-decimal (BCD) form.

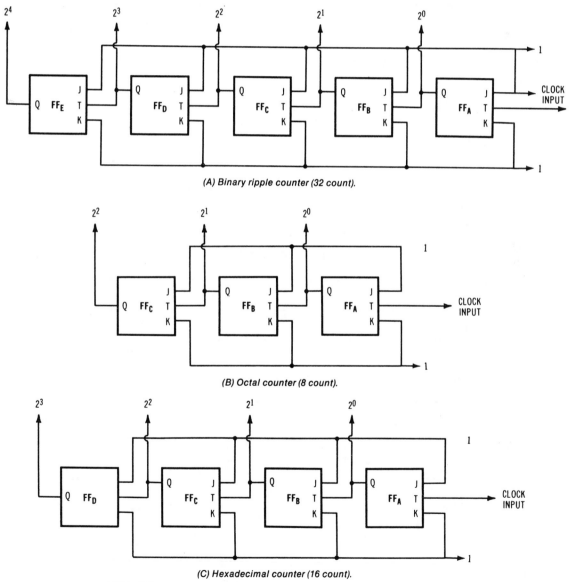

(A) Binary ripple counter (32 count).

(B) Octal counter (8 count).

(C) Hexadecimal counter (16 count).

FIGURE 5-13 Digital counter achieved with *J-K* flip-flops.

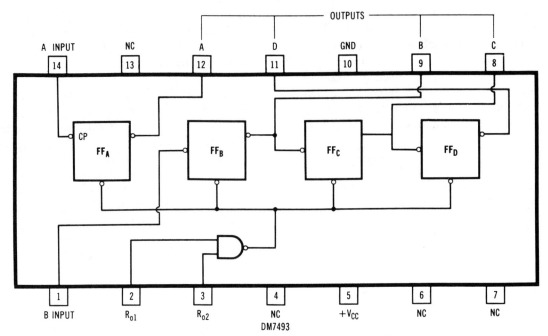

FIGURE 5-14 Four-bit binary counter IC.

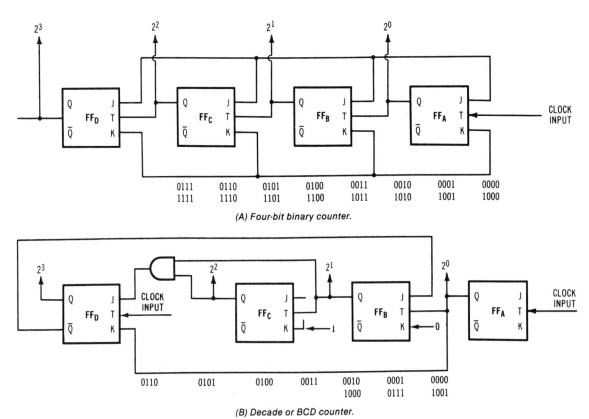

(A) Four-bit binary counter.

(B) Decade or BCD counter.

FIGURE 5-15 Binary counter to decade or BCD counter.

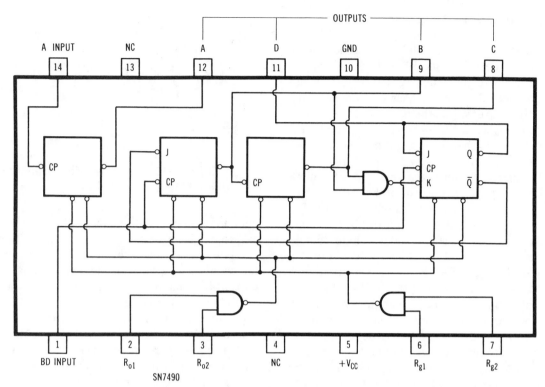

FIGURE 5-16 IC BCD decade counter.

Consider now the four-bit binary counter of Fig. 5-15A. In this counter, 16 natural counts are achieved by the four flip-flops. To convert this counter into a decade counter it must be made to skip some of its natural counts. Notice the 16 natural counts listed below the binary counter.

A method of converting a binary counter into a decade counter is shown in Fig. 5-15B. In this configuration the first seven counts occur naturally as shown. Through these steps FF_D stays at 0. The $\overline{Q}$ output of FF_D, therefore, remains at 1 during these counts. This 1 is applied to the J input of FF_B, which permits it to trigger with each clock pulse.

At the seventh count, 1s appearing at the Q outputs of FF_B and FF_C are applied to an AND gate. This action produces a logic 1 and applies it to the J input of FF_D. Arrival of the next clock pulse triggers FF_A, FF_B, and FF_C into the off state and turns on FF_D. This represents the eighth count.

When FF_D is in the on state, Q is 1 and $\overline{Q}$ is 0. This causes a 0 to be fed to the J input of FF_B, which prevents it from triggering until cleared. Arrival of the next clock pulse causes FF_A to be set to a 1. This registers a 1001_2, which is the ninth count. Arrival of the next count clears FF_A and FF_D instantly. Since FF_B and FF_C were previously cleared by the seventh count, all 0s appear at the outputs. The counter has therefore cycled through the ninth count and returned

to zero ready for the next input pulse. BCD counting of this type can be achieved in a number of ways. This method is quite common in IC devices today.

Figure 5-16 shows a complete BCD counter built on an IC. The operation of this IC is essentially the same as the one just described. When used as a BCD counter the FF_A output must be connected to the BD input. Omitting this connection and applying the clock pulses to the BD input produces a five count. With this option the counter is somewhat more versatile. The two NAND gates of this IC are used to set or reset the four flip-flops from an outside source. This counter triggers only on the negative going part of the clock pulse.

DIGITAL ELECTRONIC DISPLAYS

The average person looking at the display of a digital electronic instrument will not necessarily be familiar with binary numbers, the BCD method of display, octal, or hexadecimal systems. A digital instrument must therefore change its information into something that can be readily used without causing confusion. A function of nearly all digital instruments is the changing of signal information into decimal values that can be suitably displayed. The electronic display section of a digital instrument is responsible for this function.

Electronic Numerical Display

Numerical displays are available in two general formats. These are described as the individual numeral type and the seven-segment display. The individual numeral display uses the glow discharge principle in its operation. Figure 5-17 shows the principle of operation for this type of display. When voltage is applied to the respective electrodes it will cause gas in the enclosure to ionize. A characteristic glow will appear around the cathode. The cathode is formed in the shape of a number. The glow conforms to the area of the cathode thus forming a number. Displays of this type have cathodes shaped for the numbers 0 through 9 in the same enclosure. With an appropriate cathode–anode combination energized, a number can be viewed from the end of side of the tube. Figure 5-18 shows an internal view of the electrode structure of an individual numerical display tube. This particular tube requires a voltage of 150 V or more to produce ionization of neon gas. When the gas ionizes a characteristic orange glow is produced by the representative numeral.

The seven-segment method of display seems to be used more widely in instruments than the individual numeral display. In this type of display a numeral is divided into seven segments or slits. Figure 5-19 shows a display of this type. Illumination of two or more of these segments in an appropriate combination will produce the numbers 0 through 9. If, for example, segments *f, g, b,* and *c* are energized, the number 4 will be displayed. Energizing all seven segments will produce the number 8. Displays of this type are generally housed in a 14-pin dual in-line IC package.

The illumination of a seven-segment display is achieved in a variety of ways. An LED type of display has several discrete LEDs commonly attached to a segment. A common anode type of display has the anode of each LED in the segment and all segments

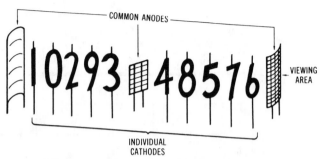

FIGURE 5-18   Exploded view of an individual numerical display tube.

commonly connected together. The positive side of a 5-V source is applied to this connection. The common-cathode connection of each segment is connected to the negative or ground of the power supply through a current-limiting resistor. A decoder IC such as an SN7447 is used to complete the path to an appropriate segment to produce illumination. Figure 5-20 shows a diagram of the segments of an LED display.

Liquid-crystal seven-segment displays are also used in industrial instruments today. The configuration of this display is primarily the same as that of the LED unit. Illumination by the liquid-crystal method is quite unique when compared with other displays. Most LCDs operate by the so-called twisted-nematic effect and are powered through a special segment addressing technique that reduces the number of connections to the display. The twisted-nematic type of LCD has a layer of liquid-crystal sandwiched between two pieces of polarized glass and a set of electrodes with seven-segment columns, as shown in Fig. 5-21. The polarized glass plates are designed to be 90° out of phase with each other. Normal light going through the top polarizer plate and into the liquid crystal undergoes a 90° twist as it passes through the material. This 90° twist permits light to pass through the polarizer electrode segments, liquid crystal, and to the bottom polarizer

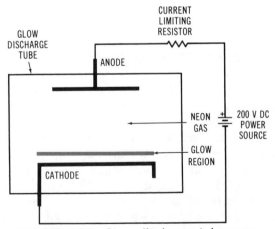

FIGURE 5-17   Glow discharge tube operation.

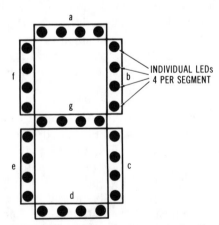

FIGURE 5-19   Seven-segmented display with four diodes per segment.

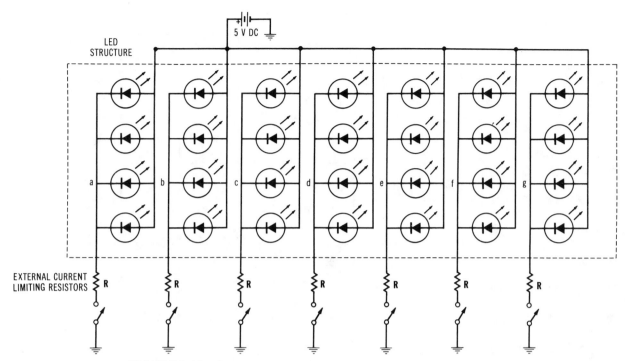

FIGURE 5-20 Segment structure of a seven-segmented LED display.

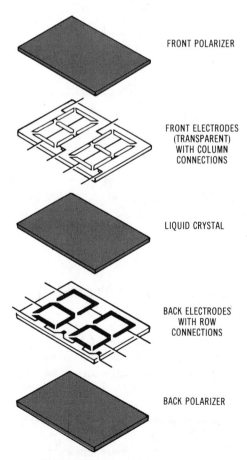

FRONT POLARIZER

FRONT ELECTRODES (TRANSPARENT) WITH COLUMN CONNECTIONS

LIQUID CRYSTAL

BACK ELECTRODES WITH ROW CONNECTIONS

BACK POLARIZER

FIGURE 5-21 LCD structure.

plate. However, applying an electric field to specific segments of the front and back electrodes will destroy the 90° twisting effect. The second polarizer then blocks light passing through the first polarizer. The end result is a dark area on a light background. The dark area is representative of the selected segments of the display. In effect, the display or number appears as a dark area on a light background. Using this operating principle, displays can be formed to produce letters, punctuation marks, and symbols, as well as numbers. As a rule, LCDs consume power in the microwatt range and are energized by low-voltage dc. A major disadvantage of the LCD is that it must operate in an area that has a reasonable level of ambient light to produce a suitable contrast in the display area.

Alphanumeric Displays

Letters, numbers, punctuation marks, and symbols for communication with a computer are frequently produced by alphanumeric displays. This display uses the 5 × 7 dot matrix type of construction. Figure 5-22 shows the layout of this display.

The 5 × 7 dot matrix display uses 35 discrete LEDs in its construction. A specific LED is controlled by a combination of two switches. If the switches in row 4 and column 5 are both turned on at the same time, diode 25 will be energized. For a complete vertical row to be energized would require one column

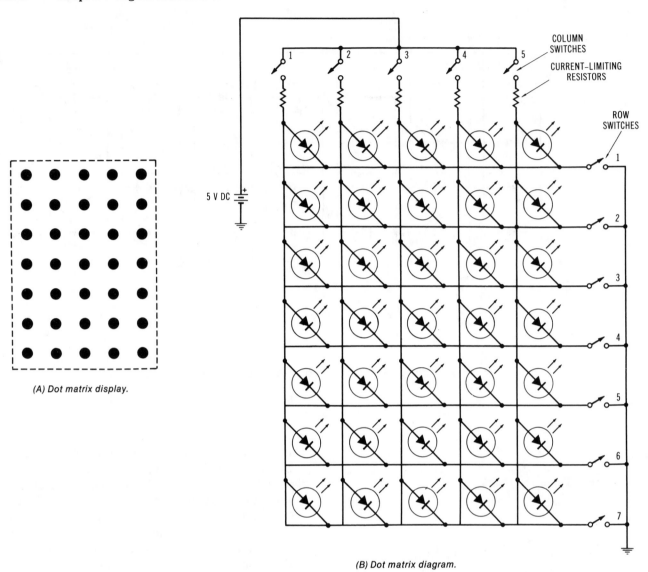

(A) Dot matrix display.

(B) Dot matrix diagram.

FIGURE 5-22 5 × 7 LED display device.

switch and all seven row switches to be turned on. A complete horizontal row would be energized by one row switch and all five column switches. Through 12 different switching combinations 96 characters can be produced by the alphanumeric display. Today these switching combinations are transmitted to the display unit through a seven-bit ASCII code. The term *ASCII* is an acronym for *American Standard Code for Information Interchange*. This code has a 3-bit group and a 4-bit group. The first 3 bits of data in the leftmost position of the 7-bit number are called the *column select* group. The next 4 bits of data represent the *row select* data. Figure 5-23 shows the ASCII code. Note that the column numbers are 000 through 111. The letter A, for example, would be 100 0001. The letter B would be 100 0010. An ASCII code is generated by pressing the letters and numbers of a typewriter key-

board. Integrated circuits accept the ASCII code and translate it into a switching combination that will energize the appropriate segments of the dot matrix display.

DIGITAL ELECTRONIC DECODING

Before a display device can be effectively used to develop a digital number, it must receive an appropriate signal from the counter circuit. The counter signal usually contains information in binary form. This information must be decoded so that it will energize the display device when a specific number occurs. Decoding is achieved by a number of four-input gates connected to the *A, B, C,* and *D* outputs of a BCD counter. When an appropriate binary number signal

	000	001	010	011	100	101	110	111
0000	NULL	DC_0	b	0	@	P		
0001	SOM	DC_1	!	1	A	Q		
0010	EOA	DC_2	"	2	B	R		
0011	EOM	DC_3	#	3	C	S		
0100	EOT	DC_4	$	4	D	T		
0101	WRU	ERR	%	5	E	U		
0110	RU	SYNC	&	6	F	V		
0111	BELL	LEM	.	7	G	W		
1000	FE_0	S_0	(	8	H	X		
1001	HT / SK	S_1	)	9	I	Y		
1010	LF	S_2	*	:	J	Z		
1011	V_{tab}	S_3	+	;	K	[		
1100	FF	S_4	,	<	L	\		ACK
1101	CR	S_5	–	=	M	]		②
1110	SO	S_6	*	>	N	↑		ESC
1111	SI	S_7	/	?	O	–		DEL

Definitions of control abbreivations:

ACK	Acknowledge	LEM	Logical end of media
BELL	Audible signal	LF	Line feed
CR	Carriage return	RU	"Are you....?"
DC_0-DC_4	Device control	SK	Skip
DEL	Delete idle	SI	Shift in
EOA	End of address	SO	Shift out
EOM	End of message	S_0-S_7	Separator (space)
EOT	End of transmission	SOM	Start of message
ERR	Error	V_{tab}	Vertical tabulation
ESC	Escape		
FE	Format effector	WRU	"Who are you?"
FF	Form feed	②	Unassigned control
HT	Horizontal tabulation	SYNC	Synchronous idle

Example of code format:

B_7 B_1 $\underbrace{100}\underbrace{0100}$ is the code for D

three-bit group   four-bit group

FIGURE 5-23 ASCII code.

appears at the input of the decoder, it will energize the display device. In an actual circuit, a decoder connects the ground or common of the power source to specific bar segments to produce a display. In a sense, this device completes the manual switching operation of Fig. 5-20 electronically. The decoder completes this operation automatically when it receives a suitable signal.

Two distinct types of decoders are available today for driving display devices. The discrete number display requires a decoder that has 10 distinct output signal possibilities. Figure 5-24 shows the circuitry of a decimal decoder-driver. In this type of decoding operation, only one output is energized at a time. The seven-segment type of decoder is uniquely different. It produces two or more output signals when it energizes the display. When the number eight is displayed, all seven outputs must be energized to actuate the display. The basic method of actuating specific gates in the decoder is very similar for both decoder types.

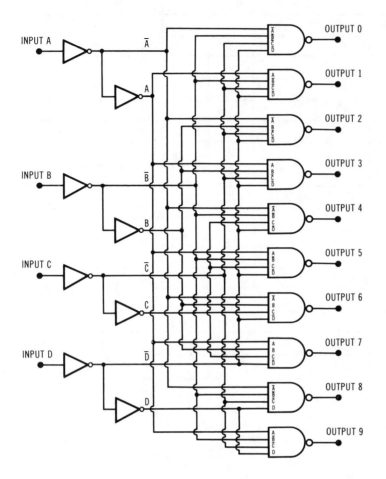

(A) Logic and connection diagrams.

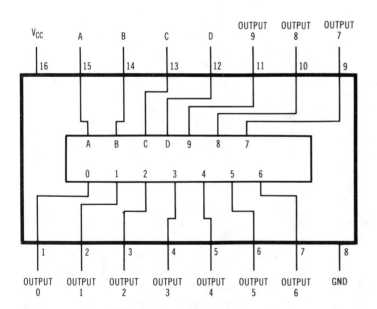

(B) Dual in-line and flat package.

INPUTS				OUTPUTS									
D	C	B	A	0	1	2	3	4	5	6	7	8	9
0	0	0	0	0	1	1	1	1	1	1	1	1	1
0	0	0	1	1	0	1	1	1	1	1	1	1	1
0	0	1	0	1	1	0	1	1	1	1	1	1	1
0	0	1	1	1	1	1	0	1	1	1	1	1	1
0	1	0	0	1	1	1	1	0	1	1	1	1	1
0	1	0	1	1	1	1	1	1	0	1	1	1	1
0	1	1	0	1	1	1	1	1	1	0	1	1	1
0	1	1	1	1	1	1	1	1	1	1	0	1	1
1	0	0	0	1	1	1	1	1	1	1	1	0	1
1	0	0	1	1	1	1	1	1	1	1	1	1	0
1	0	1	0	1	1	1	1	1	1	1	1	1	1
1	0	1	1	1	1	1	1	1	1	1	1	1	1
1	1	0	0	1	1	1	1	1	1	1	1	1	1
1	1	0	1	1	1	1	1	1	1	1	1	1	1
1	1	1	0	1	1	1	1	1	1	1	1	1	1
1	1	1	1	1	1	1	1	1	1	1	1	1	1

(C) Truth table.

FIGURE 5-24 BCD-to-decimal decoder IC. (Courtesy of National Semiconductor Corp.)

BCD-to-Decimal Decoders

The logic diagram of a BCD-to-decimal decoder is shown in Fig. 5-25. This entire circuit is built on a single IC chip and housed in a dual in-line package. This IC has A, B, C, and D inputs and 10 independent outputs. The inputs are inverted once or twice before being applied to their respective NAND gates. When an appropriate input number combination is applied, it causes the corresponding NAND gates to be actuated. A zero or ground will then appear at its output. This, in turn, energizes a selected cathode of the display by forming a completed path to ground. Digital signals other than the selected number combination applied to each NAND gate will cause the output to be at a high state or generate a one output. This output does not complete the ground path. As a result, it causes the cathodes of the display to see an open circuit. The respective elements of the display will not be illuminated by this action.

Assume now that the output of a BCD counter produces a binary signal of 0101_2 and applies it to the input of a decoder. The truth table of Fig. 5-24 shows that this signal represents the decimal number five. The decoder input sees A as 1, B as 0, C as 1, and D as 0. Directing the 1 inputs of A and C into two inverters causes these two signals to remain at the same level when applied to the NAND gate of number 5. Inputs $\bar{B}$ and $\bar{D}$, by comparison, are directed through only one inverter. This causes an inversion of the two

input signals and they appear as B and D at the input of the NAND gate of number 5. Since B and D were originally low or 0, they will now appear as a 1 or high level at the NAND gate input. This means that the four inputs of the NAND gate now represent a 1 or high level. The NAND gate shows this as a 0 or low level at output number 5. This represents a ground at the cathode of number 5 of the display. The circuit path to ground for the cathode of number 5 is completed through the NAND gate, thus producing illumination. A similar action would be achieved by each of the decoder outputs. Through this action, the decoder will generate a single output for each number combination applied to its input.

BCD-to-Seven-Segment Decoders

The logic block of a BCD to seven-segment decoder is shown in Fig. 5-26. The entire circuit of this functional block is built on a single IC chip. Inputs A, B, C, and D are applied to either one or two inverters as in the decimal decoder. The decoding process is very similar in nearly all respects to that of the decimal decoder. The seven-segment decoder, however, necessitates two or more outputs generated at the same time according to the number being displayed. In this decoder each NAND gate output will produce a zero or ground when it receives four 1s at the input. This must be done for each of the respective seven-segment number combinations.

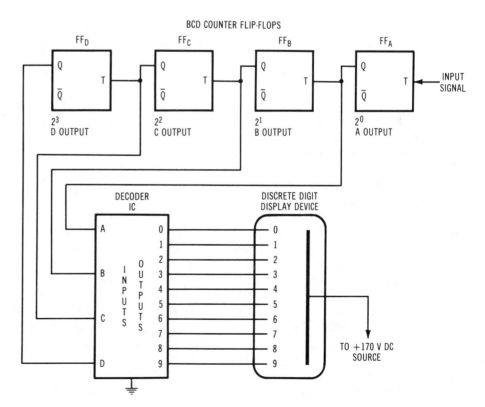

FIGURE 5-25 Logic diagram of BCD-to-decimal decoder.

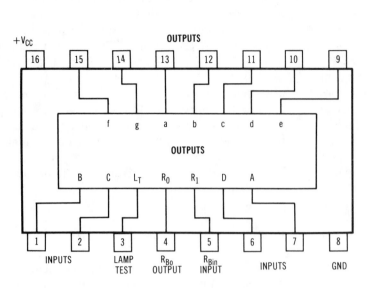

(A) IC connection diagrams.

(B) Numbers and corresponding segments.

(C) Truth table.

BCD INPUTS				7-SEGMENT OUTPUTS						
A	B	D	D	a	b	c	d	e	f	g
0	0	0	0	0	0	0	0	0	0	1
0	0	0	1	1	0	0	1	1	1	1
0	0	1	0	0	0	1	0	0	1	0
0	0	1	1	0	0	0	0	1	1	0
0	1	0	0	1	0	0	1	1	0	0
0	1	0	1	0	1	0	0	1	0	0
0	1	1	0	0	1	0	0	0	0	0
0	1	1	1	0	0	0	1	1	1	1
1	0	0	0	0	0	0	0	0	0	0
1	0	0	1	0	0	0	0	1	0	0

"0" OUTPUT=GROUNDED ELEMENT
"1" OUTPUT=OPEN CIRCUIT

FIGURE 5-26   BCD-to-seven-segment decoding.

The different decoding combinations needed to display the decimal numbers of a seven-segment display are shown in the truth table of Fig. 5-26C. The A, B, C, and D inputs and corresponding a, b, c, d, e, f, and g outputs reflect the possible decoding combinations needed to produce a specific number output. The output of this decoder can be used to actuate either light emitting diode or liquid-crystal displays. As a rule, decoders with a seven-segment display are more commonly used in digital electronic instruments than is the decimal decoder type of unit.

DIGITAL ELECTRONIC INPUT

Digital electronic instruments must employ some type of input device that changes real-world data into digital signals that can be processed by the circuit. Most of the real-world data that are of concern to a digital instrument are either of a continuous changing value type or of the on–off variety. Industrial process values such as pressure, temperature, liquid level, and fluid flow tend to change value rather gradually or on a continuous basis. This type of change produces a large number of discrete values. Data of this type are described as *analog values*.

Input data that switch abruptly from one state to another without any in-between value are described as *digital values*. A number of objects passing through a production-line counter produces a signal of this type. Any two-state operation is classified as a digital quantity. Holes or no holes in a punched card, light or no light in an optical system, or the on–off status of a switch can be used to produce digital input signals. In actual circuit operation, digital signals are produced rather easily when compared with analog inputs.

The primary responsibility of the input of a digital system is to accept real-world data and prepare it for system processing. For digital signals the process is very simple. The data are applied to the appropriate input for counting and ultimately appear on the display for evaluation. For analog data the process is more complicated. Analog data must first be converted to a digital signal before they can be processed by the system. The part of an instrument responsible for this operation is called an *analog-to-digital converter* (ADC).

Analog-to-digital conversion is achieved in a variety of different ways in digital instruments today. As a rule, the conversion process is achieved by an integrated circuit which has most of its circuitry on a single chip. The operating principle of the chip varies a great deal among different manufacturers. There is also a great deal of difference in the methodology of achieving analog-to-digital conversion. Three different methods of conversion are discussed in the next section of this presentation. In general, the ADC compares an electrical value, such as the voltage being measured, with a reference value. The resulting output is a binary signal that is applied to the digital part of the instrument for processing.

DIGITAL ELECTRONIC INSTRUMENTS

Practically any value that needs to be measured in industry today can be achieved by a digital electronic instrument. The quantity being measured is primarily limited to the capabilities of the input device. Fluid flow rates, pressure, electrical quantities, temperature, time speed, distance, light level, radiation, and sound level are typical digital instrument applications.

With such a wide range of digital electronic instrument applications today, one would immediately think that these devices must be extremely complex. To the contrary, however, all instruments contain a number of common elements. The digital display unit, for example, is basic to practically all instruments. In addition to this, the counter is nearly always of the BCD type. Its output is generally used to drive a decoder of either the seven-segment or discrete number type. This basic unit is designed to achieve the counting function of a digital instrument.

The input part of a digital instrument is undoubtedly the most unique section of the entire system. Special devices such as analog-to-digital converters are used to achieve this operation. These devices are designed to change analog information such as pressure, temperature, vibration, sound, or light level into equivalent digital signals. When these signals enter the system their output must be capable of driving the counting unit. Figure 5-27 shows a block diagram of a digital instrument with these parts.

Digital Voltmeters

One of the most widely used digital instruments in operation today is the digital voltmeter (DVM). This instrument is simply designed to change variable voltage values into signals that can be displayed in digital form. The measurement of such things as resistance, current, temperature, and pressure can then be achieved by simply changing these quantities into different voltage values. The operating principle of a DVM is basic to all digital instruments.

The input of a DVM represents the most unusual part of the entire instrument. This particular part is primarily responsible for converting an unknown voltage into a usable digital signal. It must do this accurately, be capable of rejecting extraneous electrical noises to some extent, and be reasonably stable over a suitable operating range. Three rather common conversion techniques are used to achieve this function. These are the voltage controlled oscillators, ramp function instruments, and the dual-slope method of conversion. The input part of a DVM is best described as being an analog-to-digital converter.

Voltage-Controlled Oscillator Input

When the frequency of an oscillator changes with voltage it is commonly called a voltage-controlled oscillator (VCO).The output frequency of this oscillator can then be counted during a fixed interval of time and displayed as a digital output.

Figure 5-28 shows a block diagram of a digital voltmeter with a VCO. The frequency of the VCO is commonly scaled so that it has a certain output for each volt of input. One volt should appear as a count of 1000 on the display. For a DVM with a power-line time-base generator, the VCO would need a frequency range of 0 to 60 kHz per volt sensitivity.

The time-base generator of a VCO digital volt-

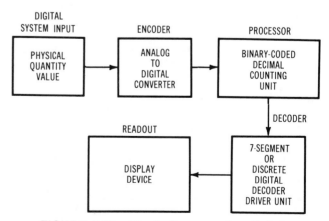

FIGURE 5-27 Block diagram of a digital instrument.

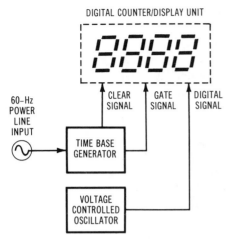

DIGITAL COUNTER/DISPLAY UNIT

CLEAR SIGNAL   GATE SIGNAL   DIGITAL SIGNAL

60-Hz POWER LINE INPUT

TIME BASE GENERATOR

VOLTAGE CONTROLLED OSCILLATOR

FIGURE 5-28   Block diagram of a VCO digital voltmeter.

meter may come from the ac power line or a separate crystal-controlled oscillator. Power-line frequency time-base units are very common and rather easy to achieve. The accuracy of the 60-Hz line frequency is quite adequate for most industrial applications of the DVM. The accuracy of this section could be improved by employing a crystal-controlled oscillator. Typically, a 100-kHz frequency is generated and scaled down to the base values of 10, 100, 1k, and 10k, according to system needs. A time-base generator is primarily responsible for turning on or gating the output of the VCO for a specific frequency count.

The basic operation of a digital voltmeter with a VCO is really quite simple. Assume that an unknown voltage of 1 V is applied to the input of the VCO. An instrument with a 60 kHz per volt sensitivity would immediately advance in frequency to 60 kHz. The power-line time-base generator would then turn on the output of the VCO for 1/60 of a second. As a result of this action, 1/60 of 60,000 Hz would produce a count of 1000. A decimal-point control circuit would show this value as 1.000 or 1 V on the display. After a short delay, the counter would clear, and reset to zero where it starts the next count. The signal processing section of the instrument simply counts the VCO output, decodes it, and ultimately drives the display device.

Ramp-Type DVM Input

The ramp type of DVM is a rather unique circuit that employs a gate, ramp generator, and a comparator attached to its input. The output of this unit drives the same counter, decoder, and display devices found in other DVMs. Basically, this input technique compares an applied voltage to that developed by the linear ramp generator. When the ramp voltage is less that the input, the comparator produces an output signal. This signal,

along with that of a clock pulse generator, is then applied to the input of the AND gate. These two inputs produce an output of clock pulses that are counted and ultimately displayed.

When the ramp generator voltage of this DVM rises in value and exceeds the input voltage, it causes the comparator to change polarity. This, in turn, causes a 0 to appear at the input of the AND gate. As a result, the clock pulse signal stops and the readout holds its display. The duration of the count is therefore accurately controlled by the linear rise time of the ramp generator. Figure 5-29 shows a block diagram of a ramp type of DVM with a representative voltage-to-time comparison circuit.

The ramp voltage of a DVM is commonly designed to rise at a rate of 1 V per millisecond. This means that the input voltage can be exceeded within a prescribed amount of time. As a result, this action causes a change in comparator polarity, which generates a 0 output. This actuates the AND gate. The clock pulses immediately stop passing into the counter. The number of counts in the display is then proportional to the value of the input voltage.

Assume now that 2 V is to be measured by the DVM. With this voltage applied, the 1.0 V per millisecond rise time of the ramp generator would also be started immediately. The output of the op amp comparator produces a 1 and applies this to the AND gate. At the same time, 1s from the clock pulse generator applied to the AND gate cause it to produce an output. After 2 ms, the ramp voltage exceeds the value of the input voltage. When this occurs, the op amp comparator changes to the 0 state and turns off the AND gate. The pulse repetition rate of the clock accurately generates 1,000,000 Hz during this period of operation. As a result, 2 ms, or 0.002 or 1,000,000 is 2000. A decimal-point control circuit would cause this number to appear as 2.000 on the display, which indicates a measured value of 2V.

DVMs of the ramp type are subject to many things that produce inaccurate measurements. The nonlinearity of the ramp, stability of the clock generator, and noise rejection are some of the more critical problems. The accuracy of this type of DVM is somewhat limited because of these problems. In general, a DVM of this type is much better than the typical 3% accuracy of a portable analog instrument.

Dual-Slope DVM Input

The dual-slope input of Fig. 5-30 has improved accuracy and is somewhat more sophisticated than the ramp or VCO circuits. This type of DVM is designed to measure a true average of the input over a given period. Nonlinear effects are canceled and the entire input is rather immune to noise.

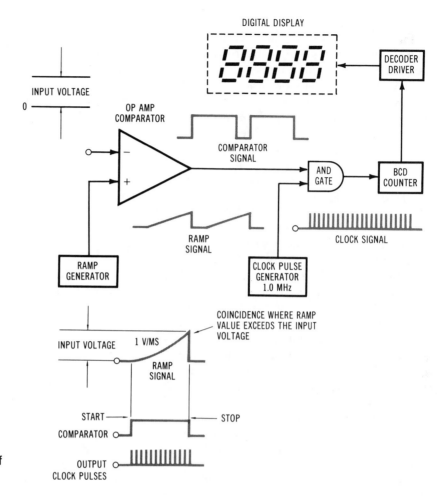

DIGITAL DISPLAY

FIGURE 5-29 Block diagram of a ramp-type DVM.

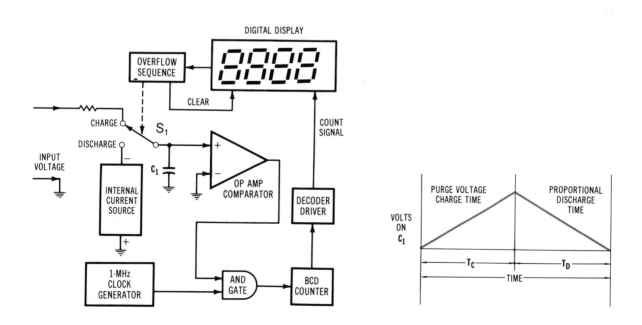

(A) Dvm circuit.

(B) Capacitor charge and discharge.

FIGURE 5-30 Dual-slope DVM.

The operation of a dual-slope DVM occurs in two steps, as shown in Fig. 5-30A. The first step occurs when an unknown voltage is applied to the DVM input through switch S_1. This action causes C_1 to charge to a ramp type of slope that is proportional to the applied input, as shown in Fig. 5-30B. During this time, a positive voltage applied to the + or noninverting input of the comparator causes a positive value, or 1, to be applied to the AND gate. Positive clock pulses applied to this gate at the same time cause the clock signal to be transferred to the BCD counter. This count is then applied to the display device.

The second operational step of the DVM occurs when the display produces an overflow signal. This takes place on the next count after 9999 appears. The overflow detector turns on, clears the display, and switches S_1 to the discharge position. At this point in time, C_1 begins to discharge into the internal current source. The positive input of the comparator continues to apply a 1 to the AND gate. As a result, a new count is now applied to the display unit. This count continues until C_1 reaches zero charge or to its original value. This, in turn, causes the comparator to produce a 0 and apply it to the AND gate. As a result, the AND gate changes to a zero state and the clock signal will not pass. The number indicated on the display now shows capacitor discharge voltage in digital form. After a short time delay the sequencer will clear the display and return S_1 to the charge position to repeat the operational cycle. The sequencer is an IC timer used as an astable multivibrator to clear the display at prescribed intervals.

SUMMARY

Any electronic device that employs a numerical signal in its operation is classified as a digital electronic system. Computers, pocket calculators, digital voltmeters, process control instruments, and timing devices are included in this classification. The operational speed at which digital signals can be processed is of prime importance in industry.

A variety of different numbering systems are in use today. The base or radix of a specific system refers to the number of discrete digits it contains. Practically all digital systems in operation today are of the binary or base 2 type. This type of system has a two-state condition, such as on or off. Quite frequently binary numbers are changed into binary-coded-decimal outputs for processing. A BCD output can be used to display a decimal number. Octal and hexadecimal systems are also used in some digital systems.

There are a number of binary logic functions in use today. Three primary functions are the AND, OR, and NOT. Electronic circuits that achieve these functions are commonly called *gates*. Combinations of primary gates account for two other important gates called NAND and NOR. Most logic gates are built on IC chips. Utilizing a specific gate function is simply a matter of interconnecting logic symbols together electrically.

To achieve different system operations, a number of discrete logic devices have been developed. Flip-flops are very important logic devices designed to generate signals, shape waves, achieve division, and have a memory function. Reset–set, trigger or T, and J-K flip-flops are in common usage. Binary counters use flip-flops to count numerical signals. Four J-K or T flip-flops can be used to achieve BCD counting when feedback is employed.

A digital system must be capable of changing information into a suitable method of display in order for it to be practical. Single-digit and seven-segment displays are used for this function. These displays may be energized by gas ionization, light-emitting diode biasing, and liquid-crystal illumination.

Before a binary signal can be effectively applied to a display device, it must be decoded. Both seven-segment and discrete number decoders are available today on IC chips. BCD to seven-segment decoders are used in nearly all display units today.

Practically all digital instruments contain a counter, decoder, driver, and display device. The input of a basic counter is by far the most unique part of a digital system. Pressure, temperature, time, light, sound, and vibration can be measured if an appropriate analog to digital converter is employed.

One of the most widely used digital electronic instruments today is the digital voltmeter (DVM). This instrument is designed to change variable voltage values into signals that can be processed and displayed in digital form. The input of a DVM can be achieved by voltage-controlled oscillators, ramp function converters, or dual-slope techniques. These conversion techniques are considered to be analog-to-digital converters. DVMs are very accurate, easy to read, and have reduced loading effects compared with conventional analog instruments.

ACTIVITIES

SWITCH-LAMP GATES

1. Connect the AND gate of Fig. 5-5.
2. Prepare a truth table showing the relationship of the two inputs and the output.
3. Connect the OR gate of Fig. 5-6. Test the circuit and prepare a truth table.
4. Connect the NOT gate of Fig. 5-7. Prepare a truth table of its operation.
5. Connect the NAND gate of Fig. 5-8. Test the circuit and prepare a truth table.
6. Connect the NOR gate of Fig. 5-9. Test the circuit and prepare a truth table.

LOGIC GATE TESTING

1. Look up the pin connections for a DM7400 IC logic gate.
2. Make a drawing of the pin connections.
3. Connect $+5$ V and the ground to the IC's $+V_{cc}$ and GND pins.
4. Select one gate to be evaluated.
5. Connect this gate so that each input is attached to the power source ground.
6. With a VOM measure the input and output voltage of the gate.
7. Develop a truth table that shows the relationship of the input and output of the gate using voltage values. Let no voltage indicate 0 and voltage indicate 1.
8. Change the input of the gate so that it conforms with the 0–1, 1–0, and 1–1 alternatives. Record the resulting output for each input alternative in the truth table.
9. Follow the preceding evaluation procedure for an SN7404, SN7408, SN7402, and an SN7432.

QUESTIONS

1. What is meant by the term *radix*?
2. What is the radix of a decimal system and a binary system?
3. What is the base of an octal numbering system?
4. What is the base of the hexadecimal numbering system?
5. The hexadecimal number 2A equals what decimal value? _____
6. The octal number 23 equals what decimal value? _____
7. The binary number 1110 equals what decimal value? _____
8. What is the BCD equivalent of the decimal number 254? _____
9. What is the largest number that can be displayed by any group of a BCD number? _____
10. What mathematical function is achieved with an AND gate?
11. What mathematical function is achieved with an OR gate?
12. What mathematical function is achieved with a NOT gate?
13. What is the name of the inputs and outputs of an *R-S* flip-flop?
14. How does an *R-S* flip-flop differ from an *R-S-T* flip-flop?
15. Describe the input and output relationship of a *J-K* flip-flop?
16. What bar segments of a seven-segment display are needed to indicate the number 5?
17. Why is decoding needed for a binary counter with a seven-segment display?
18. What is the function of the input of a digital electronic instrument?

CHAPTER 6

ELECTRONIC RECORDING INSTRUMENTS

OBJECTIVES

Upon completion of this chapter, you will be able to:

1. Identify the major parts of an electronic recording instrument.
2. Explain the functional role of the power supply of an electronic recording instrument.
3. Given a schematic diagram of a bipolar transistor amplifier, indicate the direction of current flow in each element.
4. Given a schematic diagram of a bipolar transistor amplifier, indicate how the element voltage and current values change with the application of an input signal.
5. Plot a load line for a bipolar transistor amplifier.
6. Evaluate the operation of a bipolar transistor amplifier graphically from a family of characteristic curves.
7. Explain how a power amplifier drives a servomotor.
8. Explain how dc-to-ac conversion occurs in a chopper circuit.
9. Define commonly used amplifier terms, such as *beta, dynamic gain, static operation*, and *coupling*.

IMPORTANT TERMS

In this chapter we investigate some of the basic principles of electronic recording instruments. You will have a chance to become familiar with a number of basic instrument functions. New words, such as *controller recorder*, and *servomotor*, will begin to have some meaning for you. A few of these terms are singled out for study. As a rule, the chapter will be more meaningful if these terms are reviewed before proceeding with the text.

Beta: A designation of bipolar transistor current gain determined by I_C/I_B.

Chopper: A device or circuit that is used to interrupt the flow of current, light, or other energy at a regular rate. A chopper is often called a converter.

Constant-voltage source: An electronic circuit designed to supply voltage that is of a constant value. The zener diode of a power supply is commonly used to achieve this function.

Controller: An electronic instrument that is responsible for automatic control of industrial process variables.

Converter: An electronic circuit that is used to change the dc input developed by a measuring circuit into an ac voltage. A chopper is often used as converter.

Current gain: The ratio of output current divided by input current, identified by the letters A_i.

Cutoff: In a bipolar transistor the condition in which there is no collector current flow with voltage applied to the appropriate elements.

Delta: A Greek letter designation used to denote a changing condition or value.

Driver: An electronic amplifying circuit or device that supplies an input to another amplifier. An amplifier stage preceding the power amplifier.

Dynamic: Electronic device operation that shows how the device responds to ac or changing voltage and current values.

Electronic recording instrument: An instrument designed to provide a graphic display of variations in a particular quantity being measured.

Impedance (Z): The total opposition to current flow in an ac circuit which is a combination of resistance (R) and reactance (X) and is measured in ohms.

Load line: A line drawn on a family of characteristic curves that shows how the device will respond in a circuit with a specific load resistor value.

Power amplifier: An amplifying circuit that delivers maximum output power to a load instead of providing maximum voltage gain.

Preamplifier: An amplifier intended to operate with low-level signals to provide gain and impedance matching to a level that can be handled by another amplifier. The term *preamp* is often used to describe this device.

Servomotor: The electromechanical device of a closed-loop instrument that converts an electrical signal to a mechanical position change.

Small-signal amplifier: An amplifying circuit that employs a low-level signal voltage or current that can be reproduced linearly.

Split-phase motor: A single-phase induction motor that employs an auxiliary winding that is connected in parallel with the run winding and is used to produce a rotating magnetic field for starting.

Static condition: An amplifier operating condition in which element voltage is supplied but no signal is applied for processing.

Time constant (RC): The time required for the voltage across a capacitor in an RC circuit to increase to 63% of its maximum value or decrease to 37% of its maximum value. Time = RC.

INTRODUCTION

The first instrument to be discussed is the electronic recorder. This type of instrument, like others, must employ a number of basic electronic circuits in order to operate. Power supplies, voltage amplifiers, dc-to-ac converters, and servomotor control circuits are some of the representative electronic circuits that make this instrument operate.

In general electronic recording instruments are designed to provide a graphic display of variations in a particular quantity being measured. In some applications this may be a graphic instrument that places

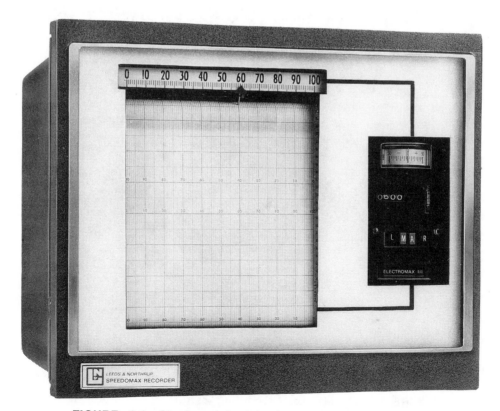

FIGURE 6-1 Single-point strip-chart recording instrument. (Courtesy of Leeds & Northrup Co.)

information on a circular chart or a strip chart that responds to the control of a servomotor unit. In addition, recorders may also display the measured quantity on a hand-deflection instrument. Figure 6-1 shows a single-point chart recording instrument that records on a 6½-in. (165-mm) calibrated strip chart. A similar instrument designed for recording on a 3-in. (76-mm) round chart is shown in Fig. 6-2. A hand-deflection indicator is shown in Fig. 6-3 for comparison. This type of indicator displays the measured variable on a 23-in. (584-mm) scale.

Electronic recording instruments usually have the capability of performing some type of control function in addition to their information gathering function. Instruments of this type are normally described as controllers or recording controllers. In this regard, a controller is ultimately used to alter a process variable or control a system function. This chapter is primarily concerned with the recording-indicator function, while some of the later chapters are concerned with controller operation.

Electronic recording instruments are basically either of the null type or of the galvanometer type. A null instrument generally responds to comparisons between the measured input and a set-point input applied to the instrument. Balance may be achieved manually or it may be self-balancing. Self-balancing instruments seem to be universal throughout the industry today.

FUNCTIONAL UNITS

Figure 6-4 shows a block diagram of the functional units of an electronic recorder of the servomotor type. The major blocks of this diagram that are of concern in this chapter are the power supply, converter or chopper, and voltage amplifiers. These functions are discussed in rather general terms that could apply to the basic circuitry of nearly any recording instrument.

Power Supply

The power supply of an electronic recording instrument is primarily responsible for providing the voltages needed to make the unit operate. A very important part of the power supply is used to develop the dc voltage that is supplied to all of the active components of the recorder. Rectifiers, filters, and voltage regulators are utilized to achieve this function. In addition to this, some instruments employ a constant-voltage unit that provides stable dc voltage values for reference standards. This unit commonly replaces the standard cell.

FIGURE 6-2 Circular chart recording instrument. (Courtesy of Leeds & Northrup Co.)

FIGURE 6-3   Hand-deflection recording instrument. (Courtesy of Leeds & Northrup Co.)

The ac voltage developed by the power supply is applied to the chart motor of instruments that employ one. Low-voltage ac is also supplied by a transformer to the chopper or converter components. This voltage is generally developed by a separate winding on the power supply transformer.

Converter Circuits

The primary purpose of the converter circuit is to change any dc input developed by the measuring circuit into an ac voltage. This conversion is necessary to use a stable, high-gain voltage amplifier. These circuits also help form the ac signal that is to be input to the voltage amplifier.

Amplifier Unit

The amplifier unit of a typical recording instrument achieves combined functions of voltage and power amplification. In many cases, the entire amplifier assembly is described as a servomotor control amplifier unit. It is generally built on a single printed-circuit board or card for easy installation and replacement. This unit is specifically responsible for receiving an error signal, which is the difference between the measuring input

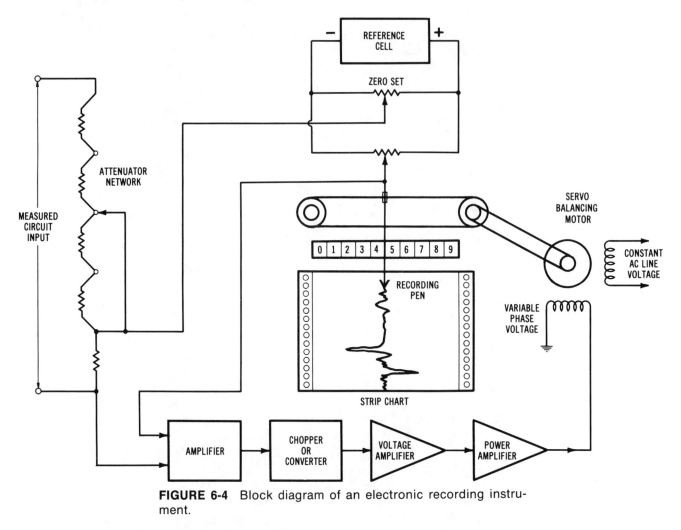

FIGURE 6-4 Block diagram of an electronic recording instrument.

and the feedback circuit, and amplifying it to drive the motor unit.

A major part of the amplifier unit involves several stages of voltage amplification. This part of the instrument is primarily responsible for amplifying the input signal to a level that is large enough to drive the power amplifier.

The power amplifier of the recorder could more accurately be called a motor control circuit. It produces a signal with a current high enough to drive the servomotor. In a sense the power amplifier is a rather low-resistance transistor that permits a larger current than can be achieved by a small-signal amplifier. A power amplifier requires a high-amplitude input signal to control the current to the motor.

POWER SUPPLY CIRCUITS

Figure 6-5 shows a simplified schematic diagram of a representative power supply used in an electronic recording instrument. In this particular circuit, 120 V rms, 60 Hz serves as the power supply input. Through

normal transformer action this voltage is stepped down to a desired value that is used to energize the rectifier. In this case, two separate secondary windings are employed.

The red-wire secondary winding of the transformer is used to supply the rectifier circuit. A secondary voltage of 24 V rms is applied to the bridge rectifier. The rectifier develops $+20$ V dc with respect to the common chassis ground. This specific voltage supplies the pen actuating motor. In addition to this, regulated voltages of $+15$ V and $+7.5$ V are developed across zener diodes D_6 and D_7. Resistor R_5 serves as the series dropping resistor for these voltages. In this case, the zener diodes are connected in series and have the same V_Z rating. These two low voltages are used to supply the transistors of the circuit. Capacitor C_4 serves as an input filter for the $+20$-V supply, and C_5 filters the $+15$-V supply.

The orange-wire secondary windings of the transformer serve as a supply for the pen amplifier of the recorder. In this particular unit, the amplifier has a split power supply of $+15$ V and -15 V for the IC voltage source built on the amplifier circuit board. Through

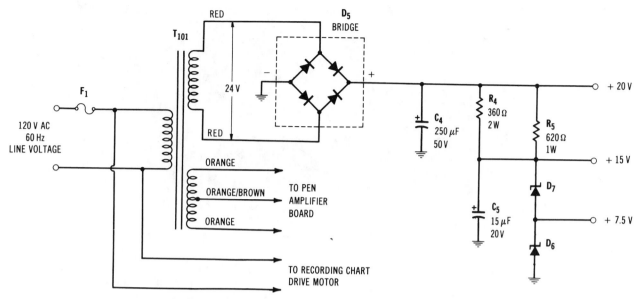

FIGURE 6-5 Simplified schematic of a representative power supply used in an electronic recording instrument.

this alternate power source, common power supply noise problems are minimized. The split dc power supply will be discussed in a later chapter in which IC operational amplifiers are utilized.

The recording chart drive motor of this unit is also remotely located but derives its source voltage from the power supply. In this particular circuit the chart motor is energized by the 120-V line voltage. The motor and transformer primary winding are both fused by F_1 for circuit protection.

A Constant-Voltage Source

Figure 6-6 shows a constant-voltage source that is used in Honeywell ElectroniK 15 recorders for the potentiometer standard of a bridge circuit. Transformer T_1 has 120-V, 60-Hz line voltage applied to it, and a dc voltage of 1.029, 4.2, or 5.064 V is developed at the output according to the values of selected resistors R_m and R_c. The supply is actually a half-wave rectifier, with C_1 forming a capacitive input filter. Regulated

output voltage is achieved by zener diodes D_1 and D_2. This specific power supply is built on a compact circuit board and is housed in a metal container for shielding purposes. The current output of this supply ranges from 6 to 8 mA at its rated output voltage.

DC-TO-AC CONVERTERS

Today, dc-to-ac converters are a common part of nearly all electronic recording instruments. The responsibility of this circuit is simply one of changing the measured input signal, which is dc, into a usable form of ac. A mechanical converter, such as the one shown in Fig. 6-7, is often used to achieve this function.

Components

Only three functional components are utilized in the converter circuit of a recorder. These are the converter itself, which is sometimes called a *chopper*, the input

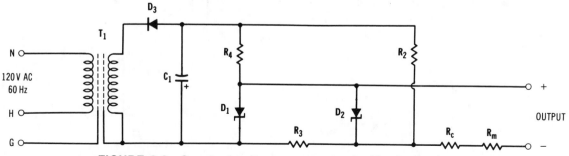

FIGURE 6-6 Constant-voltage source used with electronic recording instruments.

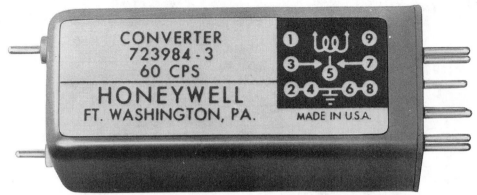

FIGURE 6-7 Mechanical dc-to-ac converter. (Courtesy of Honeywell, Inc.)

transformer, and the filter. The chopper and input transformer work together to change dc to ac. The filter is used to suppress unwanted noise and voltages that may be superimposed on the dc input signal.

Operation of the Chopper

The chopper consists of a drive coil, a vibrating reed, and a set of contacts. The chopper is shown schematically in Fig. 6-8. Terminals 4 and 5 connect the drive coil of the chopper to a 6.3-V ac, 60-Hz secondary winding of the power transformer. This ac voltage produces a changing magnetic field in the core of the drive coil. Let us assume that during the positive alternation of voltage across the drive coil terminal 4 is negative and terminal 5 is positive. Current will then conduct through the drive coil from terminal 4 to terminal 5. This direction of current conduction produces a north magnetic pole at the left end of the core of the drive coil. The right end of the core becomes a south magnetic pole. During the negative alternation of an ac voltage applied to the drive coil the direction of current is reversed. This reversal of current produces a south magnetic pole on the left end of the drive coil. Since the frequency of the voltage connected to the drive coil is 60 Hz, the polarity of the magnetic field also reverses 60 times per second.

The vibrating reed is actually the center contact of a single-pole, double-throw switch. Attached to the end of this thin metal strip is a permanent magnet. This permanent magnet is placed near one end of the drive coil. In Fig. 6-8, if this end of the drive coil is the north pole of the electromagnet produced by the coil, the vibrating reed will move upward. This movement is the result of the attraction between the south pole of the permanent magnet and the north pole of the drive coil, and the repulsion of the two north poles. This movement will also close the circuit between terminals 1 and 2. When the current to the drive coil reverses direction, a south pole is produced at this end of the coil. This causes a downward motion of the vibrating reed. The action is again produced due to the reaction between the pole of the drive coil and the poles of the permanent magnet. When the reed moves downward, contact is made between terminals 2 and 3. Since the drive coil voltage has a frequency of 60 Hz, the vibrating reed contacts both terminals, 1 and 3, 60 times per second.

DC-to-AC Conversion

The conversion of the dc voltage produced by the measuring circuits into an ac voltage incorporates the action of the chopper and the primary of the input trans-

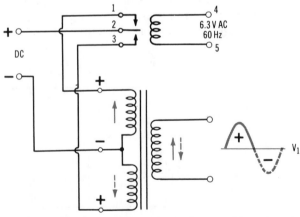

FIGURE 6-9 Conversion of a positive dc voltage to an ac voltage.

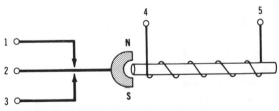

FIGURE 6-8 Diagram of a chopper.

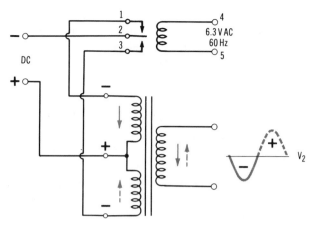

FIGURE 6-10 Conversion of a negative dc voltage to an ac voltage.

former. The connections of these two components are shown in Fig. 6-9.

This schematic indicates the situation when the output of the measuring circuits is positive. Under these conditions the voltage at terminal 2 (the vibrating reed) is positive and the center tap of the input transformer is negative. As the vibrating reed moves, terminals 1 and 2 are connected. The circuit is closed through the upper half of the primary winding of the input transformer. The dc voltage across this part of the transformer produces a current through this part of the primary. A voltage is induced into the secondary of the transformer with the polarity (positive) indicated due to this primary current. The vibrating reed then moves downward. Terminals 2 and 3 are now connected. This supplies a voltage across the lower half of the primary. The current is now conducting in the primary in the opposite direction to that described before. This current induces a negative voltage across the transformer secondary. The action of the vibrating reed produces a signal which closely approximates a square wave. The inductive action of the transformer, however, shapes this waveform so that the transformer output now closely approximates the sine wave.

Figure 6-10 shows the polarity of the voltages

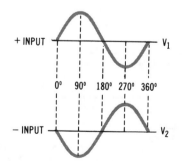

FIGURE 6-11 Phase comparison between positive and negative inputs.

when an unbalance occurs in the measuring circuit to produce a negative input to the amplifier. The vibrating reed moves upward, closing contacts 1 and 2. A voltage is then applied across the upper half of the transformer primary. Current is produced in this half of the primary that induces a negative voltage into the secondary winding. A voltage is provided across the lower half of the transformer primary when the vibrating reed moves downward. The current through this part of the primary winding produces the positive alternation across the secondary.

The difference between the ac voltage output under the two conditions should be noted. Figure 6-11 shows this relationship. You will notice that during the positive alternation produced by a positive dc input a negative alternation is produced by a negative dc input. Any time the positive maximum voltage occurs at the same time that the negative maximum occurs, the voltages are said to be 180° out of phase. Remember that a conductor moving in a circular path (360° rotation) between a pair of poles of a magnet produces one cycle of an alternating voltage. The time reference of an ac voltage, therefore, can be measured in degrees. The time difference between two corresponding points on two ac voltages measured in degrees is the phase difference or phase shift. You will notice that the time difference between the positive peak of V_1 and V_2 is one-half cycle or 180°. This is a 180° phase shift.

When the measuring circuits are balanced in the null position, the zero-voltage input will produce no ac signal as an amplifier input. The filter circuit is of particular importance under the zero-input conditions. This filter is made up of a parallel combination of a resistor and capacitor. This circuit is shown in the complete schematic of the converter circuits in Fig. 6-12. The capacitor offers a low-impedance path to very high frequencies. Any noise that might be picked up and introduced as a signal into the amplifier should be bypassed through this low-impedance path to ground.

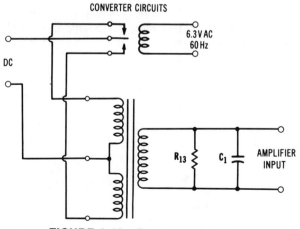

FIGURE 6-12 Converter and filter.

VOLTAGE AMPLIFIERS

The block diagram of the recorder shown in Fig. 6-4 contains several distinct voltage amplifiers that are not shown specifically. These amplifiers process the ac signal from the converter through several stages of *RC*-coupled amplifiers. Terms such as preamplifiers, signal amplifiers, and driver amplifiers are often used to describe this part of the recorder. Amplifiers of this type are primarily responsible for most small-signal amplification applications. This type of amplification increases the amplitude of the applied signal voltage to a rather high level at the output. The most important component of a voltage amplifier is its active device, which is either a bipolar or a field-effect transistor.

Instead of discussing a particular amplifier produced by a specific manufacturer, we will approach the subject of amplification in rather general terms. The principle of operation can then apply to any one of several recorders available today. Through this approach you should be able to apply the amplification function to nearly any specific circuit application.

NPN Transistor Circuit

Figure 6-13 shows the necessary connections of a simple npn silicon transistor connected into a common-emitter amplifier circuit. The operation of this circuit is based on the conventional biasing of a transistor, which was discussed previously. The emitter–base junction is forward biased, with the collector–base junction reverse biased. As noted, forward biasing of the emitter–base junction is accomplished by connecting the negative side of the battery (V_{CC}) to the emitter and the positive side, through resistor R_B, to the base. Collector current through load resistor R_L is controlled by variations in the forward bias voltage of the emitter–base junction.

A transistor amplifier of this type is designed to control the amount of current that passes through it. Current from the battery enters the emitter, passes through the base, and exits through the collector. Variations in output, or collector current (I_C), can be made by changing the base current (I_B). A small change

in I_B causes a rather substantial change in I_C. This relationship is commonly described as transistor current gain or beta (β). Expressed mathematically,

$$\beta = \frac{\Delta I_C}{\Delta I_B}$$

The Greek capital letter delta (Δ) in the formula indicates a change in value.

All of the current entering a transistor at the emitter is referred to as emitter current, or I_E. The output current, I_C, is always somewhat less than I_E because of the base current. Mathematically, this is expressed by the formula

$$I_C = I_E - I_B$$

PNP Transistor Circuit

The transistor amplifier circuit of Fig. 6-14 is a pnp counterpart of the previous npn circuit. The battery of this circuit is connected in a reverse direction to achieve proper biasing. Performance is basically the same as that of the npn circuit. Currents I_C, I_B, and I_E are represented in this diagram by arrows. The emitter current of this circuit still provides the largest current value. The composite of I_C plus I_B also equals I_E in this circuit.

Static Transistor Circuit Conditions

When a transistor is developed, the manufacturer produces operating specifications and characteristics that are used to predict its performance. A typical family of collector curves is shown in Fig. 6-15. The collector–emitter voltage (V_{CE}) is plotted on the *x*-axis of this graph, while the collector current (I_C) is plotted on the *y*-axis. The individual curves of the graph represent different base-current (I_B) values. The zero-base-current line is omitted because it represents the cutoff condition of the transistor.

A family of collector curves indicates a great deal about the operation of a transistor. If the V_{CE} voltage, for example, is held at a constant value of 9 V, an increase in base current of 5 to 10 μA will cause a change in I_C from 0.5 to 1 mA. The resulting current

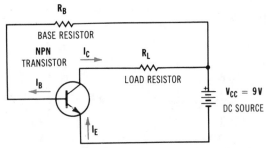

FIGURE 6-13 Basic npn transistor circuit.

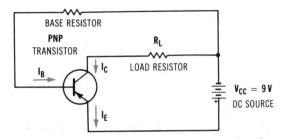

FIGURE 6-14 Basic pnp transistor circuit.

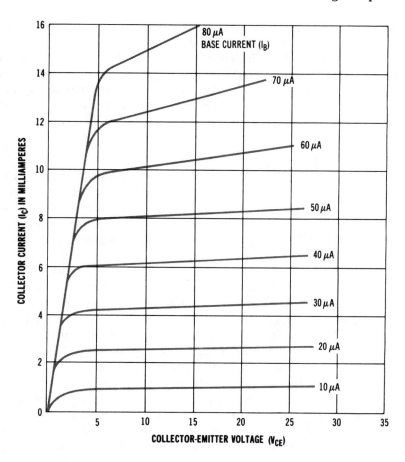

FIGURE 6-15 Family of V_{CE}/I_C characteristic curves for a transistor.

gain, or beta, produced by this action would be

$$\beta = \frac{\Delta I_C}{\Delta I_B}$$

$$= \frac{0.5 \text{ to } 1.0 \text{ mA}}{5 \text{ to } 10 \text{ } \mu\text{A}}$$

$$= \frac{500 \text{ } \mu\text{A}}{5 \text{ } \mu\text{A}}$$

$$= 100$$

The same family of curves also shows that an increase in V_{CE} from 1.0 to 9.0 V, with I_B held at 10 μA, will only cause an increase in I_C of 0.5 to 1.0 mA. This indicates that collector current can be more effectively controlled by base current changes than by V_{CE} voltage changes.

When a particular transistor is placed into a circuit with voltage supplied and no signal applied, it is considered to be in a *static* condition of operation. The npn transistor of Fig. 6-16 is used to show this condition. As you will note, this circuit has specific component values assigned. The source voltage is 30 V, and a 2200-Ω load resistance, R_L, is used. This means that the maximum collector current (I_C) that can occur can be determined by Ohm's law. Mathematically, this

is expressed as

$$I_C = \frac{V_{CC}}{R_L}$$

$$= \frac{30 \text{ V}}{2200 \text{ } \Omega}$$

$$= 0.01364 \text{ A}$$

$$= 13.64 \text{ mA}$$

This means that the total range of collector cur-

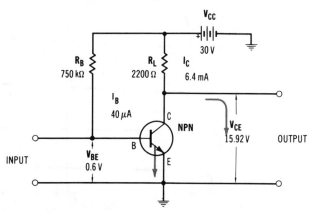

FIGURE 6-16 An npn transistor circuit.

rent that can be controlled by the transistor is from 0 to 13.64 mA. The actual static condition of operation of this circuit is determined by the selection of a value of base current. In this circuit, the 750-kΩ base resistor determines the operating base current. Mathematically, this is determined by dividing the source voltage by the base resistance, or

$$I_B = \frac{V_{CC}}{R_B}$$

$$= \frac{30 \text{ V}}{750 \text{ k}\Omega}$$

$$= 0.00004 \text{ A}$$

$$= 40 \text{ }\mu\text{A}$$

In this case, with 40 μA of base current the transistor will only be conducting 6.4 mA of collector current.

In the circuit conditions just described, the base–emitter voltage would be approximately 0.6 V dc measured across the input, or V_{BE}. The dc voltage measured across the output terminals, which in this case is the collector–emitter voltage, or V_{CE}, would be 15.92 V and is determined by the collector current, I_C, passing through load resistor R_L. An I_C of 6.4 mA through R_L produces a voltage drop of 14.08 V across R_L (0.0064 A × 2200 Ω = 14.08 V). Subtracting 14.08

V from the V_{CC} supply voltage of 30 V gives us 15.92 V for V_{CE}. This means that with no signal applied to the transistor circuit, it will have 15.92 V dc appearing as output voltage. Since the input voltage of 0.6 V and the output voltage of 15.92 V are of a constant value, the amplifier is in a static condition or dc state of operation.

Static Load-Line Analysis

By using the family of V_{CE}–I_C curves of a transistor, its static operating conditions can be determined graphically. Figure 6-17 shows the collector curves of the transistor being used in the circuit of Fig. 6-16. The diagonal line drawn across the curves is called a *load line*. It represents the extreme conditions of transistor operation for this specific circuit application. In this application, we will assume that the maximum V_{CE} voltage will be the value of the source, or 30 V. This point of the load line is located at the 30 V_{CE} mark. This means that when 30 V of V_{CE} appear across the transistor, there is no I_C current. The cutoff or nonconductive state of operation is represented by this condition.

The other operating extreme occurs when the transistor is conducting the maximum I_C through R_L. We determined this value previously by dividing V_{CC} by the value of R_L (30 V/2200 Ω = 13.64 mA). This

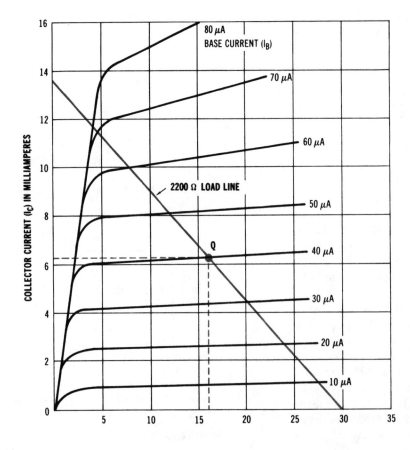

FIGURE 6-17 Family of V_{CE}/I_C characteristic curves with a load line.

point located on the I_C part of the graph indicates the maximum conduction through R_L. When this occurs there will be 0 V_{CE} voltage across the transistor. A straight line connecting these two points together indicates the two extreme operating conditions of the transistor with a 2200-Ω load resistor and a 30-V source.

Using the load line drawn on the I_C curves of Fig. 6-17, we can now display the static operating conditions of the transistor of Fig. 6-16. The intersection of the 40-μA I_B curve and the load line is labeled point Q. This base current, you will recall, was established previously by the value of R_B and V_{CC}. Projecting a straight line to the left of the graph from point Q indicates that an I_B of 40 μA produces an I_C of 6.4 mA. The resulting V_{CE} that occurs is determined by projecting a line from point Q down to the V_{CE} indicating line. A value of 15.92 V of V_{CE} is indicated. This means that the voltage across R_L is $V_{CC} - V_{CE}$ or $30 - 15.92 = 14.08$ V. These values are indicative of the static operating conditions determined previously.

Dynamic Transistor Operation

We will now take a look at the results of applying an ac signal to the base of a transistor in its static state. When this occurs, the transistor has a changing voltage gain or is in a *dynamic* state of operation. Primarily this means that a small ac voltage applied to the input of the transistor cause a variation in the emitter–base voltage (V_{BE}). This change in voltage causes a corresponding change in base current. Variations in base current will, in turn, cause a change in collector current. Ultimately, this action will cause a variation in the I_C passing through R_C which will result in a changing output voltage. Figure 6-18 shows a representative ac amplifier using the same transistor as that of the static circuit.

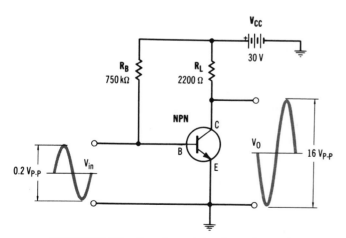

FIGURE 6-18 Npn transistor in a dynamic state of operation.

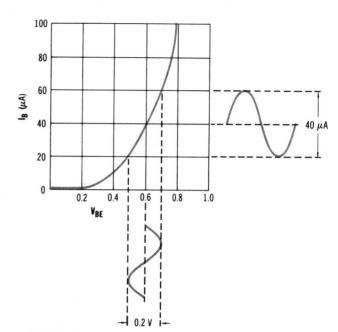

FIGURE 6-19 Input characteristic curve for a transistor.

To demonstrate dynamic transistor gain, we will first look at the input characteristic of an npn transistor. The input characteristic curve of a typical transistor is shown in Fig. 6-19. Note that the vertical axis of this graph indicates base current (I_B) and the horizontal axis displays emitter–base voltage (V_{BE}). This graph is representative of the input characteristics of the transistor used in Fig. 6-16.

In the static operation of a transistor, you will recall that with no input signal applied there was an emitter–base voltage of 0.6 V. This, in turn, caused approximately 40 μA of I_B. Locate these two points on the input characteristic curve.

Assume now that a 0.2-V peak-to-peak input signal is applied to the transistor circuit as indicated. During the positive alternation of the input signal the V_{BE} voltage will rise from 0.6 V to 0.7 V. This, in turn, will cause a change in base current from 40 μA to 60 μA. In the same manner, the negative alternation of the input signal will cause a corresponding reduction in V_{BE} from 0.6 V to 0.5 V, and a change in I_B from 40 μA to 20 μA. Primarily this means that an input voltage change of 0.2 V peak-to-peak causes a 40-μA peak-to-peak change in base current.

Transferring the changes in base current to the family of I_C curves is shown in Fig. 6-20. The load line drawn on the graph is the same as that achieved with the static circuit of Fig. 6-16. Notice particularly that the total change in I_B as a result of the ac input signal is from 60 μA to 20 μA. This represents a change in I_B of 40 mA. The corresponding change in I_C is from 10 mA to 2.8 mA. This indicates a total change in I_C

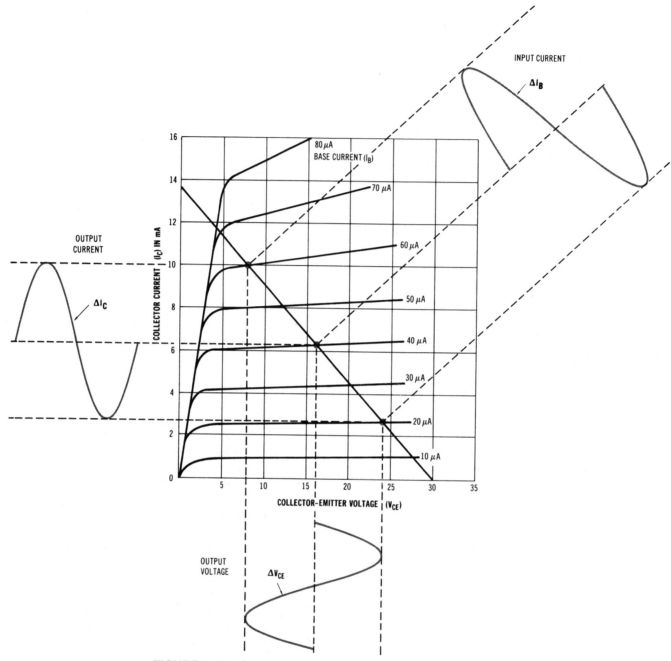

FIGURE 6-20   Graphical illustration of the amplifying action of a transistor.

of 7.2 mA. The current gain, or beta, is, therefore,

$$\beta = \frac{\Delta I_C}{\Delta I_B}$$

$$= \frac{7.2 \text{ mA}}{40 \text{ }\mu\text{A}}$$

$$= \frac{7200 \text{ }\mu\text{A}}{40 \text{ }\mu\text{A}}$$

$$= 180$$

The resulting output voltage variation of V_{CE} is

from approximately 8 to 24 V, or 16 V peak to peak. The dynamic or ac voltage gain (A_V) of this circuit is the output voltage (V_o) divided by the input voltage (V_{in}). Mathematically, this is expressed as

$$A_V = \frac{V_o}{V_{in}} = \frac{\Delta V_{CE}}{\Delta V_{BE}}$$

$$= \frac{16 \text{ V}_{p\text{-}p}}{0.2 \text{ V}_{p\text{-}p}}$$

$$= 80$$

RC Amplifier Coupling

When two or more amplifiers are connected together, the signal being processed must pass from one amplifier to the next without altering the bias voltage of the second amplifier. In ac voltage amplifiers, such as those following the converter, *RC* coupling is commonly used. Figure 6-21 shows a single-transistor amplifier, similar to the one just discussed, with an added *RC* coupling network.

The coupling components of the *RC* network are R_L and C_C. To understand the operation of this coupling network, we shall discuss it first from a capacitive reactance point of view and second in terms of a time constant. The capacitive reactance (X_C) of C_C is infinite when the voltage across it is of a constant value or dc. We therefore say that it blocks dc or responds as an open circuit to direct current. Since this is true, as can be seen from the capacitive reactance formula, $X_C = 1/2\pi fC$, then no dc will pass through R_2. The reactance of C_C to an ac such as 60 Hz, however, is quite small. Mathematically, this would be

$$X_C = \frac{1}{2\pi fC}$$

$$= \frac{1}{6.28 \times 60 \times 10 \times 10^{-6}}$$

$$= \frac{1}{3.768 \times 10^{-3}}$$

$$= 265.39 \ \Omega$$

A capacitive reactance of this value to ac is ob-

viously quite small. This means that ac of this frequency will pass very easily through C_C. When this occurs, ac will be developed across R_2. Essentially, this means that ac will pass through C_C, appear across R_2, and dc will be blocked. As a result of this action, ac signals can be passed to the next stage of amplification without the dc circuit voltages of one amplifier interfering with those of the next amplifier.

The time constant (T) of the *RC* network is equal to $R_2 \times C_C$. When the circuit is first turned on, and no signal is applied, a collector voltage of 30 V appears across C_C. This, of course, will take some time since C_C must charge through R_2 and R_C. The value of R_C is quite small compared with the value of R_2 and is considered negligible. The time constant of C_C and R_2 is, therefore,

$$T = R_2 \times C_C$$

$$= 100 \ \text{k}\Omega \times 10 \ \mu\text{F}$$

$$= 1 \times 10^5 \times 10 \times 10^{-6}$$

$$= 10^6 \times 10^{-6}$$

$$= 1 \ \text{s}$$

Complete charge of C_C will thus occur in 5 s.

When an ac signal is applied to the base of Q_1, it will cause an amplified ac signal to appear at the collector as indicated. The positive alternation of the input signal will produce a corresponding decrease in V_{CE} voltage. With C_C charged to the $+V_{CC}$ voltage, it will attempt to discharge through R_2, to ground, and through Q_1 as indicated by the arrows. Neglecting the internal resistance of Q_1, it would require at least 1 s for C_C to discharge. With the total time of the positive alternation only half of 1/60, or 1/120 of a second, C_C does not have adequate time to discharge. The amount of discharge is actually equivalent to the amount of voltage drop due to X_C. For all practical purposes, C_C remains charged to nearly the value of V_{CC}.

During the negative alternation of the input signal, the V_{CE} voltage rises in value. When this occurs, C_C attempts to charge to the increased V_{CE} voltage through R_2. In the short time that this voltage is up in value, C_C charges and a current conducts through R_2.

The charge and discharge action of C_C causes a corresponding current conduction through R_2. In effect, this means that an ac voltage appears across R_2. This voltage is representative of the V_{CE} signal of the transistor and is passed through C_C. The input of a second stage of amplification connected across R_2 would therefore receive the resulting ac signal through *RC* coupling. This signal is an excellent reproduction of the original input signal, with suitable gain, and it is inverted 180°. Today *RC* coupling is commonly used in electronic recording instruments.

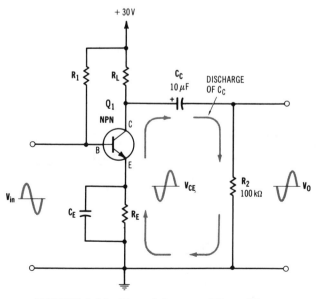

FIGURE 6-21 Transistor amplifier with an *RC* coupling network.

POWER AMPLIFIER (MOTOR-CONTROL CIRCUIT)

The last stage of amplification in an electronic recording instrument is primarily responsible for manipulation of the recording pen motor. This amplifier, depending on the design of the recorder, drives the motor in such a way that it causes the pen to move on a chart according to variations in the input signal. As a general rule, a feedback signal for the pen location is returned to be compared with the measured input signal.

Typically, one stage of power amplification is used in the motor-control amplifier. In the discussion of power amplifiers that follows, we will use two power transistors in a parallel configuration. Each transistor controls half of the resulting output signal.

Figure 6-22 shows a representative motor-control power amplifier circuit that is used in several different recorders today. An *RC* network is used to couple the voltage amplifier output signal to the bases of both transistors. An emitter resistor, R_E, is commonly connected to both transistor emitters. Collector voltage for each transistor is supplied through a special secondary winding of the power transformer. The winding is center tapped with equal amounts of voltage supplied to each transistor. These voltages are of equal amplitude and are 180° out of phase.

Amplifier Operation (No Signal Input)

No signal input to the power amplifier is provided when the measuring circuit and pen location feedback signal are properly balanced. Figure 6-23 shows a schematic of the motor-control circuit in this condition of operation. With no signal applied, the base of neither transistor receives a signal. The transistors are, however, forward biased by R_B and in a static condition of operation. With ac applied to the collector of each transistor, it means that the transistors will be alternately conductive. For example, during the first alternation of the transformer input, the collector of Q_1 is positive, while the collective of Q_2 is negative. You will recall that conduction of a transistor only occurs when the collector–base junction is reverse biased. This means that Q_1 is conductive and Q_2 is nonconductive during this alternation of the ac collector voltage. Q_2 is nonconductive because its collector–base junction is improperly biased. As you will note, current, indicated by the solid arrows, passes through R_E, Q_1, L_1, the servomotor winding, and returns to ground.

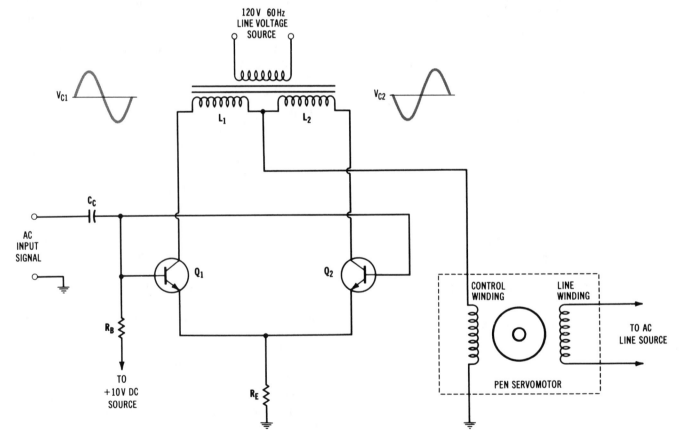

FIGURE 6-22 Power-amplifier motor-control circuit.

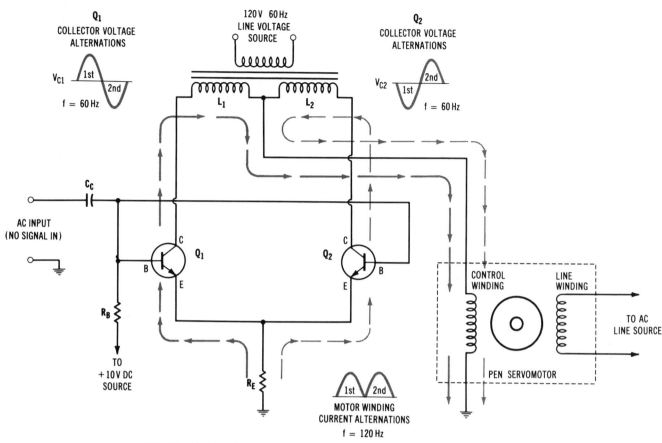

FIGURE 6-23   Operation of motor-control circuit with no ac input signal.

During the next alternation, the conduction of Q_1 and Q_2 will switch. Q_2 now becomes conductive because its collector is positive. This current is indicated by the broken arrows. The path is through R_E, Q_2, L_2, the servomotor winding, and ground. Since the resulting current through both transistors is in the same direction, the voltage drop across the motor winding has the same polarity for each half-cycle of the input. Two resulting pulses of motor winding current are produced for each cycle of the applied collector voltage input. The frequency of these pulses is 120 Hz or twice the 60-Hz line voltage. The servomotor therefore remains in a stationary position. Ideally, this represents a condition of balance. The pen should be at its center resting position, or at its zero reference location.

Amplifier Operation (with Signal Input)

Assume now that the circuit of Fig. 6-24 has an ac signal applied to its input. This condition would take place when an input signal is applied to the recorder input. When this occurs, the pen location and input signal are out of balance and an error signal is gen-erated, converted to ac, amplified, and applied to the motor-control power amplifiers.

With the bases of power amplifiers Q_1 and Q_2 commonly connected together, any input signal will appear the same at both bases. The signal of V_{C1} will, for example, appear as indicated in Fig. 6-24. For the positive input alternation, note that Q_1 is conductive and that Q_2 is not conductive. This is due primarily to the polarity of the collector voltage and the base voltage. The base of Q_1 is forward biased and the collector is reverse biased. At the same time, Q_2 is nonconductive because, although its base is forward biased, its collector is improperly biased because of the polarity of the ac line voltage. The resulting current path for this alternation is therefore through R_E, Q_1, L_1, the servomotor winding, and ground.

For the next alternation of the input signal, the bases of both Q_1 and Q_2 will swing negative. This, of course, reverse biases the bases of both transistors regardless of the polarity of the collector voltages. In this condition, no resulting current conducts through either transistor. An output, as indicated, only occurs during the positive alternation of the base input signal. The frequency of the output is 60 Hz.

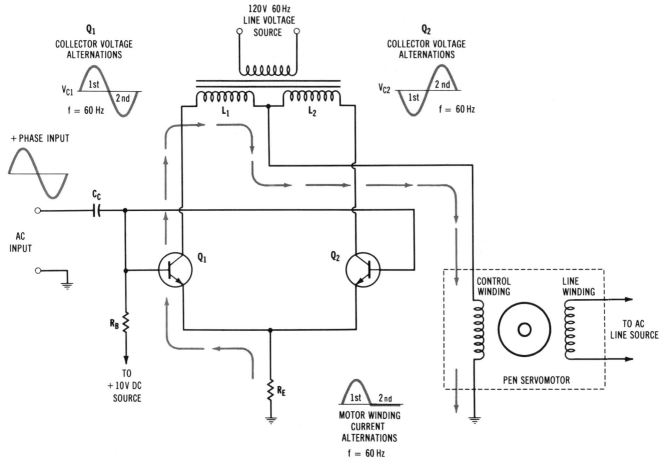

FIGURE 6-24 Operation of motor-control circuit with positive phase input signal.

If the polarity of the input signal is reversed, as indicated in Fig. 6-25, so that the first alternation is negative, the resulting output will shift. In this case, the bases of both Q_1 and Q_2 will be reverse biased because of the negative alternation. No current will conduct through either transistor in this condition of operation regardless of the collector polarity. During the next alternation of the input, which is positive, the bases of both Q_1 and Q_2 will be forward biased. The collector of Q_1 will now have the wrong polarity while the collector of Q_2 will have the correct polarity. Therefore, conduction will occur through R_E, Q_2, L_2, the motor winding, and ground. The resulting output of this sine-wave input is one pulse of output. This occurs at a frequency of 60 Hz and is shifted 180° with respect to the previous input signal.

The input signal to the power amplifier may be in phase with the line voltage supplied to the transformer, or it may be 180° out of phase with it. This is essentially determined by the direction of unbalance in the measuring circuits.

POWER AMPLIFIER OUTPUT EFFECT ON THE BALANCING MOTOR

In this section the reactions of the balancing motor due to amplifier unit outputs will be discussed. The balancing motor itself will be treated as a "black box." Since most instruments use similar balancing motors, the split-phase motor will be dealt with as a separate unit.

Figure 6-26 shows the phase relationship of the amplifier unit inputs, outputs, and the line phase. The amplifier output is the current conducting through one phase winding of the balancing motor. The relationship between the currents in the windings of the motor determines the torque produced by the motor. A phase shift of about 90° is necessary for motor rotation. The current through the winding supplied by the power supply is always 90° out of phase with the line voltage. The inputs of the amplifier units are developed by the converter circuits.

The power amplifier outputs are not pure sine

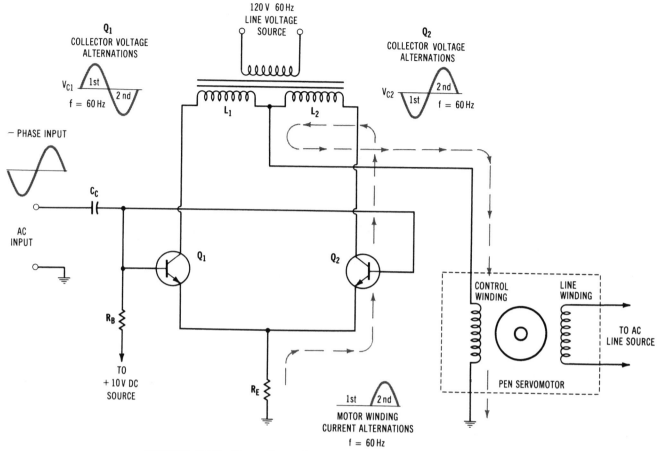

FIGURE 6-25 Operation of motor-control circuit with negative phase input signal.

waves. They are a mixture of a dc component, a 60-Hz ac component, and higher frequency components called harmonics. The only part of this output that will produce motor rotation is the 60-Hz component. When the input is zero, there is no motor rotation. The amplifier output has a frequency of 120 Hz. There is no 60-Hz ac component. Therefore, no motor rotation results. You will notice that the output developed from V_1 does have a 60-Hz ac component. This 60-Hz ac voltage is shown in Fig. 6-26. This voltage leads the line phase by 90°. Rotation of the motor occurs to re-balance or null the measuring circuits. The output resulting from V_2 is 180° out of phase with the voltage just described. The phase shift was determined by the measuring circuits. Its ac component lags the line phase by 90°. This is the condition necessary to produce rotation of the motor in the reverse direction.

We will assume that the input voltage of Fig. 6-24 is in phase with the line voltage. The resulting output will occur during the first alternation. In Fig. 6-25 the reverse is true. The input signal is 180° out of phase with the line voltage. The resulting current output, therefore, occurs during the second alternation of the

input signal. This means that the output pulse will shift 180° in respect to the phase polarity of the input signal. The shifting of this signal with respect to the voltage applied to the line winding ultimately determines the direction of motor rotation.

SUMMARY

Electronic recording instruments in general are designed to provide a type of graphic display of a particular quantity being measured. Instruments of this type record on strip charts, circular charts, or by hand deflection on a meter scale.

The major parts of a recorder are the null balancing mechanism, power supply, amplifiers, converters, power amplifier, servomotor, and recording mechanism.

The power source of a recorder is designed to supply the operating voltages needed to make the recorder function. Rectification, filtering, and regulation are common parts of the power supply.

Dc-to-ac converters are a common part of the

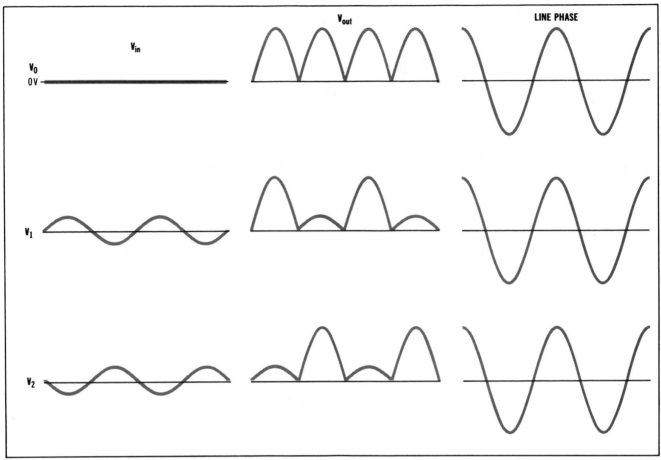

FIGURE 6-26   Phase relationships in amplifier motor-control circuit.

electronic recorder. This circuit simply changes dc into ac with a vibrating reed device known as a chopper.

A number of voltage amplifiers are needed to increase the amplitude level of the signal to a value that will drive the power amplifier. Transistors are commonly used today. They are designed to have ac gain by plotting a load line on a family of IC curves. Beta is an expression of transistor gain. The input amplifiers of a recorder are small-signal voltage amplifiers. The final output signal of these amplifiers is used to drive the power amplifier.

Power amplifiers in a recorder are commonly used to drive the servomotor mechanism. The mechanism moves the pen on a chart and actuates the balancing mechanism. The servomotor moves the pen and balancing mechanism according to the phase of the measured ac signal compared to the ac line voltage reference source.

ACTIVITIES

STATIC TRANSISTOR OPERATION

1. Construct the bipolar transistor amplifier of Fig. 6-16. The transistor should be a 2N3053 or equivalent.

2. Apply power from the dc source.

3. Measure and record the dc emitter–collector voltage (V_{CE}) and the base–emitter voltage (V_{BE}).

4. With the common side of the meter attached to the emitter (ground), note the polarity of V_{BE} and V_{CE}.

5. Measure and record the collector current (I_C) and base current (I_B).

6. Retain the circuit of this activity for the dynamic transistor operation activity that follows.

DYNAMIC TRANSISTOR OPERATION

1. Construct the bipolar transistor circuit of Fig. 6-18. Note that this is the same circuit as Fig. 6-16. The transistor should be a 2N3053 or equivalent.

2. Apply power to the circuit and measure the V_{CE} and V_{BE} voltage with a dc voltmeter.
3. Connect a low voltage ac signal of 0.1 V rms to the input of the amplifier. With an ac voltmeter measure the input voltage appearing at V_{BE}.
4. Measure and record the ac output voltage across the collector–emitter.
5. Determine the voltage gain of the amplifier.
6. If an oscilloscope is available, prepare it for operation. Connect the vertical probe to the base and the common probe to the ground. Observe the phase and

value of the ac input signal. Adjust the input voltage to a value of 0.2 V_{p-p}.
7. Move the vertical probe from the base to the collector. Observe the phase and value of the output. Determine the voltage gain of the amplifier.
8. Adjust the voltage to a value that produces the greatest output without distortion. Measure and record the output and input voltage values. Calculate the voltage gain.
9. Turn off the power source and disconnect the circuit.

QUESTIONS

1. What is an electronic recording instrument?
2. What is the function of a converter in an electronic recording instrument?
3. What is the conventional method of biasing an npn transistor?
4. Explain the meaning of the term *static operation* of a bipolar transistor.
5. What are the two extreme conditions of operation of a bipolar transistor shown by a load line?
6. What is meant by the term *dynamic operation* of a bipolar transistor?
7. Explain why the pen servomotor is balanced when no signal is applied to the input of the motor control circuit of Fig. 6-23.
8. What causes the motor control circuit of Fig. 6-24 to produce an imbalanced condition?
9. What is the function of *RC* coupling in bipolar transistor operation?
10. What is a power amplifier?

CHAPTER 7

COMPUTER-BASED SYSTEMS

OBJECTIVES

Upon completion of this chapter, you will be able to:

1. Identify the functional blocks of a basic computer and show how they are interconnected.
2. Explain the function of firmware, software, and programming.
3. Identify the functional blocks of a basic microprocessor.
4. Define computer memory.
5. Use the vocabulary associated with microprocessor terms.
6. Use the vocabulary associated with memory terms.
7. Explain the meaning of flowchart, instructional set, opcode mnemonics, and memory address.

IMPORTANT TERMS

In this chapter we investigate some of the basic principles of computer systems. In this study you have a chance to become familiar with a number of computer functions. New words, such as *memory, microprocessor, programming,* and *opcode,* will begin to have some meaning for you. A few of these terms are singled out for study. As a rule, the chapter will be more meaningful if these terms are reviewed before proceeding with the text.

Accumulator: A temporary register that is designed to store data that is to be processed.

Address: A number that identifies the location of data placed in memory.

Address register: A microprocessor function that temporarily stores the address of a memory location that is to be accessed for data.

Arithmetic/logic unit (ALU): A circuit capable of performing a variety of arithmetic or logic functions.

Bit: A contraction of the term *binary digit.*

Bus lines: A group of conductor paths that are used to connect data words to various registers.

Byte: A special term used for an 8-bit word. A byte usually consists of 8 bits.

Clear: A flip-flop input that resets the Q output to the logic 0 state.

Clock: A flip-flop input that controls timing or a pulse generator that controls computer operations.

Data register: A group of flip-flops that store data.

Enable: To make a computer operation possible, such as permitting data to be transferred from one location to another by applying a signal to the enable input.

Encoding: The process of converting an input signal into binary data or to change one digital code to another.

Erasable programmable read-only memory (EPROM): A read-only memory in which information has been entered by applying electrical pulses to selected cells. This information can be erased by applying ultraviolet light to the EPROM through a quartz window.

Execute: One of the operational cycles of a central processing unit.

Fetch: An operational cycle of the central processor unit.

Firmware: Permanently installed instructions in computer hardware that control operation.

Flowchart: A symbol diagram used to aid in program development.

Instruction: In a processor or computer, the information that tells the machine what to do. Program information.

Instructional set: A unique set of instructions that are used to control the operation of a microcomputer.

Instruction code: An exclusive number code that identifies a specific operation of the MPU.

Instruction decoder: An operation that examines an instruction code and decides if it will be performed by the ALU.

Interface: A process or piece of equipment in a computer that brings things together.

Load: A microprocessor register input that permits data to be applied to the bus lines.

Memory: A process that can store logic 1 and logic 0 bits in such a manner that they can be accessed or retrieved.

Microprocessor: An integrated circuit or set of ICs that can be programmed to process data.

Mnemonic: An abbreviation that resembles a word or reminds one of a word.

Operand: A quantity being operated on in a processing system.

Operational code (opcode): The part of a computer instruction word that designates the task to be performed.

Program: A list of tasks in an order that are used to control the operation of a processor or computer system.

Program counter: A CPU or computer counter that keeps up with the steps or operational sequence of a program.

Programmable read-only memory (PROM): A memory device that can be programmed by the user.

Read-only memory (ROM): A form of stored data that can be sensed or read. It is not altered by the sensing process.

Read/write (R/W): A memory operation that can be read from or written into with equal ease.

Registers: A digital circuit that is capable of storing or moving binary data.

Sequence controller: An electronic device or circuit function that is responsible for maintaining the operational sequence of events in proper order.

Software: Instructions, such as a program that controls the operation of a computer.

Ultraviolet (UV) light energy: Light energy used to erase data from a programmable read-only memory chip.

INTRODUCTION

Industry now has a number of instruments and machines that are classified as *intelligent* or *smart*. A smart machine is unique because it has a built-in computer. The computer permits the equipment to make decisions that may control or alter its operation. Intelligent equipment may also be described as a computer-based system. The computer is an integral part of the machine or instrument that it controls. It may be built directly into the machine or be part of a network that connects several machines to a central computer. The computer-based system is now an important tool of industry.

COMPUTERS

Computers are primarily responsible for the rapid expansion of the digital electronics field that we are experiencing today. This technology has had a decided impact on industrial instrumentation equipment. Computers make it possible to perform control operations automatically. Calculations can be made quickly, data can be manipulated, deductions can be made, and a variety of operational processes can be performed with the computer. Inventory, accounting, programmable control, manufacturing operations, and process control instrumentation have caused industry to be more dependent on the computer. Computer applications in the future will obviously be more significant than they are today.

There is a great deal more to the operation of a computer than the average person realizes. When we look at a computer-based system, we generally only see its hardware. The *hardware* includes such things as typewriter keyboards, CRT display terminals, disk drive units, counters, flip-flops, decoders, signal generation, and electronic displays. As a rule, the hardware by itself is rather useless. It cannot function or perform an operation unless it is told what to do. It must be instructed before it can be of any value in the operation of a machine. Firmware and software are needed to communicate with the computer. Essentially, these two things tell the computer what it can and cannot do. *Firmware* is a permanently installed set of instructions placed in computer memory. It tells the computer circuitry what functions to perform and in the sequence that it must follow. The manufacturer of a computer is responsible for firmware and its installation in memory. *Software* is used to instruct the computer how to do a specific operation. After one operation has been completed, the computer may be instructed to perform another task. The software function of a computer is usually called a *program*. In general, a computer can utilize an unlimited number of programs. The computer operator is responsible for program use, its installation, and in some cases its development. This, in general, accounts for the versatility function of the computer.

Computer Hardware

The hardware of a computer can take on a variety of different shapes and sizes depending on the system. Most industrial computer-based systems are now of the microcomputer type. This type of system has the computer assembly built on a small printed circuit board or card. Other systems may have the computer function built on a single integrated circuit chip called a *microprocessor*. This chip can perform the basic functions of a computer. As a rule, microprocessors are responsible for most industrial applications of the computer.

Regardless of a computer's design, it has a number of basic operational functions that must be performed by the hardware. These functions are arranged in a specific sequence or pattern. In general, all computer-based systems must conform to this basic operational pattern. The end result of each computer is primarily the same. A block diagram of basic computer functions is shown in Fig. 7-1. Notice that this includes input/output, an arithmetic logic unit, control, memory, and a power source.

The input of a computer is often considered as an interface between the real world and the machine. *Interface* refers to the process of bringing two or more things together. The keyboard of a computer frequently serves as an interface device. This device accepts real-world input and converts it into something

that the computer understands. Basically, the input of a computer is responsible for encoding. This refers to the process of changing decimal numbers and/or alphabetic characters into a coded signal. The resulting signal is generally some type of binary data. Typewriter keyboards and calculator keyboards serve as the input for most digital electronic systems. In addition to this, data may also be supplied by punched cards, magnetic disks, and analog-to-digital converters.

Coded data developed by the computer input are applied to the arithmetic/logic unit (ALU). The ALU manipulates these data according to a prescribed set of instructions in the firmware. Normally, the incoming data are placed in a data register. This part of the computer serves as a short-term storage circuit for binary data. Registers are capable of storing one data word. A word may be either 8 or 16 bits of data in a group. Registers consist of a number of flip-flops. Each flip-flop will store one bit of data. These data can be applied to all flip-flops at one time or may be shifted into the register one bit at a time.

To demonstrate the operation of the ALU, assume that the computer needs to add two numbers. Initially, the first number is applied to the input. It is transported to the ALU and applied to register A. The second number is then entered into register B. A command from the control section tells the ALU to add the values in registers A and B. The sum is then placed in register A, replacing the first number. In performing this operation the ALU responds only to binary data words. The ALU is responsible for all computer math functions and logic decisions.

The control unit of a computer is responsible for directing the operation of the ALU. In a sense, all functional operations are directed by the control unit. Control is basically achieved by two operational cycles called *fetch* and *execute*. During the fetch cycle instructional data are retrieved from the memory unit. The execute cycle tells how to carry out the instruction. Computer operation is based on repeated fetch and execution cycles. The operational time of each cycle is directed by the control unit.

After the ALU has completed an operational cycle its accumulated data are transferred to the computer output. These informational data must then be translated into something that the real world understands. Normally, this operation is called *decoding*. Essentially, decoding is responsible for changing data into alphabetical or digital information. This information is then used to actuate the output device. Printers, typewriters, cathode-ray tube terminals, and digital displays are typical computer output devices. For an industrial computer-based system, the output could be used to actuate motors, relays, solenoids, lamps, or heating elements. The computer output serves as the control element for these load devices.

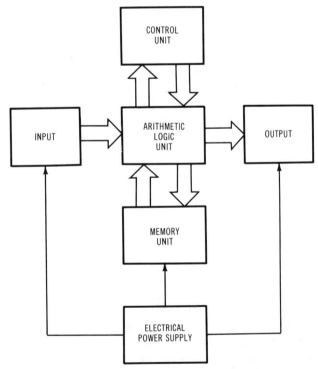

FIGURE 7-1 Block diagram of a basic computer.

MICROPROCESSORS

A microprocessor (MPU) is the arithmetic/logic unit and control section of a computer scaled down so that it fits on a single IC chip. Typical chip sizes are 0.16 in. × 0.16 in. and contain thousands of transistors, resistors, and diodes. Several dozen U.S. companies are now manufacturing these chips in a multitude of different variations. Most of these devices are of the metal-oxide-semiconductor (MOS) technology, with some being of the complementary-MOS (CMOS) type. A rather substantial number of these devices are found in dedicated industrial computer-based systems.

A microprocessor is essentially a digital device that is designed to receive binary data. It may then store these data for future processing, perform arithmetic and logic operations in accordance with previously stored instructions, and deliver the results to an output device. In a sense, a microprocessor is responsible for performing the operational functions of a basic computer.

A block diagram of a typical microprocessor shows that it contains a number of basic computer components connected in a rather unusual manner. Figure 7-2 shows a diagram of the internal functions of a microprocessor. Its construction includes an arithmetic/logic unit, an accumulator, a data register, address registers, program counter, instruction decoder, and sequence controller.

Arithmetic/Logic Unit

All microprocessors have an arithmetic and logic unit (ALU). The ALU performs mathematical and logical operations on the data words supplied to it. It is made to work automatically by control signals developed by the instruction decoder.

The ALU simply combines the contents of its two inputs, which are called the *data register* and *accumulator*. As a general rule, addition, subtraction, and logic comparisons are the primary operations performed by the ALU. The specific operation to be performed is determined by a control signal supplied by the instruction decoder.

The data supplied to the inputs of an ALU are normally in the form of 8-bit binary numbers. Upon

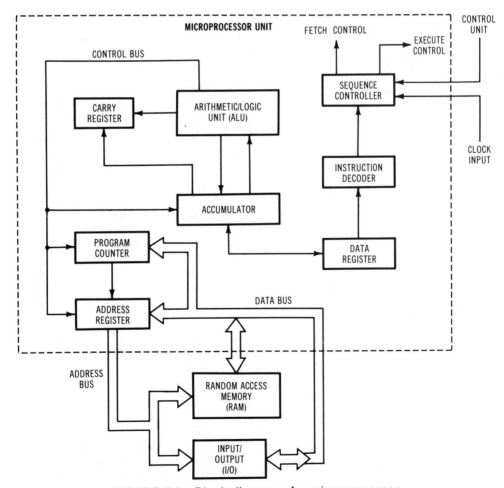

FIGURE 7-2 Block diagram of a microprocessor.

being received at the input, these data are combined by the ALU in accordance with the logic of binary arithmetic. Since a mathematical operation is ultimately performed on the two data inputs, the latter are often called *operands*.

To demonstrate the operation of the ALU, assume now that two binary numbers are to be added. In this case, let us consider the addition of the numbers 6_{10} and 8_{10}. Initially, the binary number 00000110_2 is placed in the accumulator. The second operand, 00001000_{10}, representing the number 8_{10}, is then placed into the data register. When a proper control line to the ALU is activated, binary addition is performed, producing an output of 00001110_2, or 14_{10}, which is the sum of the two operands. This value is then stored in the accumulator, where it replaces the operand that appeared there originally. The ALU only responds to binary numbers.

Accumulators

The accumulators of a microprocessor are temporary registers that are designed to store operands that are to be processed by the ALU. Before the ALU can perform, it must first receive data for the accumulator. After the data register input and accumulator input are combined, the logical answer or output of the ALU appears in the accumulator. This particular function is essentially the same for all microprocessors.

In microprocessor operation, a typical instruction would be to "load the accumulator." This instruction enables the contents of a particular memory location to be placed into the accumulator. A similar instruction might be "store accumulator." In this operation the instruction causes the contents of the accumulator to be placed in a selected memory location. Essentially, the accumulator serves in one capacity as an input source for the ALU, and then as a destination area for its output.

Data Registers

The data register of a microprocessor serves as a temporary storage location for information applied to the data bus. Typically, this register will accommodate an 8-bit data word. An example of a function of this register is operand storage for the ALU input. In addition, it may be called on to hold an instruction while the instruction is being decoded, or it may temporarily hold data prior to the data being placed in memory.

Address Registers

Address registers are used in microprocessors to temporarily store the address of a memory location that is to be accessed for data. In some units this register

may be programmable. This means that it permits instructions to alter its contents. The program can also be used to build an address in the register prior to executing a memory reference instruction.

Program Counter

The program counter of a microprocessor is a memory device that holds the address of the next instruction to be executed in a program. As a general rule, this unit simply counts the instructions of program in sequential order. In practice, when the MPU has fetched instructions addressed by the program counter, the count advances to the next location. At any given point during the sequence, the counter indicates the location in memory from which the next information will be derived.

The numbering sequence of the program counter may be modified so that the next count may not follow a numerical order. Through this procedure, the counter may be programmed to jump from one point to another in a routine. This permits the MPU to have branching capabilities should the need arise.

Instruction Decoders

Each specific operation that the MPU can perform is identified by an exclusive binary number known as an instruction code. Eight-bit words are commonly used for this code. Exactly 2^8 or 256_{10} separate or alternative operations can be represented by this code. After a typical instruction code is pulled from memory and placed in the data register, it must be decoded. This instruction decoder simply examines the coded word and selectively decides which operation is to be performed by the ALU. The output of the decoder is first applied to the sequence controller.

Sequence Controller

The sequence controller performs a number of very vital functions in the operation of a microprocessor. Using clock inputs, the circuitry maintains the proper sequence of events required to perform a processing task. After instructions are received and decoded, the sequence controller issues a control sign that initiates the proper processing action. In most units the controller has the capability of responding to external control signals.

Buses

The registers and components of most microprocessors are connected by a bus-organized type of network. In a computer-based system, the term *bus* is defined as a group of conductor paths that are used to connect data words to various registers. A simplifi-

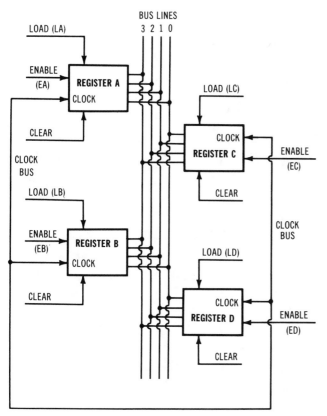

FIGURE 7-3 Registers connected to common bus lines.

cation of registers connected by a common bus is shown in Fig. 7-3.

The utility of bus-connected components is the ease with which a data word can be transmitted or loaded into registers, In operation, each register has inputs labeled clock, enable, load, and clear. When the load and enable input lines are low or at 0, each register is isolated from the common bus line.

When transferring a word from one register to another, it is necessary to make the appropriate inputs high or at the 1 state. For instance, to transfer the data of register A to register D, enable A (EA) and load D (LD) inputs must both be in the 1 state. This will cause the data of register A to appear on the common bus line. When a clock pulse arrives at the common inputs, the transfer process is completed.

The word length of a bus is based on the number of conductor paths that it employs. Buses for 4, 8, and 16 bits are commonly used in microprocessors. New MPUs may use 8-, 16-, or 32-bit address buses.

MEMORY

Industrial computer-based systems range from a number of single-chip microprocessors to some rather complex networks that employ several auxiliary chips con-

nected together in a massive system. In general, the primary difference in this broad range of hardware is in the memory capabilities of the system. Single-chip microprocessor units are generally limited in the amounts of memory that they can have owing to the large number of essential logic functions needed to make the unit operational. Additional memory can be achieved much more economically through the use of auxiliary chips. The potential capabilities of a microprocessor type of system are primarily limited by the range of memory that it employs.

Memory refers to the capability of a device to store logical data in such a way that a single bit or group of bits can be easily accessed or retrieved. In practice, memory can be achieved in a variety of different ways. Computer-based systems are usually concerned with read/write memory and read-only memory. These two classifications of memory are accomplished by employing numerous semiconductor circuit duplications on a single IC chip.

Read/Write Memory

Read/write semiconductor memories are the most widely used form of electronic memory found in computer-based systems today. Read/write chips of the large-scale integration or LSI type are capable of storing 16,384 or 16K bits of data in an area less than one-half of a square centimeter. New technologies are responsible for 64K, 128K, 256K, and 512K units. The actual structure of a chip includes a number of discrete circuits, each having the ability to store binary data in an organized manner. Access to each memory location is provided by coded information from the microprocessor address bus. The read/write function indicates that data can be placed into memory or retrieved at the same rate.

A simplification of the memory process is represented by the 8×8 memory unit of Fig. 7-4. As shown here, the memory units of an IC are organized in a rectangular pattern of rows and columns. This particular chip employs eight rows that can each store 8-bit words or a total of 64 single bits of memory. To select a specific memory address, a 3-bit binary number is used to designate a specific row location and three additional bits are used to indicate the column location. In this example, the row address is 3_{10}, or 011_2, and the column address is 5_{10}, or 101_2. The selected memory address is at location 30_{10}.

Many read/write memory chips employ a single MOS transistor for each memory cell. Figure 7-5 shows the circuitry of a discrete MOS transistor memory cell. Electronic data in the form of a binary signal are stored in the transistor as a charge on a small capacitor. The gate and source–drain electrodes of the transistor serve as the plates of a capacitor. No charge

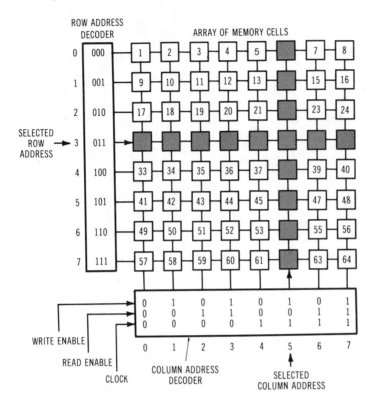

FIGURE 7-4 Simplification of the memory location process.

across the two electrodes represents a 0 or no memory condition. A charge appearing across the two electrodes indicates a one (1) or active memory state. When a row select line is activated, it energizes the gate of each transistor in the entire row. When a column line is selected, it energizes the source–drain electrodes of each transistor in the column. Simultaneous

activation of a row and column energizes a specific transistor memory cell.

The charge placed on an MOS transistor memory cell must be restored periodically to overcome component leakage. Charge restoration is achieved by a special transistor circuit outside the memory cells. The charge-restoring transistor is called a *thresholding amplifier*. In actual circuit operation, charge regeneration occurs every few milliseconds.

Eight-bit word storage is achieved in the MOS memory unit by energizing one row and all eight columns simultaneously. The row and column decoders are responsible for this operation. Some of the common read/write memory units available today have capacities of 8×8, 32×8, 128×8, 1024×8, and 4096×8.

To write a word into memory, a specific address is first selected according to the data supplied by the address bus. See the block diagram of the read/write memory unit of Fig. 7-6. The address decoder, in this case, selects the appropriate row and column select lines. A high or 1 write-enable signal applied to the control unit causes the data bus signal to be transferred to the selected memory address. These data then charge the appropriate memory cells according to the coded 1 or 0 values. Removing the write-enable signal causes the data charge accumulations to remain at each cell location. The output is disconnected from the data bus after the write operation has been completed.

To read the charge accumulation appearing at

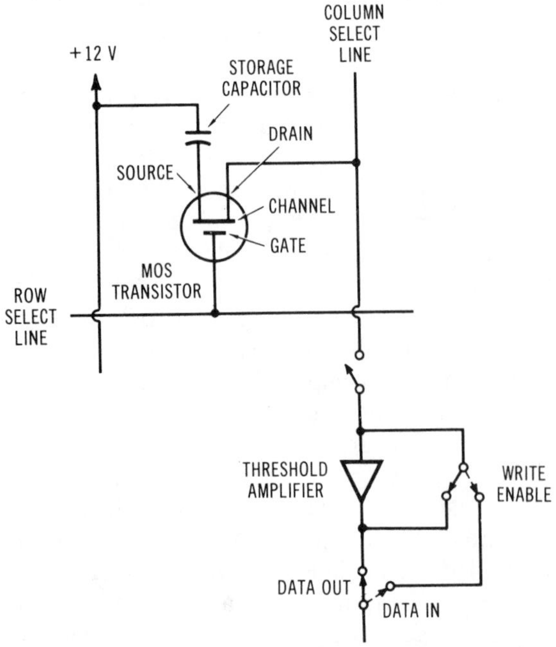

FIGURE 7-5   MOS transistor memory cell.

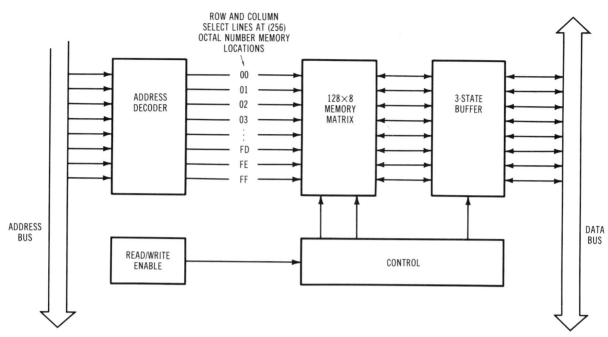

FIGURE 7-6 Read/write memory block diagram.

each memory cell, the read-enable control line must be energized. Selecting a specific memory location causes charge data to appear at the data bus as memory output signals. Charge restoration from the thresholding amplifier continuously maintains the same charge at each cell. This means that reading from memory does not destroy the charge data at each cell. All this takes place as long as the memory unit is energized electrically. A loss of electrical power or turning the unit off destroys the data placed at each memory location. Essentially, solid-state read/write memory is classified as volatile because of this characteristic.

Read-Only Memory

Most computer-based systems necessitate memories that contain permanently stored or rarely altered data. A prime example of this would be math tables and firmware program data. Storage of this type is provided by read-only memory (ROM). Information is often placed in this type of memory unit when the chip is manufactured. ROM data are considered to be *nonvolatile*. This means that the data will not be lost when the power source is removed or turned off.

Read-only memory is achieved in a variety of ways. One very common process employs fusible links built into each memory cell. Data can be placed in memory by melting the necessary fusible links. This type of procedure is used to open interconnecting conductors, to place a diode between conductors, or place a small capacitor between the two electrodes. Obviously, a fusible link cannot be re-formed after it has

once been altered. This principle of ROM development can only be used once to alter the chip.

Programmable read-only memory (PROM) chips can be altered or erased by exposure to an ultraviolet energy source. Exposing a PROM to a UV energy source causes all memory cells of the entire unit to change to the zero state. Altering the chip electrically will cause a new program to be initiated.

Figure 7-7 shows a cross-sectional view of a UV erasable ROM. The floating gate of the MOS transistor

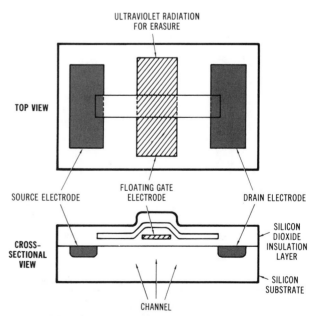

FIGURE 7-7 Optically erasable ROM.

ACCUMULATOR AND MEMORY INSTRUCTIONS

Legend for column sub-headers: OP = Operation Code (Hexadecimal); ~ = Number of MPU Cycles; ≏ = Number of Program Bytes. Condition code register bits: 5=H, 4=I, 3=N, 2=Z, 1=V, 0=C.

Operations	Mnemonic	IMMED OP	~	≏	DIRECT OP	~	≏	INDEX OP	~	≏	EXTND OP	~	≏	IMPLIED OP	~	≏	Boolean/Arithmetic Operation	H	I	N	Z	V	C
Add	ADDA	8B	2	2	9B	3	2	AB	5	2	8B	4	3				$A+M \rightarrow A$	↕	•	↕	↕	↕	↕
	ADDB	CB	2	2	DB	3	2	EB	5	2	FB	4	3				$B+M \rightarrow B$	↕	•	↕	↕	↕	↕
Add Acmltrs	ABA													1B	2	1	$A+B \rightarrow A$	↕	•	↕	↕	↕	↕
Add with Carry	ADCA	89	2	2	99	3	2	A9	5	2	B9	4	3				$A+M+C \rightarrow A$	↕	•	↕	↕	↕	↕
	ADCB	C9	2	2	D9	3	2	E9	5	2	F9	4	3				$B+M+C \rightarrow B$	↕	•	↕	↕	↕	↕
And	ANDA	84	2	2	94	3	2	A4	5	2	B4	4	3				$A \cdot M \rightarrow A$	•	•	↕	↕	R	•
	ANDB	C4	2	2	D4	3	2	E4	5	2	F4	4	3				$B \cdot M \rightarrow B$	•	•	↕	↕	R	•
Bit Test	BITA	85	2	2	95	3	2	A5	5	2	B5	4	3				$A \cdot M$	•	•	↕	↕	R	•
	BITB	C5	2	2	D5	3	2	E5	5	2	F5	4	3				$B \cdot M$	•	•	↕	↕	R	•
Clear	CLR							6F	7	2	7F	6	3				$00 \rightarrow M$	•	•	R	S	R	R
	CLRA													4F	2	1	$00 \rightarrow A$	•	•	R	S	R	R
	CLRB													5F	2	1	$00 \rightarrow B$	•	•	R	S	R	R
Compare	CMPA	81	2	2	91	3	2	A1	5	2	B1	4	3				$A-M$	•	•	↕	↕	↕	↕
	CMPB	C1	2	2	D1	3	2	E1	5	2	F1	4	3				$B-M$	•	•	↕	↕	↕	↕
Compare Acmltrs	CBA													11	2	1	$A-B$	•	•	↕	↕	↕	↕
Complement, 1's	COM							63	7	2	73	4	3				$\overline{M} \rightarrow M$	•	•	↕	↕	R	S
	COMA													43	2	1	$\overline{A} \rightarrow A$	•	•	↕	↕	R	S
	COMB													53	2	1	$\overline{B} \rightarrow B$	•	•	↕	↕	R	S
Complement, 2's	NEG							60	7	2	70	6	3				$00-M \rightarrow M$	•	•	↕	↕	①	②
(Negate)	NEGA													40	2	1	$00-A \rightarrow A$	•	•	↕	↕	①	②
	NEGB													50	2	1	$00-B \rightarrow B$	•	•	↕	↕	①	②
Decimal Adjust, A	DAA													19	2	1	Converts Binary Add. of BCD Characters into BCD Format	•	•	↕	↕	↕	③
Decrement	DEC							6A	7	2	7A	6	3				$M-1 \rightarrow M$	•	•	↕	↕	④	•
	DECA													4A	2	1	$A-1 \rightarrow A$	•	•	↕	↕	④	•
	DECB													5A	2	1	$B-1 \rightarrow B$	•	•	↕	↕	④	•
Exclusive OR	EORA	88	2	2	98	3	2	A8	5	2	B8	4	3				$A \oplus M \rightarrow A$	•	•	↕	↕	R	•
	EORB	C8	2	2	D8	3	2	E8	5	2	F8	4	3				$B \oplus M \rightarrow B$	•	•	↕	↕	R	•
Increment	INC							6C	7	2	7C	6	3				$M+1 \rightarrow M$	•	•	↕	↕	⑤	•
	INCA													4C	2	1	$A+1 \rightarrow A$	•	•	↕	↕	⑤	•
	INCB													5C	2	1	$B+1 \rightarrow B$	•	•	↕	↕	⑤	•
Load Acmltr	LDAA	86	2	2	96	3	2	A6	5	2	B6	4	3				$M \rightarrow A$	•	•	↕	↕	R	•
	LDAB	C6	2	2	D6	3	2	E6	5	2	F6	4	3				$M \rightarrow B$	•	•	↕	↕	R	•
Or, Inclusive	ORAA	8A	2	2	9A	3	2	AA	5	2	BA	4	3				$A+M \rightarrow A$	•	•	↕	↕	R	•
	ORAB	CA	2	2	DA	3	2	EA	5	2	FA	4	3				$B+M \rightarrow B$	•	•	↕	↕	R	•
Push Data	PSHA													36	4	1	$A \rightarrow M_{SP}, SP-1 \rightarrow SP$	•	•	•	•	•	•
	PSHB													37	4	1	$B \rightarrow M_{SP}, SP-1 \rightarrow SP$	•	•	•	•	•	•
Pull Data	PULA													32	4	1	$SP+1 \rightarrow SP, M_{SP} \rightarrow A$	•	•	•	•	•	•
	PULB													33	4	1	$SP+1 \rightarrow SP, M_{SP} \rightarrow B$	•	•	•	•	•	•
Rotate Left	ROL							69	7	2	79	6	3				M 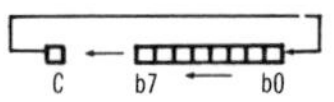 (C ← b7…b0 rotate)	•	•	↕	↕	⑥	↕
	ROLA													49	2	1	A	•	•	↕	↕	⑥	↕
	ROLB													59	2	1	B	•	•	↕	↕	⑥	↕
Rotate Right	ROR							66	7	2	76	6	3				M 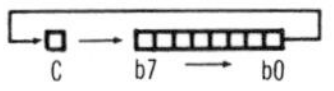 (C → b7…b0 rotate)	•	•	↕	↕	⑥	↕
	RORA													46	2	1	A	•	•	↕	↕	⑥	↕
	RORB													56	2	1	B	•	•	↕	↕	⑥	↕
Shift Left, Arithmetic	ASL							68	7	2	78	6	3				M 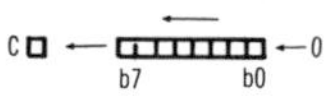 (C ← b7…b0 ← 0)	•	•	↕	↕	⑥	↕
	ASLA													48	2	1	A	•	•	↕	↕	⑥	↕
	ASLB													58	2	1	B	•	•	↕	↕	⑥	↕
Shift Right, Arithmetic	ASR							67	7	2	77	6	3				M 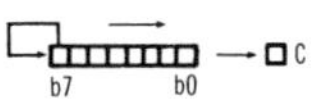 (b7…b0 → C)	•	•	↕	↕	⑥	↕
	ASRA													47	2	1	A	•	•	↕	↕	⑥	↕
	ASRB													57	2	1	B	•	•	↕	↕	⑥	↕
Shift Right, Logic	LSR							64	7	2	74	6	3				M 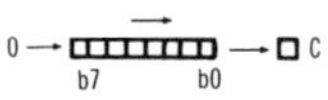 (0 → b7…b0 → C)	•	•	R	↕	⑥	↕
	LSRA													44	2	1	A	•	•	R	↕	⑥	↕
	LSRB													54	2	1	B	•	•	R	↕	⑥	↕
Store Acmltr	STAA				97	4	2	A7	6	2	B7	5	3				$A \rightarrow M$	•	•	↕	↕	R	•
	STAB				D7	4	2	E7	6	2	F7	5	3				$B \rightarrow M$	•	•	↕	↕	R	•
Subtract	SUBA	80	2	2	90	3	2	A0	5	2	B0	4	3				$A-M \rightarrow A$	•	•	↕	↕	↕	↕
	SUBB	C0	2	2	D0	3	2	E0	5	2	F0	4	3				$B-M \rightarrow B$	•	•	↕	↕	↕	↕
Subtract Acmltrs	SBA													10	2	1	$A-B \rightarrow A$	•	•	↕	↕	↕	↕
Subt with Carry	SBCA	82	2	2	92	3	2	A2	5	2	F2	4	3				$A-M-C \rightarrow A$	•	•	↕	↕	↕	↕
	SBCB	C2	2	2	D2	3	2	E2	5	2	7D	6	3				$B-M-C \rightarrow B$	•	•	↕	↕	↕	↕
Transfer Acmltrs	TAB													16	2	1	$A \rightarrow B$	•	•	↕	↕	R	•
	TBA													17	2	1	$B \rightarrow A$	•	•	↕	↕	R	•
Test, Zero or Minus	TST																$M-00$	•	•	↕	↕	R	R
	TSTA													4D	2	1	$A-00$	•	•	↕	↕	R	R
	TSTB													5D	2	1	$B-00$	•	•	↕	↕	R	R

Condition code register columns (bottom): H I N Z V C

LEGEND

OP Operation Code (Hexadecimal)
→ Number of MPU Cycles
≏ Number of Program Bytes.
+ Arithmetic Plus.
− Arithmetic Minus
· Boolean AND.
M_{SP} Contents of memory location pointed to be Stack Pointer

+ Boolean Inclusive OR.
⊕ Boolean Exclusive OR.
$\overline{M}$ Complement of M.
• Transfer Into.
0 Bit Zero.
00 Byte Zero.

CONDITION CODE SYMBOLS

H Half carry from bit 3.
I Interrupt mask
N Negative (sign bit)
Z Zero (byte)
V Overflow 2's complement
C Carry from 7

R Reset Always
S Set Always
↕ Test and set if true, cleared otherwise
• Not Affected

Note: Accumulator addressing mode instructions are included in the column for IMPLIED addressing

FIGURE 7-8 MC6800 Instructional set. (Courtesy of Motorola Semiconductor Products, Inc.)

of this device is not connected to anything. Data to be stored in the cell are written into the transistor by applying 25 V dc between the gate and drain, while the source and substrate are at a ground potential. This condition causes a static field to appear between the gate and source, which causes electrons to move with a great deal of velocity. Electrons that move through the thin silicon dioxide layer become trapped on the floating gate. The charged gate–drain electrodes serve as the plate of a small capacitor. A charged condition represents a 1 or high state, while an uncharged state represents a 0 condition. The charged condition of each transistor of memory cell is nonvolatile.

Erasing the charged data of each MOS cell is achieved by exposing the chip to a source of ultraviolet energy. This action causes the silicon dioxide layer to be temporarily conductive. As a result, excessive leakage causes the charge formed on the gate to dissipate. During the exposure operation, each cell is discharged at the same time. Memory can be restored by writing data back into each transistor cell. PROMs can be altered while attached to a circuit board if the need arises.

An alternative to the UV erasable ROM is the electrically alterable read-only memory (EAROM). This type of memory chip permits individual cells or memory words to be erased instead of the entire chip. Cell structure is primarily the same as that of the UV erasable ROM. The floating gate structure of each cell is altered by having a discrete insulation strip between it and the drain. Selective charge and discharge of cells can be achieved by electrical signals applied to the gate. Applications of the EAROM are not very common today. This chip is more costly than others, has slow access time, and some long term memory retention problems.

A Programming Example

To demonstrate the potential capabilities of a computer-based system we discuss its ability to solve a straight-line computation problem. In this problem the system is used simply to add two numbers and then indicate the resulting sum. This type of problem is obviously quite simple and could easily be solved without the help of a microcomputer system. It is, of course, used in this situation to demonstrate a principle of operation and to show a plan of procedure more so than for its problem-solving capability.

In practice, a microcomputer system could not solve even the simplest type of problem without the help of a well-defined program that works out everything right down to the smallest detail. After the program has been developed, the system simply follows this procedure to accomplish the task. Programming

is an essential part of nearly all computer system applications.

Before a program can be effectively prepared for a microcomputer system the programmer must be fully aware of the specific instructions that can be performed by the system. In general, each microcomputer system has a unique list of instructions that are used to control its operation. The *instruction set* of a microcomputer is the basis of all program construction. Figure 7-8 shows an instruction set for the Motorola MC6800 microprocessor system.

Assume now that the programmer is familiar with the instruction set of the system being used to solve our straight-line computation problem. The next step in this procedure is to decide on what specific instructions are needed to solve the problem. A limited number of operations to perform can generally be developed without the aid of a diagrammed plan of procedure. Complex problems, by comparison, usually require a specific plan to reduce confusion or to avoid the loss of an important operational step. Flowcharts are commonly used to aid the programmer in this type of planning. Figure 7-9 shows some of the flowchart symbols that are commonly used in program planning.

The first step in preparing a program to solve our problem is to make a flowchart that shows the procedure to be followed. In this case we set up the program to accomplish the following:

Find: the number N (a decimal value)
Use equation: $x + y = N$
Let: $x = 0A_{16}$ (operand)
$y = 07_{16}$ (operand)

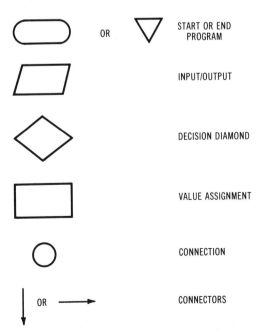

FIGURE 7-9 Flowchart symbols.

STEPS

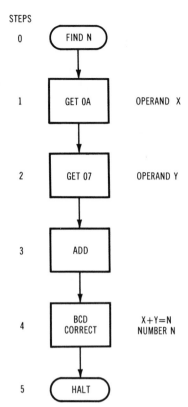

FIGURE 7-10 Flowchart for problem solving.

to energize an output to produce a decimal readout. The fifth and final step of the program is an implied halt opcode that stops the program.

A programming sheet for our problem example is shown in Fig. 7-11. Notice that this sheet indicates the step number, a representative memory address in hexadecimal values, instruction bytes in hexadecimal values, opcode mnemonics, binary equivalents, addressing mode, and a description of each function. The memory address locations employed in the program begin with 66_{16}. In practice it is a common procedure to reserve the first 100 memory locations, or 00 to 63_{16}, for branch instructions. We have arbitrarily selected location 66_{16} or 102_{10} to avoid those addresses being reserved for branching operations.

The program example used here is only one of literally thousands of programs that can be employed by a computer-based system to perform industrial operations. We have tried to show only one simple problem that could be achieved by a computer. The potential capabilities of this type of system are virtually unlimited. The type of microprocessor employed by a system and its unique instructional set are the primary factors that govern its operation in program planning.

SUMMARY

Industry now has instruments and machinery that are classified as intelligent or smart equipment. This type of equipment is also described as a computer-based system.

The most obvious part of a computer-based system is its hardware. Typewriter keyboards, CRT display terminals, input devices, output devices, and memory storage devices are some examples of hardware.

Firmware and software are needed to communicate with a computer. Firmware is a permanently installed set of instructions in the computer's memory.

Figure 7-10 shows a simple program flowchart to be followed to solve the problem. The first step, numbered 0, is a simple statement of the problem. Steps 1 and 2 are then used as the operands of the problem and indicate the values of x and y. After these values have been obtained, each is placed into a specific accumulator. Step 3 is then responsible for the addition operation. Since this system deals with binary numbers, conversion to decimal values is also necessary. Step 4 is a conversion operation that changes binary numbers to BCD values. This value can then be used

Programming Sheet

Title: X+Y=N   Date: _____

Purpose: Find N   Time: _____

Steps	Memory Address (Hexadecimal)	Byte 1	Byte 2	Byte 3	Opcode	Addressing Mode	Operand (Binary)	Description
1	66	86	—	—	LDA A	Immediate		Load accumulator A
	67	—	0A	—			0000 1010	Use data $0A_{16}$
2	68	C6	—	—	LDA B	Immediate		Load accumulator B
	69	—	08	—			0000 1000	Use data 018_{16}
3	6A	1B	—	—	ABA	Inherent		Add the contents of accumulators A and B
4	6B	19	—	—	DAA	Inherent		Correct for bad output
5	6C	3E	—	—	HLT	Inherent		Stop all operations

FIGURE 7-11 Programming sheet.

It tells the hardware what functions to perform and in what sequence. The manufacturer of a computer is responsible for its firmware. Software is used to instruct the computer how to do specific operations. Software is generally called a program. The computer operator is responsible for program use, installation, and development.

Computer systems vary in size and type. Large mainframe computers have given way to microcomputers. This type of unit contains several ICs built on a common printed circuit board. Microprocessors are single-chip computers. Regardless of the computer system being used, it is composed of an input/output, arithmetic/logic unit, control, memory, and a power source.

A microprocessor is the primary control section of a computer scaled down to fit on a single IC chip. The arithmetic/logic unit (ALU) achieves the arithmetic function. Accumulators function as holding registers for operands that are used to temporarily store the address of a memory location that can be accessed for data. A program counter is used to hold the address of the next instruction to be executed in a program. Instruction decoders are used to decipher an instruction after it has been pulled from memory. Sequence controllers maintain the logical order in which events are performed by the microprocessor. Buses are conductor paths that supply data words to registers.

Most computer-based systems employ auxiliary memory units to extend their operating capabilities. Memory permits data to be accessed or retrieved. Read/write memory permits data to be placed or stored in specific memory cells and retrieved at a later time when it is needed. Read-only memory contains permanently stored or rarely altered data. Permanent program data are a prime example of a ROM application. A PROM is a programmable read-only memory. Electrical energy is used to store data in a PROM, and ultraviolet energy is used to erase it.

Programming refers to a series of acceptable instructions developed for a computer to permit it to perform a prescribed operation or function. A hardwired program is achieved by electrical circuit construction. Firmware systems have programmed material placed on read-only memory chips. Software programming is created on paper and transferred to the system through a keyboard, punched cards, or magnetic tape. Most computer-based systems combine firmware and software instructions in their programming material.

When programming a computer-based system, the programmer must be aware of specific unit instructions, decide on what instructions are needed to solve a problem, plan a flowchart, develop a programming sheet, initiate the program, correct it if needed, then execute its operation.

ACTIVITIES

COMPUTER STARTUP

1. This activity necessitates a computer, a startup disk, a disk drive, a display (CRT), and a keyboard.
2. Familiarize yourself with the computer operator's manual.
3. Familiarize yourself with the basic operating controls of the computer. This generally includes the power on–off switch and the CRT brightness control.
4. Insert the startup disk, label side up, into disk drive 1.
5. Turn on the CRT and wait a few minutes for warm-up.
6. Adjust the brightness control to a suitable level.
7. Turn on the computer power switch. This generally produces a beeping sound and the indicator lamp will be on. An LED on the front of the disk drive unit should light and the unit should make a whirring or clicking sound. After a few seconds the LED will turn off and the sounds will stop. This is an indication that the program on the disk has been loaded into the computer.
8. Depending on the startup disk, some information will appear on the CRT display.
9. The procedure just described is often called computer startup or "booting" the computer.

COMPUTER KEYBOARD FAMILIARIZATION

1. Boot the computer.
2. Refer to the keyboard of the computer.
3. Familiarize yourself with the keyboard.
4. A keyboard has special function keys, control keys, a return key, the space bar, uppercase and lowercase keys, the shift key, and cursor movement keys.
5. Identify the location of these keys.
6. The display should have a cursor located in the upper left corner. Find the cursor.
7. The return key moves the cursor to the beginning of the next line. Depress the return key one or two times while noting the position change of the cursor.
8. The cursor movement keys are generally identified with direction arrows. Locate these keys and press each one while observing the positional change of the cursor.
9. Depress one of the letter keys. Hold the key down for a short time while observing the display. This action should cause the letter to be repeated.

10. Press the return or enter key.

11. Insert a floppy disk that contains BASIC software language into disk drive A.

12. On the keyboard, type BASIC adjacent to the prompt > appearing on the screen and press the ENTER or RETURN key. The computer will now have BASIC software language resident in its memory and is ready to accept a BASIC program prepared by the operator.

13. Type the following: PRINT "HELLO, MY NAME IS JOHN."

14. Then press the return or enter key. Describe your findings.

15. Type PRINT 15 + 34.

16. Then press the return or enter key. What happens?

17. Other math operations can be performed using the same procedure. Subtraction uses the minus sign (−), multiplication uses the (*), and division uses (/). Try several problems using different math functions.

18. To return the computer to its initialized MSDOS state after programming or program operations are complete, type the command SYSTEM after the prompt > and strike the enter or return key. Remove any floppy disks from the disk drive. The computer may now be used for any other programming language available or be "powered off."

QUESTIONS

1. What are the fundamental parts of a basic computer?

2. Explain the differences between software, hardware, and firmware.

3. What is a microprocessor?

4. What is the functional role of an ALU?

5. What is the function of an accumulator?

6. What is meant by the term *memory* in a computer-based system?

7. Explain the meaning of the term *PROM*.

8. What is a read/write memory?

9. How are data erased from a PROM?

10. What is an instructional set for a microprocessor?

11. Define the term *programming*.

12. How is a flowchart used in programming?

CHAPTER 8

ELECTRONIC RECORDERS

OBJECTIVES

Upon completion of this chapter, you will be able to:
1. Identify the functional blocks of an electronic recorder.
2. Discuss the operation of an electronic recorder with a functional block diagram.
3. Explain how the principles of resistance, capacitance, and inductance are used in the operation of an electronic recorder.
4. Explain how diodes and transistors are used in the operation of an electronic recorder.
5. Explain the functional operation of the discrete blocks of an electronic recorder.

IMPORTANT TERMS

In this chapter we are going to investigate some of the basic operating principles of an electronic recorder. You will have an opportunity to become familiar with a number of the basic functions of this instrument. New words, such as *feedback, vector, phasing*, and *torque*, will begin to have some meaning for you. A few of these terms are singled out for study. As a rule, the chapter will be more meaningful if these terms are reviewed before proceeding with the text.

Balance motor: The servomotor of an electronic recorder. It is responsible for moving the pen and altering the slide-wire balancing mechanism.

Direct coupling: A method of connecting two amplifying circuits together without capacitors or inductors so that both ac and dc signals can pass through the coupling path.

Dither signal: A small ac voltage applied to the servomotor amplifier that causes a slight position change when the balance state is reached.

Feedback: The return of a portion or all of the output of a circuit or a device to its input.

Instantaneous value: A value of ac voltage that occurs at any instant or time along the waveform.

Kirchhoff's voltage law: In any current loop of a circuit the sum of the voltage drops is equal to the voltage supplied to the loop.

Null detection: An electronic circuit that determines a condition of balance or zero output.

Phase: The position of a point on a periodic waveform with respect to the start of the cycle and expressed in electrical degrees.

Phase angle: An angle, expressed in electrical degrees or radians, that indicates the amount by which the current in an electrical circuit leads or lags the voltage.

Points: The marking tip or stylus of a recorder.

Pythagorean theorem: A geometry proof stating that the square of the hypotenuse of a right-angled triangle is equal to the sum of the squares of the other two sides.

Reactance (X): An opposition to ac current flow due to inductive reactance (X_L) or capacitive reactance (X_C).

Reactive component: A device that offers capacitive or inductive reactance in a circuit or to a signal path.

Recorder: An instrument designed to provide a graphic

display of variations in a particular quantity being measured.

Resonance: A circuit condition in which the inductive reactance (X_L) equals the capacitive reactance (X_C) at a specific frequency.

Resonant circuit: An ac circuit that exhibits resonance.

Torque: Mechanical energy in the form of rotary motion.

Vector: A straight line that indicates a quantity that has magnitude and direction.

INTRODUCTION

There are numerous commercial electronic recording instruments available today for use in industrial control applications. As a general rule, each manufacturer tries to do something unique with its instrument to improve performance or make it more competitive than the others. Such things as multipoint recorders, different recording techniques, physical size, and electronic circuitry account for some of the differences in these instruments.

A person working with electronic recording instruments should have some understanding of the circuitry needed to make the instrument perform. Due to the number of instruments available, it is difficult to single out one particular instrument and refer to it as an industrial standard. A typical electronic recorder is discussed in this chapter. It is only one example of a number of instruments available today. The circuits discussed are all solid state. The operation of this instrument could, however, apply to a number of other commercial instruments. When working with a specific recorder, the manufacturer's operational and instructional literature should always serve as the primary reference source.

An electronic universal multipoint recorder is shown in Fig. 8-1. This particular recorder has the capability of being adjusted so that it can record information from 2 to 24 points. Converting the number of points being measured can be done within a matter of seconds by altering a shorting plug. This type of recorder uses a syncro-balancing motor-control unit.

FUNCTIONAL UNITS

The functional units of an electronic recorder are shown in block diagram form in Fig. 8-2. A recorder of this type contains two major divisions with several discrete electronic functions in each division. The amplifier unit contains a power supply, filter, chopper, voltage amplifier, and a driver stage. The motor unit contains a power amplifier and a balancing motor.

The operation of each discrete functional block of the recorder is discussed briefly in this chapter. The components making up each block involve the prin-

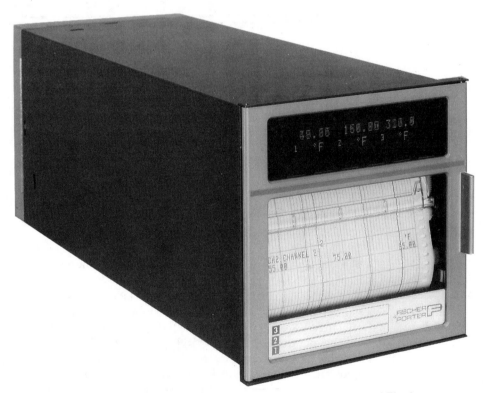

FIGURE 8-1 Universal multipoint recorder. (Courtesy of Fischer & Porter Co.)

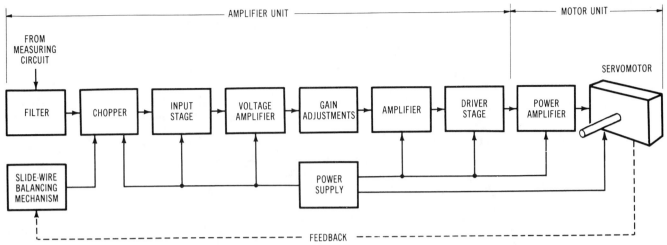

FIGURE 8-2 Block diagram showing the functional units of a recorder. (Courtesy of Honeywell, Inc.)

ciples of resistance, capacitance, and inductance, as well as diode and transistor operation. These components and their principles of operation serve as the basis for the discussion. In a sense we will be using the recorder as a means of discussing specific component applications. Through this approach, you will be able to increase significantly your understanding of electronic theory.

GENERAL RECORDER OPERATION

Using the block diagrams of Fig. 8-2 as a reference, we first discuss the recorder in general operational terms. Initially, the input to the amplifier is a dc signal that varies according to the measured value of the variable under consideration. It may be of either positive or negative polarity. This signal then passes through the filter, which removes the extraneous noise. The chopper then changes dc into a series of pulses that have a characteristic ac shape. This signal is amplified and ultimately applied to the driver stage. It is then added to the rectified sine wave that drives the motor unit. Motor rotational direction is based on the phase difference between this ac signal and a fixed phase signal from the power line. A positive dc input causes clockwise motor rotation until balance is achieved, and a negative input causes counterclockwise rotation. A "dither" signal applied to the input is a small voltage that causes a slight oscillation in motor position when it reaches a state of balance.

Power Supply

The power supply of an electronic recorder provides ac to the motor as a line reference source and also develops dc voltage for component operation. A simplified schematic of the power supply is shown in Fig.

8-3. The main rectifier employing CR_5 and CR_6 is a full-wave circuit with two zener diode regulators. Three different dc voltage values are developed for circuit operation. Capacitor C_{13} is part of the RC filter. Diode CR_3 is added to the circuit to prevent reverse voltage spikes from entering the dc supply circuit. It is forward biased, which permits full conduction capabilities during normal operation.

Diode CR_1 and its associated components serve as a half-wave rectifier to develop a negative dc voltage. This voltage is regulated by VR_3 and filtered by C_2, R_6, and R_2. Note that CR_1 is placed in the power supply in a reverse polarity. The negative dc voltage developed by this rectifier is ultimately used to bias the collector of the waveshaping amplifier of the chopper circuit.

Input Filter

The input filter of an electronic recorder is shown in Fig. 8-4. A three-section RC network is formed by R_9–C_4, R_{10}–C_5, and R_{12}–C_6. A dc signal applied to the input charges each section of the filter in succession. Through this action, the filter removes any noise or extraneous ac voltages that may appear at the input. Resistor R_{11} is used to couple the input to the output so that any ac passing through the filter will be of equal amplitude but 180° out of phase. As a result of this action, any resulting ac is canceled by phase inversion. A discussion of phase shifting and component influence on ac phase is presented later in this chapter.

Chopper/Waveshaper

The chopper of an electronic converter is a circuit that changes dc to ac. This circuit supersedes the mechanical vibrating chopper discussed in Chapter 6. A simplified schematic of the electronic chopper is shown

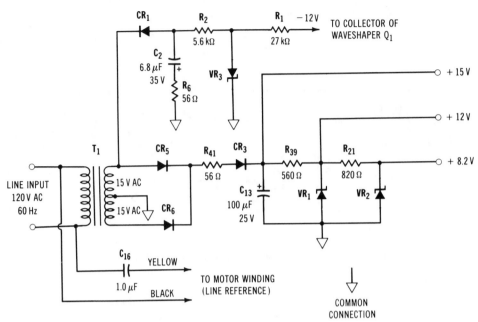

FIGURE 8-3 Simplified schematic of the power supply used in a recorder. (Courtesy of Honeywell, Inc.)

in Fig. 8-5. As you will note, this circuit employs a metal-oxide-semiconductor field-effect transistor (MOSFET) and a bipolar transistor. Q_1 is a waveshaping transistor that changes the ac secondary voltage of transformer T_1 into a negative-going square wave. The base circuit of Q_1 is preceded by an RC low-pass filter consisting of R_4, R_5, and C_3. This filter bypasses the high frequencies to ground and passes low frequencies, such as 60 Hz. As a result of this, 60 Hz from T_1 is applied to the base of Q_1.

Diode CR_1 is likewise connected to the secondary voltage of T_1. In this case, CR_1 is a half-wave rectifier with an RC filter comprised of C_2, R_6, and R_2. Zener diode VR_3 regulates the dc voltage to -12 V dc. This

voltage is used to reverse bias the collector of Q_1. As a result of this circuit, Q_1 produces a square-wave output that changes between $+6$ and -6 V.

The MOSFET, labeled Q_2 in the diagram, has three signals applied to it. The square-wave signal from Q_1 is supplied to the gate through capacitor C_1. This signal causes the source–drain resistance of Q_2 to vary between a very high value and a few hundred ohms on alternate halves of the waveform. This, in turn, causes the output of Q_2 to respond alternately to the dc slide-wire feedback signal at the source and the dc input signal at the drain. The output of Q_2 therefore corresponds to these two input signal levels. Input and feedback signals of equal value will result in zero ac output. An imbalance in the two input signals will result in an ac output signal that will shift phase in either the positive or negative direction. Through this circuit, the dc input will produce an ac output that shifts phase according to the polarity of the input voltage.

Voltage Amplifiers

The next function of an electronic recorder is voltage amplification. Figure 8-6 shows a simplified circuit of the input voltage amplifier following the chopper. Note that Q_3 is a junction field-effect transistor and that Q_4 and Q_5 are bipolar transistors.

Transistor Q_3 is a high-impedance n-channel JFET. The ac signal from the chopper is applied to its input through capacitor C_7. The gate of a JFET, you will recall, varies conduction through the channel. It

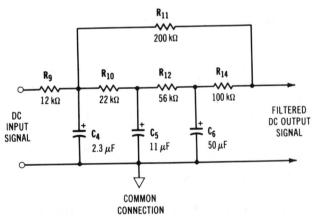

FIGURE 8-4 Input filter of the recorder. (Courtesy of Honeywell, Inc.)

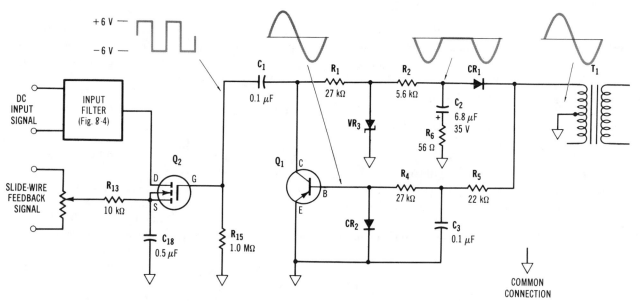

FIGURE 8-5 Chopper/waveshaper circuit of the recorder. (Courtesy of Honeywell, Inc.)

is normally reverse biased. The applied ac signal will either add to or reduce the effect of reverse biasing. As a result of this, the ac signal controls the source–drain current passing through the channel. In effect, a small change in reverse bias voltage will have a rather significant influence on the drain current passing through R_{17}. The resulting voltage drop is representative of the amplified output signal. This is then coupled to Q_4 through C_8. Resistor R_{44} is a source biasing resistor which stabilizes amplification. The gate resistor, R_{16}, serves as a discharge or return path for the ac signal passing through C_7. This amplifier has a char-

acteristically high input impedance and a rather low output impedance.

The output signal of Q_3 is applied to the input of Q_4 through capacitor C_8. Transistor Q_4 is forward biased by a voltage divider composed of R_{18}, R_{20}, R_{23}, R_{24}, and CR_4. Q_4 is connected in an emitter-follower circuit configuration. It has high input impedance and low output impedance. The output signal, which has a gain of slightly less than one, is developed across emitter resistor R_{19}. In this situation, Q_4 is used as an impedance matching device to develop maximum signal transfer to Q_5.

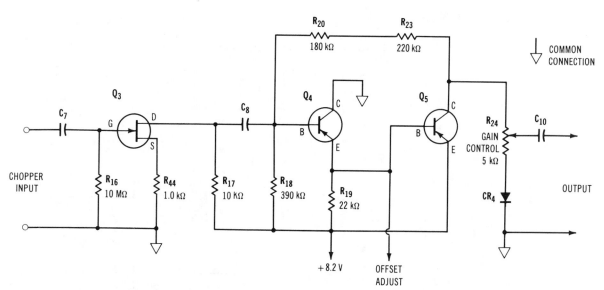

FIGURE 8-6 Input voltage amplifier stage of a recorder. (Courtesy of Honeywell, Inc.)

The emitter output of Q_4 is coupled directly to the base of Q_5. Base biasing is achieved by R_{19} and the offset adjust control (not shown). This transistor has a low input impedance. It is, however, connected in a common-emitter circuit configuration with the output developed across collector resistor R_{24}. This resistor also serves as the gain adjusting control. The output of the amplifier, which is passed through capacitor C_{10}, can be adjusted to any level of gain depending on the desired level of sensitivity. The current gain, or beta, of the composite input amplifier can, therefore, be adjusted to meet the application needs of the input signal. CR_4, called a *pedestal diode*, is used to prevent the polarity of C_{10} from developing a reverse voltage for the next amplifier during low gain settings.

Direct-Coupled Amplifiers

Transistors Q_6, Q_7, and Q_8 serve as a second block of voltage amplifiers. As you will note in Fig. 8-7, these transistors are all bipolar transistors connected in a direct-coupled circuit configuration. The input to this circuit is through C_{10}. Transistor Q_6 is biased by emitter resistor R_{27}. Resistors R_{25} and R_{26} serve as a discharge path for C_{10}. The ac signal appearing across R_{26} is applied to the base of Q_6. Resistor R_{28} serves jointly as a load resistor for Q_6 and as a base bias resistor for Q_7. This positive voltage reverse biases the collector of Q_6 and forward biases the base of Q_7 at the same time. Any variation in the collector current of Q_6 is applied directly to the base of Q_7.

The output of Q_7 is direct coupled to the base of Q_8. The emitters of these two transistors are both connected to the common or negative side of the power supply. The composite output of the three transistors appears across C_{14}. A rather substantial amount of signal gain is achieved through this block of amplifiers.

Driver Amplifier/Power Amplifier

The drive amplifier, Q_9, of Fig. 8-8 employs RC coupling between the collector of Q_8 and its base. Capacitor C_{14} serves as the coupling capacitor, and resistor R_{33} serves as the discharge resistor. Voltage changes at the top of R_{33} are applied to the base of Q_9 through R_{45}. A voltage divider made up of R_{40} and R_{33} establishes forward bias voltage for the base of Q_9. The emitter resistor, R_{42}, determines the transistor operating point and establishes the gain capabilities of the driver.

The output of the driver amplifier is direct coupled to the base of power amplifier Q_{10}. The resistance of Q_{10} serves as the load resistor for Q_9. Variations in the collector current of Q_9 and the base of Q_{10} are needed to control the high current passing through Q_{10}. The total current of the power amplifier is ultimately determined by the gain range switch. The position of this switch determines the value of the emitter resistor.

The dc voltage applied to the collector of Q_{10} is the unfiltered full-wave output of the power supply. This voltage provides the conventional reverse bias for the collector. With the ac signal applied to the input taking on either a positive or negative phase, as indicated, the output will have either a leading or lagging output across C_{17}. If the first alternation of the input

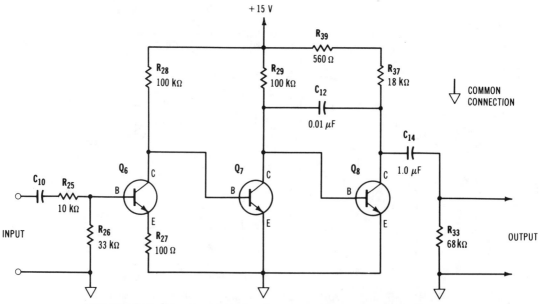

FIGURE 8-7   Direct-coupled voltage amplifiers used in the Honeywell ElectroniK 15 recorder. (Courtesy of Honeywell, Inc.)

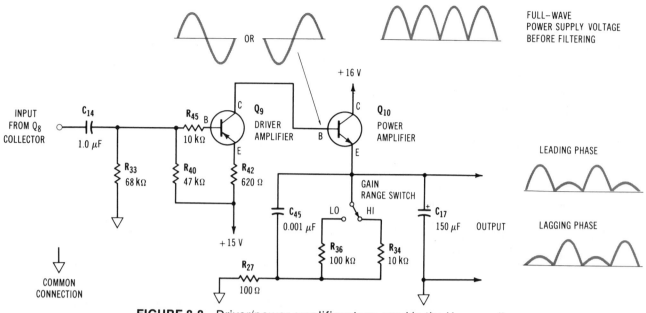

FULL–WAVE
POWER SUPPLY VOLTAGE
BEFORE FILTERING

LEADING PHASE

LAGGING PHASE

FIGURE 8-8 Driver/power amplifier stage used in the Honeywell ElectroniK 15 recorder. (Courtesy of Honeywell, Inc.)

is positive, the output will produce positive pulses or be of a leading phase as indicated. If the first alternation is negative, the output pulse will have a reduced amplitude as indicated. Through this mixing process, the output has a phase shifting capability that is dependent on the polarity of the original dc input signal.

The power amplifier of this unit is considered to be single ended. The voltage gain produced by Q_{10} is less than 1, but is still great enough to alter the control winding voltage of the servomotor. Typical voltage values are 40 V peak to peak due to the previous level of amplification. The voltage output of the power amplifier is considered to be a control voltage when it is applied to the servomotor. The resulting operation of the power amplifier with respect to the full-wave rectifier input is called *null detection*. This effect is similar to that of the two-amplifier null detector of Chapter 6.

PHASING

From the preceding discussions, it can be seen that phasing is of importance in the operation of recording instruments. Any discussion of phase relationships applies, in many cases, to all instruments since they are very much alike in circuitry. Some background information on the effect of components on phase will be necessary.

Resistive AC Circuit

Resistance is defined as that component of a circuit that causes a voltage drop that is directly proportional to the current through it. This relationship (Ohm's law)

holds true for current that varies with time as well as for constant current. Figure 8-9 shows this relationship in the resistive circuit. The current and voltage in this circuit are obviously in phase as can be seen in the time diagram of Fig. 8-9C. A vector diagram for this circuit is shown in Fig. 8-9B. A vector is a quantity that indicates not only magnitude but also direction. The length of the vector represents the magnitude of either a current or a voltage; the direction of a vector represents the phase relationships of currents and voltages. Two vectors pointing in the same direction, as in Fig. 8-9B, are in phase. Vectors are merely a method

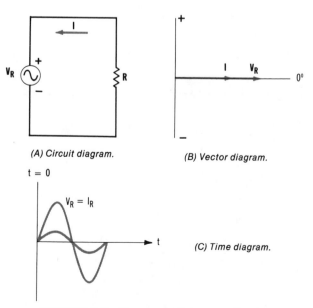

(A) Circuit diagram.

(B) Vector diagram.

(C) Time diagram.

FIGURE 8-9 Purely resistive ac circuit.

of visually representing voltages and currents and their phase relationships. A vector that is rotated counter-clockwise from a reference vector is said to have a *leading* (positive) phase angle. The amount by which it leads the reference is indicated by the amount of rotation. A vector that is rotated clockwise from a reference vector has a *lagging* (negative) phase angle.

Inductive AC Circuit

Inductance is defined as that property of a circuit which tends to oppose any change of current through it. Simply stated this means that an inductor opposes changes in current. Since voltage is proportional to the rate of change of current, the voltage across an inductor is out of phase with the current, as indicated in Fig. 8-10C. The ratio between the voltage across an inductor and the current through it is called inductive reactance ($V_L = IX_L$). Reactance implies a 90° phase shift as opposed to resistance. The vector diagram (Fig. 8-10B) indicates that voltage leads current through an inductor by 90°.

The phase shift in this circuit results from the fact that the magnetic field produced by a coil varies with the current. A changing magnetic field in turn induces a counterelectromotive force into the circuit. The voltage drop, therefore, is produced by the change in current and not by its magnitude.

Capacitive AC Circuit

Capacitance is defined as that circuit element that has a current proportional to the rate of change of voltage across it. Simply stated this means that a capacitor opposes changes in voltage. Since current is propor-

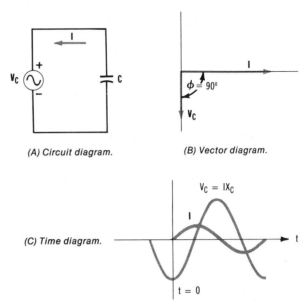

(A) Circuit diagram.   (B) Vector diagram.

(C) Time diagram.

FIGURE 8-11   Purely capacitive circuit.

tional to the rate of change of voltage, it is out of phase with the voltage, as indicated in Fig. 8-11C. The ratio between the voltage and current of a capacitor is called capacitive reactance ($V_c = IX_c$). Both inductive and capacitive reactance are measured in ohms.

The vector diagram (Fig. 8-11B) indicates that capacitive current leads the voltage by 90°. Conversely, voltage lags the current. The phase shift in this circuit results from the ability of the capacitor to store energy in an electric field (i.e., its ability to charge). As described in Chapter 6, any increase in the voltage across the capacitor results in a decrease in current, thereby decreasing its charging rate.

Series *RL* Circuit

Figure 8-12A shows a series *RL* circuit. Actually, any circuit containing an inductor involves some resistance because the wire used to wind a coil has some resistance. A fixed resistive component may also be inserted into the inductive circuit. A series circuit is defined as a circuit that provides a single path for current. Since this is true, the same current will pass through each element in the series circuit. Current, therefore, is used as the reference in the series circuit since it is the common value for each component.

If the effective current is measured in this circuit, then the voltage across R (V_R) and the voltage across L (V_L) can be computed. By Ohm's law, $V_R = IR$ and $V_L = IX_L$. As indicated in the time diagram for this circuit (Fig. 8-12C), V_R is in phase with the current, while V_L leads the current by 90°. Kirchhoff's voltage law states that in a series dc circuit, the sum of the voltage drops in the circuit equals the applied voltage (V_A). This is also true in an ac circuit at any instant

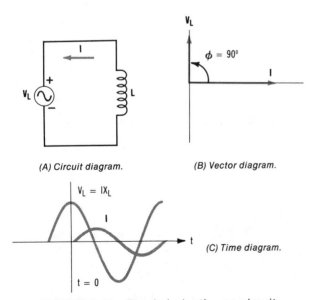

(A) Circuit diagram.   (B) Vector diagram.

(C) Time diagram.

FIGURE 8-10   Purely inductive ac circuit.

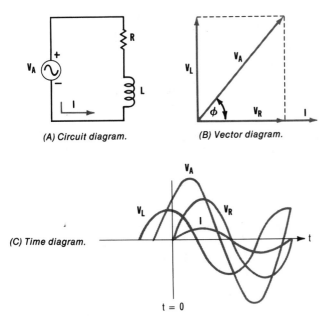

(A) Circuit diagram.

(B) Vector diagram.

(C) Time diagram.

FIGURE 8-12 Series *RL* circuit.

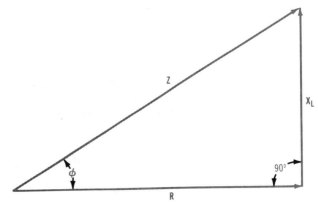

FIGURE 8-13 Impedance vector diagram.

total opposition to ac current. Its symbol is Z, and the unit of measure is the ohm.

Series *RC* Circuit

Figure 8-14 indicates the voltage and current relationships found in an *RC* circuit. We now know that the voltage across the capacitor lags the current by 90°. We can see from the vector diagram (Fig. 8-14B) that V_A also lags the current. The size of the phase angle (ϕ) depends on the relative magnitudes of V_R and V_C. The only difference between the capacitive circuit and the inductive circuit is that in the capacitive circuit current leads voltage, while in the inductive circuit, current lags voltage.

Solving a sample problem should help us to see the applications of the principles discussed. Figure 8-15A indicates the known values in the circuit. The resistance, R, the capacitance, C, the current in the circuit, and the frequency of the applied voltage are the

of time. It is obvious then that peak or effective values cannot be added directly since they do not occur at the same time. In the time diagram (Fig. 8-12C), instantaneous values of voltage have been added to produce the sine wave that represents the applied voltage (V_A). This is a most laborious process. The same result is obtained by adding V_R and V_L vectorially. The vector sum of two vectors is the diagonal of the parallelogram whose sides are composed of the vectors V_R and V_L. The phase angle (ϕ) between I and V_A is also obtained from the vector diagram (Fig. 8-12B). Mathematically, V_A can be found by the Pythagorean theorem since V_R and V_L form the legs of a right triangle, and V_A is its hypotenuse. Therefore,

$$V_A = \sqrt{V_R^2 + V_L^2}$$

and

$$\phi = \arctan \frac{V_L}{V_R}$$

By convention, reference vector I is drawn horizontally from left to right and is given a phase angle of 0°.

Resistance and inductive reactance are directly proportional to V_R and V_L in the series circuit. Since this is true, a vector diagram representing reactance and resistance can be drawn (Fig. 8-13). The vector sum of resistance and reactance in a series circuit is referred to as the *impedance* (Z) of the circuit.

The impedance value is the ratio of the applied voltage and the current ($V_Z = IZ$). The phase angle (ϕ) is the same as that found in the voltage diagram since these two triangles are similar (i.e., their sides are proportional). Impedance could be described as the

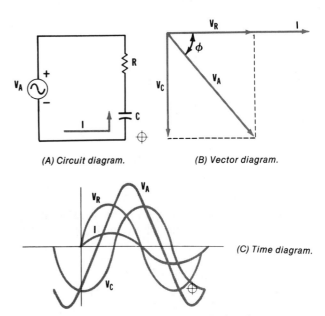

(A) Circuit diagram.

(B) Vector diagram.

(C) Time diagram.

FIGURE 8-14 Series *RC* circuit.

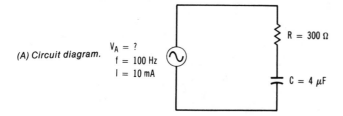

(A) Circuit diagram. $V_A = ?$ $f = 100$ Hz $I = 10$ mA $R = 300\ \Omega$ $C = 4\ \mu F$

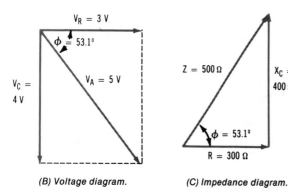

(B) Voltage diagram.   (C) Impedance diagram.

FIGURE 8-15   Series *RC* solved by vectors.

known quantities. Since frequency and capacitance are known, the capacitive reactance can be computed as follows:

$$X_c = \frac{1}{2\pi f C}$$

$$= \frac{1}{6.28 \times 100 \times 4 \times 10^{-6}}$$

$$= \frac{0.159}{4} \times 10^4$$

$$= 0.039 \times 10^4$$

$$\approx 400\ \Omega$$

Also,

$$V_R = IR$$

$$= 10 \times 10^{-3} \times 3 \times 10^2$$

$$= 3\ \text{V}$$

$$V_C = IX_C$$

$$= 10 \times 10^{-3} \times 4 \times 10^2$$

$$= 4\ \text{V}$$

From the voltage vector diagram (Fig. 8-15B), the applied voltage, V_A, and the phase angle, ϕ, could be obtained graphically. Mathematically,

$$V_A = \sqrt{V_R^2 + V_C^2}$$

$$= \sqrt{3^2 + 4^2}$$

$$= \sqrt{25}$$

$$= 5\ \text{V}$$

$$\phi = \arctan \frac{V_C}{V_R}$$

$$= \arctan \frac{4}{3}$$

$$= \arctan 1.333$$

$$= 53.1°$$

Following the same procedure with the impedance diagram (Fig. 8-15C), yields

$$Z = \sqrt{R^2 + X_C^2}$$

$$= \sqrt{300^2 + 400^2}$$

$$= \sqrt{25 \times 10^4}$$

$$= 500\ \Omega$$

$$\phi = \arctan \frac{X_C}{R}$$

$$= \arctan \frac{400}{300}$$

$$= \arctan 1.333$$

$$= 53.1°$$

This agrees with the impedance obtained by Ohm's law,

$$Z = \frac{V_A}{I}$$

$$= \frac{5\ \text{V}\underline{/53.1°}}{10\ \text{mA}}$$

$$= 500\ \Omega\ \text{at}\ 53.1°$$

Series *RLC* Circuit

Figure 8-16A shows a series *RLC* circuit. Current is still the reference vector since it is common to all components. V_R is in phase with I. V_L leads the current by 90°. To find V_A, these three vectors must be added vectorially. Any two of these vectors are added first and the sum added to the remaining vector. In the example shown in Fig. 8-16B, V_L and V_R were added to produce $V_L + V_R$. This vector was then added to V_C to produce V_A. Since V_L and V_C are in opposite directions, their sum is merely the difference in the magnitudes (i.e., $V_L - V_C$). Mathematically, $V_A = \sqrt{V_R^2 + (V_L - V_C)^2}$. You will notice that if X_L and X_C are equal, V_A is equal to V_R. Under these conditions, the circuit is said to be resonant. This will happen at one frequency for every capacitor-inductor combination. At this frequency, you will also notice that $Z = R$. Impedance at this frequency, therefore, is at its min-

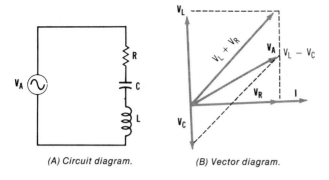

(A) Circuit diagram.

(B) Vector diagram.

FIGURE 8-16 Series *RLC* circuit.

imum. If it were possible for a circuit to have a resistance of 0 Ω, it would occur at resonance. If Z is at a minimum, then current is maximum in a series circuit at resonance.

Parallel RL Circuit

In a parallel circuit, such as shown in Fig. 8-17A, the voltage is the same across each component (i.e., $V_A = V_R = V_L$). This value, therefore, is the reference vector in a parallel circuit. I_R is in phase with V_A. I_L is 90° out of phase with the applied voltage and lags the voltage. The sum of the branch currents in a parallel dc circuit equals the total current. This is true instantaneously in the ac circuit. Therefore, these effective currents must be added vectorially. The vector I_T represents this vector sum, and ϕ represents the generator phase angle.

Parallel RC Circuit

In Fig. 8-18, an *RC* circuit is analyzed. Since this is a parallel circuit, V_Z is again the reference vector. I_R is in phase with V_A. In a capacitor, I leads V by 90°. This

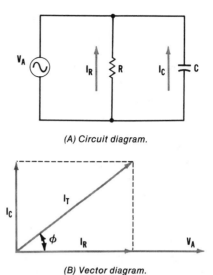

(A) Circuit diagram.

(B) Vector diagram.

FIGURE 8-18 Parallel *RC* circuit.

is indicated in the vector diagram (Fig. 8-18B). I_T is the vector sum of I_R and I_C. The phase angle is again indicated by ϕ. You will notice that in the parallel circuit, $I_T = \sqrt{I_R^2 + I_X^2}$, whether I_X is capacitive or inductive. Since this is true, you will note that an impedance vector diagram cannot be used since I_T and Z are inversely, not directly, proportional. The simplest way to find the impedance is by the use of Ohm's law (i.e., $Z = V_A/I_T$).

Parallel RLC Circuits

Figure 8-19A shows the schematic diagram of a parallel *RLC* circuit. Voltage is the reference vector in a parallel circuit. I_R is in phase with the applied voltage. I_C

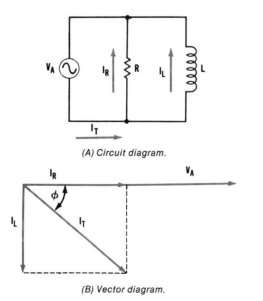

(B) Vector diagram.

FIGURE 8-17 Parallel *RL* circuit.

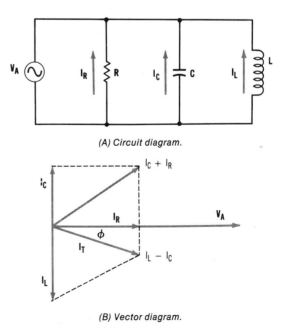

(A) Circuit diagram.

(B) Vector diagram.

FIGURE 8-19 Parallel *RLC* circuit.

leads this voltage by 90°, while I_L lags by 90°. Adding vectors I_C and I_L produces a vector that is the difference in their magnitudes and has the direction of the larger. When this vector is added to I_R, I_T is the resulting vector. In the vector diagram shown (Fig. 8-19B), vectors I_C and I_R were added first. The results, however, are the same regardless of which vectors are added first. A lagging phase angle has resulted in Fig. 8-19 since I_L is greater than I_C. If I_C were of a greater magnitude than I_L, the phase angle would lead. If I_C were equal to I_L, then I_T would equal I_R. This condition occurs at the resonant frequency of the circuit, at which current is at its minimum. If current is minimum, impedance Z is maximum. Notice that a series resonant circuit offers a minimum impedance at the resonant frequency while a parallel resonant circuit offers a maximum impedance. At frequencies other than the resonant frequency, an RLC circuit can be either capacitive or inductive. This depends on the relative sizes of L and C.

PHASE RELATIONSHIPS IN THE NULL DETECTOR

The condition for rotation of the balance motor has been stated. It has been stressed that a phase shift of 90° between the current in the line-phase winding and the control winding is necessary for rotation of the balance motor. An understanding of the action of the two-phase motor under such conditions will help us to understand the importance of "phasing" in the null detector. Figure 8-20 shows the four possible conditions that might exist in the balancing motor. Line voltage is always applied to the line winding of the balance motor. Current, therefore, is always conducting in this winding and its phase is constant, as indicated in the diagram. A thorough analysis of a two-phase motor is more complex than the discussion that will be used here. Only the basic characteristics of the balance motor need to be analyzed for an understanding of the null detector.

Diagram A in Fig. 8-20 shows only the effect of the line current on the motor. This situation would also occur if there were no outputs from the amplifier or if the amplifier output has no 60-Hz component. At the beginning of a cycle of current applied to the line winding of the motor, current is rising. We will arbitrarily assign the polarity of the magnetic field produced by this current as shown in diagram A. This changing current induces a field into the squirrel-cage rotor of the motor. This rotor field is in the same direction as the field produced by the line winding. Since the north pole of the coil and the south pole of the rotor attract, no torque is produced and no rotation results. As the line current varies, the induced rotor field varies in step with it.

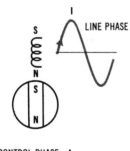

CONTROL PHASE   A

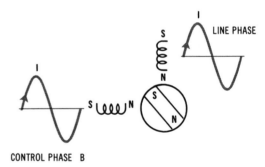

CONTROL PHASE   B

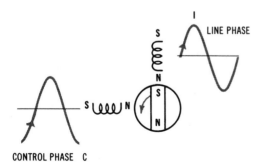

CONTROL PHASE   C

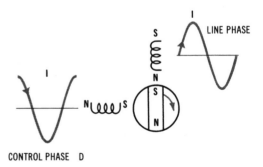

CONTROL PHASE   D

FIGURE 8-20   Effect of phase on balance motor.

In diagram B, a voltage is applied to the control winding. This current is in phase with the line current and at the same frequency. As the current in each winding rises, the same magnetic field is produced by each winding. The induced field in the rotor is the result of both of these windings and will be somewhere in between the two. Its actual location will be determined by the relative magnitudes of the two currents. Since the rotor field is equally attracted by the field of both windings, no torque is produced.

Diagram C indicates the situation that exists when the line current leads the control current by 90°. At the beginning of a cycle of current in the line winding, current is rising. The induced rotor field is the same as in situation A. Current in the control winding at this instant is at its maximum negative value. No change in this current is occurring at this instant; control current induces no field in the rotor. As the line current continues to rise, however, its rate of change decreases. The current in the control winding begins to rise at this time. Since control current is rising, its magnetic field polarity is the same as that of the line winding. Since its field strength is rising while that of the line winding is decreasing, the rotor field is attracted more and more in the direction of the control winding, resulting in a torque in the counterclockwise direction that starts rotation. As the currents vary, this rotation will continue.

Diagram D indicates the situation that exists when line current lags the control winding current by 90°. Clockwise rotation is the result of these currents. When the line current rises, the control current starts to fall. This means that the polarity of the magnetic field produced by the control winding is opposite to that of the line winding. This results in a repulsion-attraction between the field of the control winding and the rotor field induced by the line current in such a direction as to produce a clockwise torque, resulting in clockwise rotation.

It has been shown that a 90° phase shift between the two phase windings of the balance motor must be produced to create motor control, but how is this phase difference accomplished? In order to answer this question every reactive component in the null detector should be analyzed. We will find that most of these components have very little effect on the phase relationships. We will investigate the converter drive coils, input transformers, coupling capacitors, control windings, and line phase windings to find out which ones affect the phase relationships. In all cases we will use the line voltage for our reference. Since the phase characteristics are more involved in the Honeywell ElectroniK 15 recorder, we discuss it first, keeping in mind that the same thing may be accomplished in recorders made by other manufacturers.

Phase Relationship of an Electronic Recorder

Figure 8-21 shows a schematic of the two coil windings of the servomotor of the electronic recorder. The control winding has C_{17} of the power amplifier in parallel with the winding. This forms a parallel LC circuit with practically no shift in phase occurring. The line winding, by comparison, has capacitor C_{16} in series with the winding. This series combination causes the phase of the line winding to shift 90°. The shift in phase will be of a constant value and serves as the reference.

The phase of the control winding is subject to a shifting effect depending on the polarity of the dc input signal to the amplifier. If the input is of a positive dc value, the control winding will lag the line winding by 90°. This will cause the motor to rotate in a clockwise direction. When the polarity of the dc input is negative, the control winding will lead the line winding by 90°. This causes the motor to rotate in a counterclockwise direction until balance occurs. Ultimately this means that the control signal shifts phase according to the input polarity, while the phase of the line winding remains at a constant position. As a result of this action, the servomotor responds to any polarity change that takes place in the input signal.

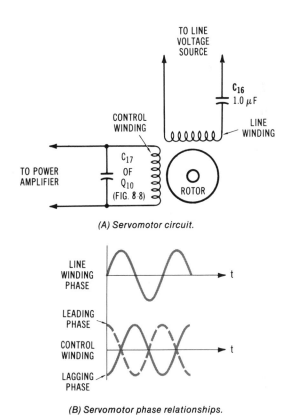

(A) Servomotor circuit.

(B) Servomotor phase relationships.

FIGURE 8-21 Servomotor phase relationships.

Phase Relationships of Mechanical Chopper Recorders

For recorders that employ a mechanical chopper, there are some other phase considerations to take into account. Figure 8-22A shows a schematic representation of the chopper drive coil, and Fig. 8-22B shows a vector diagram indicating its effect on phase relationships. An applied voltage of 68 V ac is delivered to the drive-coil circuit by the power transformer. Since this is a series circuit, current will be used as a reference. This is the reference we wish to use since it is the current through the drive coil that controls the vibrating reed. The circuit is composed of R_8 and the converter drive coil, which has an inductive reactance, X_L, and a resistance, R. The large resistor (R_8) inserted in the circuit not only drops the ac voltage applied to the circuit from 68 volts to 19 V, it also reduces the phase angle. In the simplified vector diagram (Fig. 8-22C), V_R and V_{R8} have been added. They can be added directly since these voltages are in phase. The voltage drops across the resistances are so much larger than that across X_L that the phase angle is very small (10°). An additional lag of 8° is due to contact closure time in the converter. This makes a total phase difference of 18° that can be attributed to the drive coil.

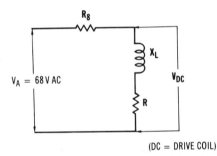

(DC = DRIVE COIL)

(A) Schematic representation of a drive coil.

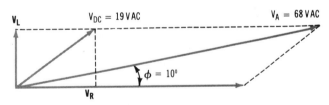

(B) Vector diagram.

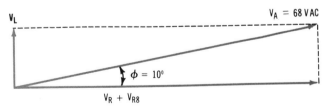

(C) Simplified vector diagram.

FIGURE 8-22 Drive-coil phase relationships.

The effect of the input transformer is negligible because of the manner in which such transformers are wound. A transformer can be connected so that its output voltage is either in phase or 180° out of phase with the primary voltage. Either connection could be used here since a 180° phase shift will not produce rotational torque. We have also noted that each transformer produces a 180° phase shift. Again, this will not produce rotational torque in a two-phase motor.

The next reactive component to be considered is the coupling capacitor. These components also have very little effect on amplifier phase relationships. Figure 8-23A shows a typical RC coupling network. With a typical coupling capacitor ($C = 0.022$ μF) used in the amplifier, at a frequency of 60 Hz, X_C will be

$$X_C = \frac{1}{2\pi f C}$$

$$= \frac{1}{6.28 \times 60 \times 0.022 \times 10^{-6}}$$

$$= \frac{0.159}{1.32} \times 10^6$$

$$\simeq 120\,000\ \Omega$$

This value is about one-eighth the size of R. Vector R will therefore be about eight times as long as X_C in the vector diagram (Fig. 8-23B), producing a phase angle of about 7°. Even when four coupling circuits are used, the total phase shift would be small. Some of the capacitors used for coupling have even less reactance than the one involved here.

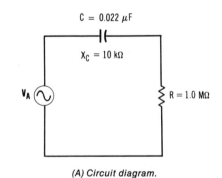

(A) Circuit diagram.

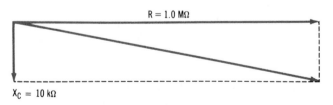

(B) Vector diagram.

FIGURE 8-23 Phase shift in a coupling network.

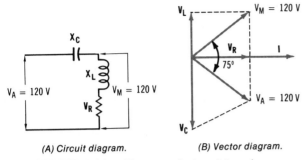

FIGURE 8-24 Phase relationships in a motor drive circuit.

The control winding of the balance motor is considered next. It is the output load of the whole amplifier, and the power amplifier in particular. A capacitor is connected in parallel with this winding. These two components form a resonant circuit at a frequency of 60 Hz. This parallel circuit then acts as a resistive load. No phase shift occurs in the control winding.

Most of the phase shift from the line voltage occurs in the line winding. This winding is connected to the line voltage, V_A, through a capacitor. The reactance of this capacitor at 60 Hz is much greater than the reactance of the coil winding. The result is that the overall effect of the circuit is capacitive. As seen in Fig. 8-24, 120 V (V_M) is applied to the line winding itself. This voltage is the vector sum of the voltage drops across the inductive reactance of the coil and the resistance of the coil. V_M leads I by about 37.5°. This voltage drop (V_M) added vectorially to the voltage drop across the capacitor (V_C) equals the applied voltage, V_A, of 120 V. This voltage lags the current by 37.5°. The effect of this circuit is that V_A lags V_M by 75°.

The combined effect of the phase shift caused by the drive coil (18°) and that of the line winding (75°) is a total of 93°—about the phase shift desired. This difference in phase in the line winding will be constant. The power amplifier output, however, changes its phase 180° with a change of polarity of the dc input voltage. The result is that the amplifier output may lead the line phase by 90° or lag by 90°. This produces the type of motor action necessary for rebalance in the measuring circuits.

SUMMARY

An electronic recorder contains two major divisions that are discussed in this chapter. The amplifier has a power supply, filter, chopper, voltage amplifier, and driver stage. The motor unit contains a power amplifier and the balancing motor.

The power supply provides ac to the line winding of the motor and it develops dc for component operation. A full-wave rectifier with a center-tapped transformer and an RC filter provides dc operating voltages for component operation. Three different dc voltages are provided. Two are regulated by zener diodes. In addition to this, a negative dc source is also developed for the waveshaping circuit.

The input filter is a three-section RC network that filters noise and extraneous ac from the dc input signal.

The chopper changes dc input into ac. This is achieved by a waveshaping transistor that changes ac from the transformer into a pulsating dc. An IGFET has a square wave applied to its gate, the dc slide-wire feedback signal to its source, and the dc input to its drain. The output corresponds to the two signal inputs and alternates between high- and low-resistance conduction due to the gate signal.

Voltage amplification is achieved first by a JFET and two bipolar transistors in the input section of the amplifier. Following this, three direct-coupled amplifiers are employed. The collector of one transistor and the base of the next transistor are commonly connected together in this circuit configuration.

The driver/power amplifier is responsible for the output of the amplifier unit. An amplified ac signal processed by the driver is used to control a large amount of current controlled by the power amplifier. The term *null detector* is sometimes used to describe this section of the amplifier.

The servomotor of a recorder is responsible for moving the pen and altering the slide-wire balancing mechanism. The servomotor is basically a two-phase induction motor. The ac applied to the line winding has a fixed phase relationship, while that applied to the control winding shifts according to the polarity of the dc input. A positive dc input will produce a lagging phase of 90° with respect to the phase of the line winding. This will cause a clockwise rotation of the motor. When the polarity of the dc input is negative, the control winding will lead the line winding by 90°. This, in turn, causes counterclockwise rotation of the motor.

When ac is applied to a resistor, the current and voltage remain in phase. In an inductive circuit with ac applied, the current lags the voltage by 90°. A capacitive circuit causes the current to lead the voltage by 90°. When R, C, or L components are connected in series, the phase relationship between I and V is dependent on the predominant component value. In a circuit containing parallel components, the voltage serves as the reference component with the current either lagging or leading, depending on the predominant component. The resulting effect of current is the vector sum of the individual component current values.

ACTIVITIES

RECORDER INVESTIGATION

1. Refer to the manufacturer's operational manual of an electronic recorder that is available for investigation. This investigation should be performed with the power disconnected or turned off.

2. Open the front door of the recorder's enclosure.

3. Indicate the type of recording performed by the instrument. This should be circular chart, fanfold, or strip chart.

4. Locate the recording point or stylus. Identify the type of point used in the system.

5. Does the instrument have single- or multipoint recording capabilities?

6. Describe the point replacement procedure or refilling operation of the instrument.

7. Remove the recording assembly to expose the circuitry. Where is the servomotor located?

8. Identify the power supply and primary circuit board of the instrument. Notice how the circuit boards are attached to the instrument.

9. Locate the input or inputs to the instrument.

10. Reassemble the instrument.

QUESTIONS

1. What are the two major divisions of an electronic recorder?

2. What is the function of the power supply of an electronic recorder?

3. Explain the functional role of the chopper of an electronic recorder.

4. Why are the voltage amplifiers of an electronic recorder directly coupled?

5. Describe the voltage and current relationship of a purely resistive ac circuit such as Fig. 8-9.

6. Describe the voltage and current relationship of a purely inductive ac circuit such as Fig. 8-10.

7. Describe the voltage and current relationship of a purely capacitive ac circuit such as Fig. 8-11.

8. In a series resonant circuit when X_L and X_C are equal, what is the resulting impedance?

9. In a parallel resonant circuit when X_L and X_C are equal, what is the resulting impedance?

10. What are the two windings of a servomotor?

11. Which winding of the servomotor is attached to the power amplifier?

12. Which winding of the servomotor has a fixed phase relationship?

13. What causes the rotor of a servomotor to have rotational torque?

CHAPTER 9

ELECTRONIC TRANSMITTERS

OBJECTIVES

Upon completion of this chapter, you will be able to:

1. Identify the major parts of an electronic transmitter that accepts a millivolt input signal.
2. Explain the functional role of the power supply of an electronic transmitter.
3. Explain how optical isolation is achieved in an electronic transmitter.
4. Given a schematic diagram of the power supply of an electronic transmitter, identify where regulation, filtering, and dc-to-ac conversion is achieved.
5. Explain how the Wheatstone bridge is used in the measuring circuit to determine an input voltage value.
6. Describe the operation of a Hartley oscillator.
7. Indicate where an oscillator is used in the operation of an electronic transmitter.
8. Identify different transducer input circuits.

IMPORTANT TERMS

In this chapter we investigate some of the basic operating principles of an electronic transmitter. You will also have a chance to become familiar with a number of basic instrument functions. New words, such as *oscillator, transducer*, and *optical isolation*, will begin to have some meaning for you. A few of these terms are singled out for study. As a rule, the chapter will be more meaningful if these terms are reviewed before proceeding with the text.

Ambient: The immediate surroundings or encompassing conditions of an environment.

Ambient temperature: The immediate temperature surrounding an object.

Bandpass: The range of frequency that will pass through a filter network or tuned circuit without a significant reduction in amplitude.

Cold junction: A junction of wires in a thermocouple that connect to the measuring instrument. This junction is generally referenced at the freezing point or at room temperature.

Darlington pair: A circuit of two like transistors that are connected to respond as a single transistor. The current gain of this configuration is the product of the two individual transistor current gains.

Differential pressure: The difference in pressure between two pressure sources, measured relative to one another.

Duty cycle: A percentage that identifies the ratio of the time a device is conducting to the total cycle time. In digital circuits, this is the time that a pulse is at its high or 1 level during an operational cycle.

Flywheel effect: The ability of a resonant circuit to maintain oscillation at a continuous frequency when fed short-duration pulses of energy.

Hartley oscillator: A transistor circuit that generates a sine wave due to feedback energy produced by a tapped inductor and a capacitor.

Indicator: The part of an electronic instrument that shows if it is on or off or indicates a specific quantity.

Interface: A process or piece of equipment in a circuit that brings two things or quantities together.

Isolation: To insulate or set apart specific circuits or electronic parts.

Light-emitting diode (LED): A semiconductor diode that gives off light when forward biased.

Optical: A science that deals with light, its production, and its response.

Optical isolation: A process that separates components or circuit parts by light energy.

Oscillator: An active electronic circuit that generates or produces an output that is a periodic function of time.

Phototransistor: A transistor-like device whose characteristics are a function of the incident light applied to its base.

Positive feedback: An output to input signal path that returns an in-phase output signal to the input to increase amplification. Excessive amounts of feedback cause distortion, instability, and oscillation.

Pressure: Force per unit area. Measured in pounds per square inch or pascal.

Process: A collection of functions performed in and by the equipment in which a variable is to be controlled.

Radio frequency (RF): Electromagnetic radiation having a frequency range of 10 kHz to 300 GHz.

Radio-frequency interference (RFI): An opposing radio-frequency energy source that occurs outside a system, sometimes called electronic noise.

Regeneration: A positive feedback process.

Saturation: A condition in which a further increase in one quantity or value produces no further increase in a dependent quantity.

Sensor: A device that detects a change in physical quantity and converts it into a usable signal that can amplify, control, or measure another quantity.

Thermistor: A resistive device whose value changes with temperature.

Transducer: Any device that converts a physical quantity of one type into a different form or new physical quantity.

INTRODUCTION

In many electronic process-control applications, the signal may go directly to the system controller, indicator, or recorder without any further signal processing. In other applications, controllers may be positioned at some remote location from the process variable being manipulated. In this case, special pieces of interface equipment are needed to receive the process signal from the sensor and give it the power needed for transmission to the controller. In achieving this function, the signal must be strong enough to travel the required distance without excessive noise or any reduction in accuracy or sensitivity to the signal. Signal transmitters are used to achieve this function in electronic instruments.

A simplified block diagram of a millivolt-to-current transmitter is shown in Fig. 9-1. This instrument has a number of unique things in its circuitry that have not been discussed previously. For example, it has electrical feedback rather than the electromechanical devices used in some of the preceding units. Input and output isolation is accomplished optically. Isolation between electrical power and the input/output circuitry is also provided on all units as standard equipment. When used in thermocouple input applications, a cold junction is self-contained in a constant-temperature oven. Transmitters of this type have high reliability and are immune to radio-frequency (RF) interference and ambient temperature variations.

POWER SUPPLY

The power supply of Fig. 9-2 shows one of the unique options of the transmitter. In this case, 24 V dc is used as the primary input source. In order to isolate the unit from the primary electrical sources and to produce higher voltages, a dc-to-ac converter is used in the primary circuit. This circuit causes the primary dc voltage to be switched on and off through the two transistors. As a result of this action, the magnetic field of the transformer expands and collapses with each switching change.

When dc power is applied to the two transistors, more current will conduct through one transistor than the other due to a slight difference in the transistors and the tolerance values of the components. Assume in this case that Q_{100} is initially conducting more heavily than Q_{101}. This will result in a base-drive voltage being induced into the lower winding of transformer T_{100} of such a polarity as to cause the collector current of Q_{100} to increase and the collector current of Q_{101} to decrease. As a result, Q_{100} is driven into saturation and Q_{101} is driven to cutoff. At the saturation point of Q_{100}, there is no further change in collector current and no further voltage induced into the lower winding of transformer T_{100}. This means that the cutoff voltage to the base of Q_{101} is removed, which allows it to begin conduction. Through transformer action, this now causes an increase in the collector current of Q_{101} and a decrease in the collector current of Q_{100}. Ultimately, Q_{101} saturates and Q_{100} goes into cutoff. The process then repeats itself with Q_{100} conducting again and Q_{101} being cut off again.

The switching action of Q_{100} and Q_{101} produces

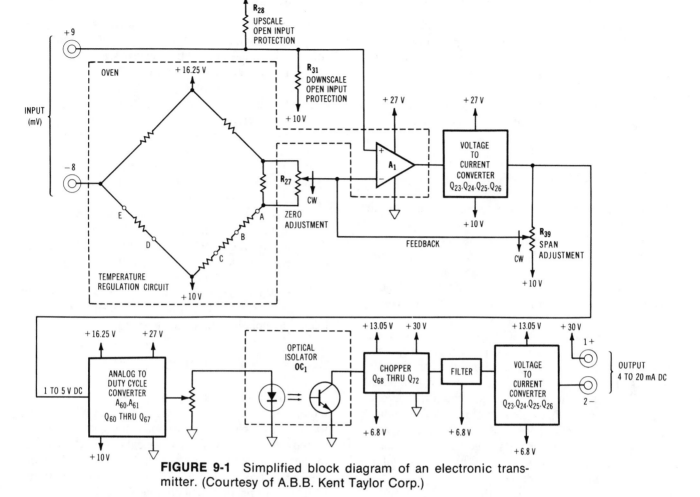

FIGURE 9-1 Simplified block diagram of an electronic transmitter. (Courtesy of A.B.B. Kent Taylor Corp.)

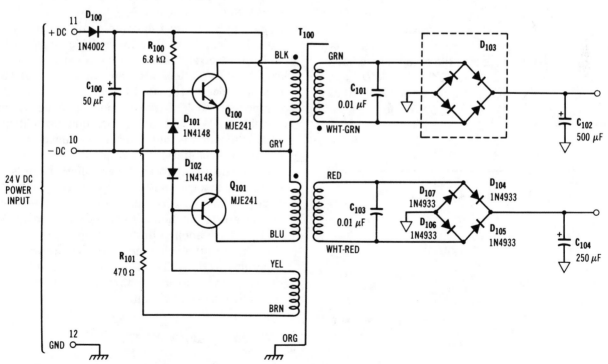

FIGURE 9-2 Power supply of a transmitter. (Courtesy of A.B.B. Kent Taylor Corp.)

a varying voltage across the primary winding of transformer T_{100} which is either stepped up or down by the secondary windings. The secondary windings of this circuit are connected to typical bridge circuits with RC filtering. Through the action of this dc-to-ac converter power supply, it becomes possible for this transmitter to work from a common dc source and still have electrical isolation.

VOLTAGE REGULATION

In conjunction with the power supply, a transmitter has a number of regulated output voltages. Figure 9-3 shows a simplified schematic of the voltage regulator for the +30-, +13.05-, and +6.8-V dc supplies. Through this type of circuitry, the dc voltage will change very little regardless of variations in load or line voltage. Regulation in this case is achieved by zener diodes.

A zener diode, as you will note, is connected in reverse-bias mode. The cathode, which is representative of the n material, is connected to the positive side of the dc source, while the anode, or p material, is connected to the negative side. When the reverse-bias voltage exceeds the zener voltage (V_Z), the diode goes into conduction. When this occurs, the voltage across the diode remains at its rated V_Z value. A series-connected resistor absorbs the remaining voltage through an increase in current.

In Fig. 9-3 it can be seen that the positive output of the power supply is connected to the cathodes of D_{64} and D_{66} through a 1000-Ω resistor and a 499-Ω resistor. When the power supply is turned on to produce voltage, 30 V appears at the output immediately. With the zener voltage of D_{64} at 6.25 V and D_{66} at 6.8 V, these voltages will appear across the respective diodes. Note also that 6.25 V plus 6.8 V equals 13.05 V,

which appears across the two diodes. The difference between +30 and +13.05 V is +16.95 V, which is the combined voltage that appears across R_{79} and R_{80}.

Any change in load or line voltage that reduces the cathode voltage of D_{64} or D_{66} reduces the total conduction current of the series circuit. This reduction in current decreases the voltage drop across R_{79} and R_{80}, which compensates for the reduction in voltage. Any rise in voltage across the circuit increases the diode current. This results in an increased voltage drop across R_{79} and R_{80}, which maintains the voltage across the diodes at the rated V_Z value.

The zener voltage of the two diodes maintains a constant value within an operating range of current depending on the wattage rating of the diodes. A 1-W 6.8-V zener diode, for example, would accommodate up to $I = P/V = 0.1471$ A, or 147.1 milliamperes of current. When diodes are connected in series, as they are in Fig. 9-3, both diodes must have a similar current-handling capability or wattage rating.

CONVERTER CIRCUITS

The transmitter block diagram of Fig. 9-1 should have indicated an additional feature of the unit with respect to converter circuitry. Previous units, for example, employed some type of mechanical dc-to-ac converter in the signal path to produce ac for easy signal processing. A converter drive coil, mechanism, and transformer were needed to accomplish this operation. The transmitter is unique in this regard because it does not employ a mechanical dc-to-ac converter in its signal path. It simply accepts a low-voltage dc signal in the millivolt range and processes it through direct-coupled amplifiers to produce a suitable dc output current. In this unit, dc is processed by not employing reactive components such as transformers or capacitors in the signal path. With the aid of operational amplifier ICs, dc amplification is much easier to achieve than it was in older transmitters.

An electronic form of signal conversion is, however, used in a transmitter for a different purpose. The circuit in this case is called an *analog-to-duty-cycle converter*. Its function is to change the variable value of dc voltage applied to its input into a square wave. The on, or duty-cycle time, of the square wave varies according to the value of the applied dc voltage. The output of the converter is then applied to the input of the optical isolator circuit.

VOLTAGE AMPLIFIERS

A great deal of the voltage amplification achieved by a transmitter is accomplished by dc operational amplifier ICs. In this circuit application, a dc voltage

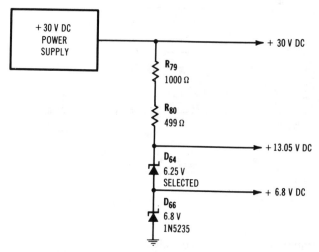

FIGURE 9-3 Simplified schematic of a voltage-regulator circuit.

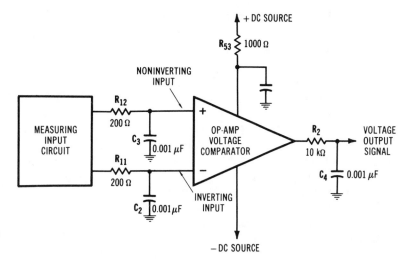

FIGURE 9-4 Op-amp voltage comparator.

value is applied to the input and an amplified version of dc appears at the output.

Figure 9-4 shows an IC op amp used as a voltage comparator. In this circuit, when the inverting input is negative with respect to the noninverting input, the output produces a positive-going dc voltage. In the same manner, a positive-going input signal will be inverted and produce a negative-going output signal. When negative or positive signals are applied to the noninverting input, the output will be an amplified version of the input with no change in polarity.

The input of the op amp shown is connected to the output of a Wheatstone bridge circuit. In this application, the amplifier responds to very small changes in dc voltage produced by the input signal. A minute change in input signal voltage will produce a high level of voltage amplification through the IC. Gain capabilities of several thousand are not unusual with this type of amplifier circuit. Op amps used in this type of circuit

do not employ an external feedback resistor between the input and output that restricts amplification. Since there is no intervening reactive component in the signal path, dc will pass very readily through the op amp and its associated circuitry.

OPTICAL ISOLATORS

Figure 9-5 shows an optical isolator that is used in a number of transmitters. The isolator is a unique circuit innovation that provides isolation between the input signal and the output or load device. Essentially, the amplified dc input signal is applied to a circuit that changes dc voltage of a variable value into a square wave. This analog to duty-cycle converter alters the on and off time of the square wave in accordance with the voltage value of the dc input signal. The square wave is then applied to a light-emitting diode (LED)

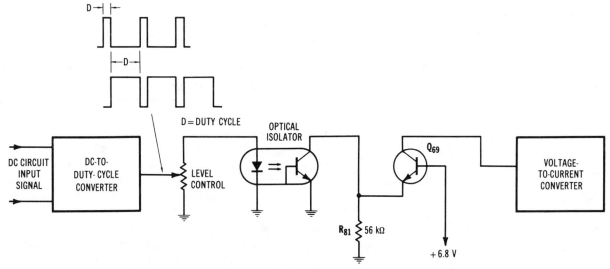

FIGURE 9-5 Simplified schematic of an optical-isolator circuit.

in the optical coupler. Variations in light intensity produced by the LED are then picked up by the base of a phototransistor housed in the same enclosure. The light-signal path from LED to phototransistor can only occur in one direction, which assures isolation between the input and output.

The output of the phototransistor of the optical isolator must now be changed from a square wave into a variable dc voltage before it is applied to the output circuit. Initially, the square-wave signal is applied to a chopper circuit. The square wave simply turns a transistor on and off according to the duty cycle of the square wave. The output of the chopper is then filtered and applied to a voltage/current converter circuit. Ultimately, the signal is applied to the current output amplifier circuit for further processing.

DARLINGTON AMPLIFIER

Figure 9-6 shows a schematic of two transistors connected into a Darlington pair circuit configuration. A circuit of this type develops the output signal of a transmitter. A Darlington amplifier has a high input impedance, a low output impedance, and very high current gain. The voltage gain of the circuit, however, is less than 1. All of these features are desirable characteristics for the current amplifier.

A Darlington pair is often referred to as a double emitter-follower circuit configuration. The input in this case is applied to the base of Q_1, and the output is taken from the emitter of Q_2. The combined transistor pair has slightly less than unity voltage gain. The total current gain, however, is the effective product of the current gains (beta, or h_{fe}) of the two transistors. The input impedance is increased because of this current multiplication and is essentially the total current gain times the value of the emitter resistor, R_E.

Many Darlington amplifiers are built on a single chip and housed in the same case. This type of configuration behaves in the same manner as a single tran-

sistor emitter-follower amplifier with a very high current gain. Devices of this type are designed to simplify circuit construction techniques.

MEASURING CIRCUITS

Figure 9-7 is a simplified schematic of the measuring circuit of a transmitter. We have in this circuit the Wheatstone bridge, zero adjustment, downscale or upscale open input protection circuits, temperature regulation, cold junction block amplifier, and the span-adjustment control. The input to the unit is a low-level signal of a few millivolts. The combination of jumper connections between terminals A through E on the bridge is used to position the range of the zero adjustment determined by the value of the input signal, from -25 to $+50$ mV dc.

The dashed line surrounding the bridge circuit components indicates those things that are temperature sensitive and located in the oven. The temperature of the oven is controlled by a temperature-regulating circuit consisting of a thermistor sensor and a transistor. Heat from the transistor is used as a thermal source for the enclosed oven. The thermistor is a sensitive resistor that changes its resistance value with changes in temperature. Through this resistance, the base current of the transistor is varied accordingly.

The voltage source for the bridge circuit is $+16.25$ and $+10$ V from the regulated source. The $+16.25$-V source is connected to the junction of resistors R_3 and R_9, while the $+10$-V source is applied to the junction of R_7 and R_8. Resistor R_{27} is a zero-adjust potentiometer located outside of the oven assembly. When the bridge is balanced to zero by R_{27}, there should be approximately 0.5 mA of current in each leg of the bridge. When the bridge is balanced, the voltage drop on one side should equal the voltage drop of the alternate side. When this occurs, there should be zero voltage input to the inverting input $(-)$ of the op amp. The cold block junction is attached to the positive input and applied to the noninverting input $(+)$ of the op amp. Variations in temperature are compensated for through this connection.

When the signal applied to the input of the bridge circuit increases, it alters the input voltage of op amp A_1. An increase in input voltage causes a corresponding increase in output voltage. The output of A_1 is then used to control the voltage signal applied to the voltage-to-current converter. The output of the converter provides from 4 to 20 mA of dc to the instrument load. This represents the full current range of the instrument. The 4 mA value rather than 0 mA identifies a zero transducer or 0-mV input signal. If a conductor carrying the 4- to 20-mA signal breaks, a 0-mA signal

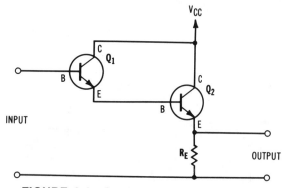

FIGURE 9-6   Darlington pair current amplifier.

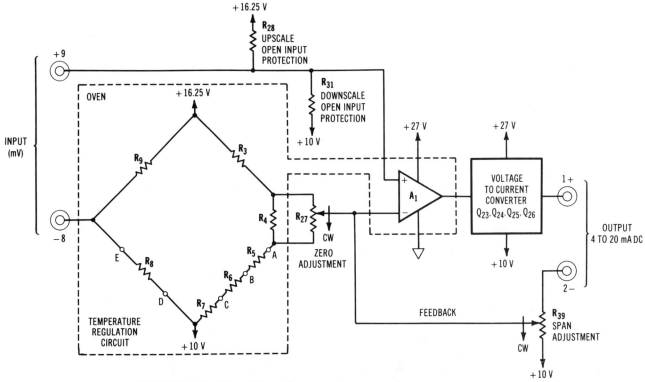

FIGURE 9-7 Simplified diagram of a transmitter measuring circuit. (Courtesy of A.B.B. Kent Taylor Corp.)

will be sensed by the receiving system. This 0-mA signal is used to identify a defective condition of operation.

A portion of the output signal forms a feedback loop that returns to the inverting input of amplifier A_1. When the value of the input signal equals the output feedback signal applied to A_1, the transmitter becomes stabilized at a new value.

The transmitter also has provision for upscale and downscale open-input protection. Upscale open-input protection is achieved by connecting an 82-MΩ resistor (R_{28}) to the +16.25-V dc source and the non-inverting input of amplifier A_1. In normal operation, the input signal and the bridge circuit are shunted across resistor R_{28}, which makes it inoperative. If by chance the input circuit becomes open, R_{28} provides a voltage signal to the input of amplifier A_1 which causes the output of the transmitter to drift upscale.

Downscale open-input protection is provided by a 22-MΩ resistor (R_{31}). This resistor is connected to the +10-V dc source and the noninverting input of A_1. In normal operation, the input signal and the bridge circuit are shunted across resistor R_{31}, which makes it inoperative. If by chance the input circuit becomes open, R_{31} provides a voltage signal to the input of amplifier A_1 which causes the transmitter output to drift downscale.

PRESSURE-TO-CURRENT TRANSMITTERS

The preceding transmitters accepted dc voltage in the millivolt range and developed a current output of 4 to 20 mA. A number of transmitters are available today that respond to other process input signals. This type of transmitter generally has some rather unique changes in its circuitry. A transmitter that responds to changes in pressure at the input and produces an output current is called a *P/I unit* (from the symbols *P* for pressure and *I* for current). The P/I transmitter will be presented here. This transmitter incorporates a circuit that uses a modified version of the Hartley oscillator. A basic version of the oscillator will be discussed first, then the modified version will be discussed, pointing out the major differences in circuit design. A simplified block diagram of a P/I transmitter is shown in Fig. 9-8.

The oscillator of a P/I transmitter controls the direct current of a power supply as its output. The sensing device used by the oscillator is a coil in the detector assembly. The inductance of this coil varies with the position of the pivoted beam. The beam position is determined by an input working against the spring that provides a method of mechanically zeroing the transmitter. Feedback repositions the pivoted beam by the magnet unit. The output of this transmitter

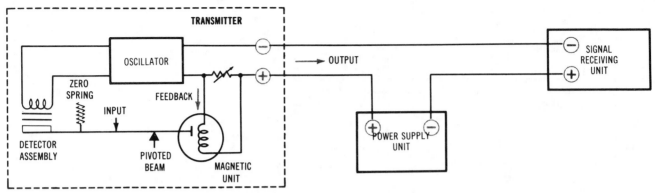

FIGURE 9-8   Simplified diagram of a P/I transmitter. (Courtesy of Honeywell, Inc.)

can be used by several different types of instruments, such as indicators, recorders, and controllers.

HARTLEY OSCILLATOR

Figure 9-9 is a schematic diagram of a series-fed Hartley oscillator. The identifying characteristic of this oscillator is the tapped coil in the base–emitter circuit. Coil L_1 and capacitor C_1 form a resonant tank circuit. The values of these components determine the frequency at which the circuit oscillates. The base capacitor, C_B, and the base resistor, R_B, provide automatic or self-bias for the oscillator. The emitter of the transistor is connected to the coil in the tank circuit. This circuit provides an output to input feedback path, which is necessary for any oscillator circuit to operate. All of the emitter current of the circuit must pass through this series part of the coil. As a result of this connection, the term series-fed oscillator is used to describe this circuit condition.

The Hartley oscillator is often found in radio receivers and transmitters. Oscillators are important in the field of electronics because they are high-frequency ac generators. Oscillators provide ac voltage at fre-

quencies at which electromechanical generators are neither practical nor possible. Four circuit characteristics are necessary before an oscillator can function. It must:

1. Be an amplifying device
2. Have a source of dc power
3. Have a frequency-determining device
4. Have positive feedback

The Hartley oscillator meets all of these requirements. When dc power is applied, the transistor starts to conduct because of its forward-biased base–emitter junction. The path of this current is from the negative side of the power supply, through the lower windings of coil L_1 to the tap, to the emitter, the collector, and back to the positive side of the power supply. As the current through the tapped portion of L_1 increases, a voltage is induced in the entire coil by transformer action. The lower portion of the coil acts as the primary of the transformer, whereas the entire coil is the secondary. When the primary and secondary of a transformer are a part of the same coil, it is called an *autotransformer*. The polarity of the voltage induced on the base of the transistor is such as to provide positive feedback or regeneration. As shown in the schematic, a rising current through the coil primary induces a positive voltage on the base side of the coil. This positive voltage is applied through C_B to the base of the transistor, which causes a rise in base current through the coil, thus inducing a more positive voltage on the base. The feedback cycle continues until the transistor collector–emitter current reaches saturation. This action takes place almost instantly.

As the base becomes more and more positive, more base current is applied to capacitor C_B in the direction shown. When the positive voltage across L_1 is removed because of transistor saturation, this leaves the base with a relatively high negative voltage from C_B. Then C_B discharges through R_B. During this time,

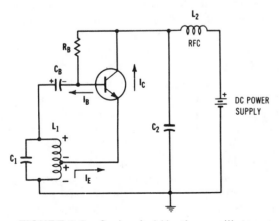

FIGURE 9-9   Series-fed Hartley oscillator.

the base is held negative, which reverse biases the base–emitter junction and cuts off the transistor collector–emitter current.

When the transistor reaches saturation, its collector–emitter current no longer rises. No transformer action can occur at this time since the magnetic field remains constant. As a result, no voltage will be induced across L_1. The base voltage then begins to go negative due to the charge on C_B. The base current decreases as a result. A decreasing current through the primary of L_1 induces a negative potential at the base side of L_1. This negative cycle of operation continues during the cutoff time of the transistor, however, because of the *flywheel effect* of the *LC* resonant tank circuit. The oscillating process is thus continuous. The frequency of oscillation is determined by the values of L and C:

$$ f_O = \frac{1}{2\pi \sqrt{LC}} $$

TRANSMITTER OSCILLATOR

Figure 9-10 is a complete schematic diagram of the transmitter oscillator circuit. A dc supply of 42 V is necessary for the proper operation of the oscillator. Current and voltage receiving units will be placed in series with the power supply. A variable series resistor adjusts the load current so it will have a value of between 4 and 20 mA. The voltage drop across the 2.5-Ω resistor is used as a test voltage. The 80-μF capacitor filters out any ac voltage that might be present across

it. A direct-current path is provided around the magnet unit by L_2, the 750-Ω resistor, and the coarse and fine span adjustments which determine the amount of feedback fed to the balance beam by the magnet unit. The 250-μF capacitor is a bypass capacitor. These components, with the exception of the magnet unit, have the primary function of controlling the current (or voltage) output.

Let us now consider the operation of the oscillator section of Fig. 9-10. Its operation is essentially the same as that of the Hartley oscillator previously discussed. One of the first noticeable differences is that it uses two transistors instead of one. In this case, Q_1 and Q_2 are connected in parallel to provide sufficient output current. Resistors R_1, R_2, and R_3 and diode D_1 provide the voltage drop that determines the emitter potential. Diode D_1 has the added responsibility of sensing the amplitude of the base signal and providing for changes in the forward bias of the transistors that are reflected in current output. Base voltage is determined by the voltage divider formed by R_5, R_6, R_7, R_8, and R_9. Both R_7 and R_9 are temperature-compensating resistors. L_1 and C_2 make up the Hartley resonant circuit and determine the frequency of operation. The feedback path is through C_1, the tapped portion of L_1, and the detector coil, which senses the position of the pivoted beam. The circuit described previously is a series-fed Hartley oscillator; this one is a modified shunt-fed Hartley oscillator.

Circuit operation will be discussed referring to the simplified schematic in Fig. 9-11. Coil L_2 represents a parallel combination of L_2 and the magnet unit from Fig. 9-10. Resistance R_E is a combination of the

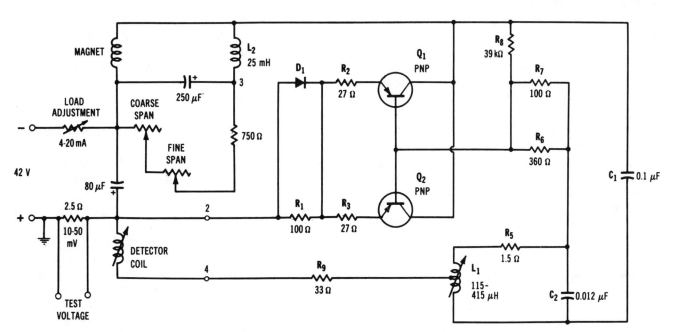

FIGURE 9-10 Transmitter oscillator circuit.

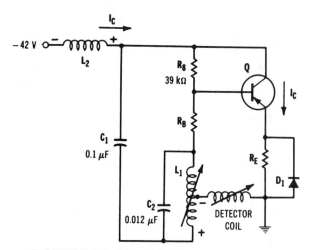

FIGURE 9-11 Simplified transmitter oscillator circuit (shunt-fed Hartley).

resistors in the emitter circuit, while resistance R_B is a combination of the resistors in the base circuit. When power is applied, collector current I_C will conduct through L_2, transistor Q, and the parallel combination of R_E and D_1. For this direction of current, D_1 has a very low resistance. The emitter is at practically ground potential. Resistor R_8 develops a potential at the base that provides the proper forward bias. The variable inductor L_1 determines the frequency of the oscillator. The bandpass of the LC circuit is somewhat broadened by R_5 (see the complete schematic in Fig. 9-10). This enables the circuit to oscillate over a broader range of frequencies than would otherwise be possible. The only major difference between the series-fed and the shunt-fed Hartley is the reference point or ground. With the series-fed Hartley (Fig. 9-9), the bottom of L_1 was grounded. This allowed the collector–emitter current to conduct through L_1. In the shunt-fed circuit, no collector–emitter current conducts through L_1 since the emitter is grounded.

As mentioned before, when power is applied, collector current conducts through D_1, Q, and L_2. A voltage drop is developed across L_2 with the polarity indicated. This voltage causes current to conduct through the detector coil and the lower portion of L_1. The current through this portion of L_1 then induces a voltage across L_1 that increases the forward bias on the transistor so that the collector current increases. This continues until the transistor saturates, the feedback cycle reverses, and oscillations are maintained in the L_1–C_2 tank circuit as described for the circuit of Fig. 9-9.

TRANSDUCERS

The purpose of a transducer is to change an input signal of one type into an output signal of a different type that is proportional to the original input. The type of

output to be produced depends on the function we wish to perform. Four types of transducers are discussed in this section. In all but one case described here, the output is current, whereas the input is some kind of pressure signal. By transforming the input pressure to a 4 to 20-mA current, efficient long-range transmission of the signal can be accomplished to a standard recorder, indicator, or controller. The fourth transducer has a current input with a pressure output.

Pressure-to-Current Transducer

Figure 9-12 illustrates the P/I (pressure-to-current) transducer. A balanced pivoted beam is the heart of this instrument. With no pressure input, the zero spring positions the beam. The position of the beam is sensed by the detector coil, the inductance of which changes as the beam changes position. The change in inductance is caused by the change in the air gap between the core of the coil and the ferrite disk. The magnet unit provides for feedback to rebalance the beam. The span-adjusting resistors control the amount of feedback current.

The operating sequence, with an increase in pressure in the input bellows, moves the right end of the pivoted beam down. This increases the air gap between the detector coil and the ferrite disk, causing a decrease in the inductance of the coil. The decrease in inductance changes the oscillator frequency, which results in an increase in current. This output current passes through the magnet unit and the span-adjusting resistors. The amount of this current depends on the set value of the parallel span-adjusting resistors. The magnetic field produced by the feedback current through the magnet unit opposes the pressure input. The upward force of the left end of the beam produced by the magnet unit balances the beam at the new operating position. Any further change in pressure provides a corresponding change in output current.

Current-to-Pressure Transducer

Figure 9-13 is a simplified schematic of the I/P (current-to-pressure) transducer. The sequence of operating steps follows. An increase in current produces a magnetic field across the magnet coil, the strength of which is determined by the position of the span-adjusting resistors. The reaction between the magnet coil and the permanent magnets within the magnet until moves the beam upward. The movement of the beam increases the pressure on the nozzle, resulting in an increase in the air pressure going to the pilot valve. The output pressure and the pressure feedback also increase. The increase in pressure in the rebalancing bellows sets up a new operating balance point for the pivoted beam. The zeroing spring sets the balance point for a mini-

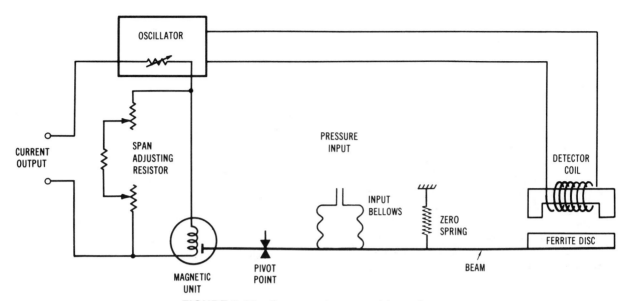

FIGURE 9-12   Pressure-to-current transducer.

mum current input. Any further change in current produces a corresponding change in pressure output.

Processed Pressure-to-Current Transducer

The PP/I (processed pressure-to-current) transducer is shown schematically in Fig. 9-14. This transducer is much like the P/I transducer. The primary differences arise from the fact that the processed pressures are relatively high. These higher pressures must be fed into the system through a high-pressure bellows or a Bourdon tube. Which of these devices is used depends on the range of pressures to be measured. The bellows positions a bellows lever that is connected to the pivoted beam through an input spring. Movement of the beam occurs when an unbalance results between the input spring and the zero spring. The repositioning of the beam causes a change of inductance in the oscil-

lator circuit. The increase results in a direct-current increase through the oscillator. A portion of this output current provides feedback to the beam through the magnet unit. The new balanced position of the pivoted beam is the operating point for this particular pressure. Any further change in pressure results in a beam unbalance. The same sequence of events is followed as the system adjusts to a new operating point.

Differential Pressure-to-Current Transducer

Figure 9-15 illustrates the action that occurs in the differential P/I transducer. The differential pressure between the high- and low-pressure bellows is sensed by the liquid-filled torque tube. A change in differential pressure rotates the takeoff arm. This motion is transferred to the cross-spring beam through the input spring to the pivoted beam. The actions of the beam,

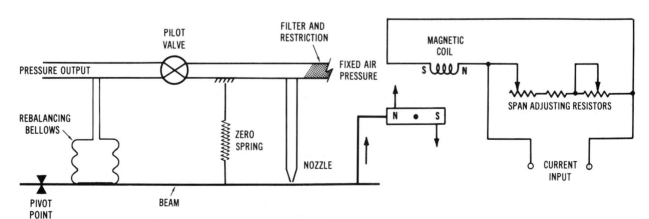

FIGURE 9-13   Current-to-pressure transducer.

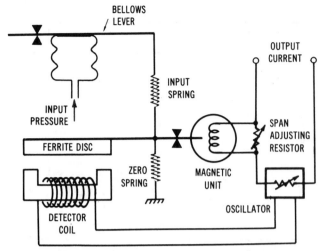

FIGURE 9-14 Processed pressure-to-current transducer.

oscillator, and magnet unit are exactly the same as for those transducers discussed previously.

SUMMARY

Signal transmitters are designed to receive the process signal from the sensor and give it the power needed for transmission to a controller located some distance away. Typically, a transmitter employs temperature regulation, voltage-to-current conversion, an analog to duty-cycle converter, optical isolators, a chopper, filtering, and a final voltage-to-current converter.

The power supply of a transmitter generally employs low-voltage dc as a primary power source. A dc-to-ac converter is commonly used in a transmitter power supply. Alternate transistors conduct which switches the dc on and off. This changing voltage is applied to a transformer that steps it up or down according to the needs of the instrument. The transformer output is then rectified, filtered, and regulated.

One form of conversion used in a transmitter is called analog-to-duty-cycle conversion. This function changes variable dc values into square waves. The on part of the square wave, or its duty cycle, is made to change according to the value of the applied dc.

The voltage amplification function of a transmitter discussed in this chapter is achieved by operational amplifiers. Gain capabilities of several thousand are typical with IC op amps. With these devices, dc can be amplified because no intervening reactive components are involved. The op-amp circuits discussed here have high amplification capabilities because no feedback resistor is used to restrict gain.

Optical isolators are used to provide isolation between the input signal and the output or load device. First, ac is changed into a square wave with a variable duty cycle. This signal is then applied to a light-emitting diode in the optical coupler. Changes in light intensity are then picked up by the base of a phototransistor housed in the same enclosure. The signal path can only occur between the LED and the phototransistor, which assures isolation between input and output. The output of the phototransistor is then changed into a variable dc value before being applied to the output circuit.

A Darlington amplifier is two transistors connected in a double-emitter-follower configuration. It has high input impedance, low output impedance, and high current gain.

The measuring circuit of a transmitter employs a Wheatstone bridge, zero adjustment, downscale or upscale open-input protection, temperature regulation, cold block junction amplification, and the span-adjustment control. The input is normally only a few mil-

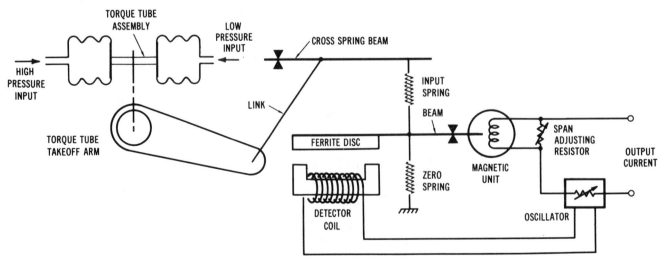

FIGURE 9-15 Differential pressure-to-current transducer.

livolts. The voltage source for the measuring circuit is supplied by a regulated section of the power supply.

An important part of a transmitter is the oscillator. Essentially, an electronic oscillator changes dc into alternating current. Through this part of the transmitter a small value dc input signal can be sent or transmitted to a distant location where it may achieve control or produce a signal indication.

A series-fed Hartley oscillator contains a tapped coil as an identifying component. This coil and a capacitor are used to form the *LC* frequency-determining components. In addition to this, there must be an amplifying device, dc power source, and a positive feedback loop.

The input to a transmitter is developed by a transducer. Essentially, a transducer changes energy of one form into a different form. Pressure-to-current (P/I), current-to-pressure (I/P), processed pressure-to-current (PP/I), and differential pressure-to-current are some examples of instrumentation transducers.

ACTIVITIES

TRANSMITTER INVESTIGATION (mV input to mA output)

1. Refer to the manufacturer's operational manual of the electronic transmitter being used in this investigation. The electrical power source should be disconnected from the transmitter or turned off for this investigation.
2. Remove the external covering or enclosure of the transmitter.
3. Describe the primary function of the transmitter with respect to its input and output.
4. Locate the input signal source of the transmitter. Describe the value range of the input signal being accepted by the transmitter.
5. Locate the power supply. Does it have measurement points that can be evaluated when the unit is energized?
6. Locate the measuring circuit of the unit. Identify its common features.
7. Locate the output feed line of the transmitter. Describe the value range of the output.
8. Describe the circuit board replacement procedure of the transmitter.
9. Reassemble the transmitter.

TRANSMITTER INVESTIGATION (P/I, PP/I, or DP/I)

1. Refer to the manufacturer's operational manual of the transmitter being used in this investigation. The electrical power source should be disconnected or turned off through this entire investigation.
2. Remove the housing or enclosure of the transmitter.
3. Describe the type of input and output being used by the transmitter of this investigation.
4. Locate where the input signal source is attached to the transmitter. Describe the unique features of the input. What is its range of operation?
5. Identify the power supply of the transmitter. Is there anything unique about the power supply? What are some of the voltage values produced by the power supply?
6. Does this unit have an oscillator? If so, where is it located?
7. Locate the output of the transmitter. Does this part of the transmitter have any distinguishing features?
8. Explain how the circuit boards are removed from the unit.
9. Reassemble the transmitter.

QUESTIONS

1. What is the primary role of an electronic transmitter?
2. Describe the power supply of an electronic transmitter?
3. Why is an op amp used to amplify dc in a transmitter?
4. What occurs when an op amp is used without an external feedback resistor?
5. What is the function of an optical isolator in a transmitter?
6. Why does the light-signal path of an optical coupler occur in only one direction?
7. What is the advantage of using a Darlington amplifier?
8. List some of the functions of the measuring circuit of an electronic transmitter.
9. Explain the fundamental operation of a Hartley oscillator.
10. What is meant by the term *transducer*?
11. What is the fundamental role of an oscillator in the operation of a transmitter?
12. What are some examples of instrumentation transducers?

pH INSTRUMENTS

OBJECTIVES

Upon completion of this chapter, you will be able to:

1. Define pH.

2. On a pH scale identify different values of acid, base, and neutral salt solutions.

3. Explain how a pH probe responds when placed in an ionized solution.

4. Draw a block diagram of a laboratory type of pH instrument.

5. Explain the primary function of each block of a laboratory type of pH instrument diagram.

6. Given a schematic diagram of a pH analyzer, be able to identify circuit parts that will achieve conversion, amplification, phase detection, and current output.

IMPORTANT TERMS

In the study of pH instruments one frequently encounters a number of new and somewhat unusual terms. These terms play an important role in the presentation of this material. Terms such as *comparator*, *ionization*, *phase detector*, and *degeneration* will begin to have some meaning for you. A few of these terms are singled out for study. As a rule, the chapter will be more meaningful if these terms are reviewed before proceeding with the text.

Acid: A water-soluble substance that is generally sour to the taste and is capable of reacting with a base solution to form a salt that contains hydrogen molecules or ions. Typical acid solutions are boric acid, orange juice, vinegar, and photographic developing chemicals.

Acidity: A measure of the hydrogen ion content of a solution.

Alkalinity: A measure of the hydroxyl ion content of a base solution.

Base: A water-soluble substance or brackish tasting compound that is capable of reacting with an acid to form a salt. An alkaline substance containing a number of free hydroxyl ions. Typical base solutions are household lye, bleach, ammonia, borax, and baking soda.

Comparator: An active device or circuit that compares an input voltage to a predetermined dc voltage reference.

Degeneration: A negative feedback process that takes an out-of-phase output signal and returns it to the input of a circuit.

Emitter follower: A high-input-impedance amplifier circuit that has unity voltage gain with its input and output signals in phase.

Extraneous voltage: Voltage that exists or comes from something outside the primary generating source.

Half-cell: A pH electrode that develops a voltage that is proportional to the hydrogen ion content of a solution in which it is placed.

Inverting input: The input lead of an operational amplifier that causes the output signal to be inverted.

Ion: An atom that has lost or gained electrons, causing it to possess a negative or positive charge.

Ionization: The process of producing charged particles or ions.

Leakage current: A small but unwanted current flow that occurs due to insulation breakdown or excessive heat.

Module: A packaged functional assembly of wired electronic components.

Negative feedback: A degenerative feedback circuit.

Noninverting input: An input lead for an integrated circuit that does not shift the phase of the output signal.

Parallax: The displacement of an object that occurs when it is not viewed in a straight line. Meter-reading parallax occurs when viewing the deflection hand at an angle.

pH: A measure of the acidity or alkalinity of a solution based on the concentration of hydrogen ions. pH values less than 7 are considered acidic with values greater than 7 considered to be basic or alkaline.

Phase detector: A circuit that provides a dc output voltage that is proportional to the phase difference of its two input signals.

Probe: The input device of a pH instrument that receives voltage for further processing by its circuitry.

Referenc half-cell: A pH electrode that is responsible for producing a stable voltage that is independent of solution properties.

Spurious ac: A false or unnatural ac signal that may be developed in a conductor or signal path.

Temperature compensation: The operational characteristics of a circuit or component that are independent of temperature changes.

INTRODUCTION

It is extremely important in many chemical process applications to know if a particular chemical solution has a predominantly acid or alkaline (base) content. Some common examples of acid solutions are vinegar (acetic acid), the citric acid in fruit juice, and dilute sulfuric acid, which is used as a battery electrolyte. Ammonia water, by comparison, is a rather weak base solution, while concentrated lye mixtures form a very strong base solution. Acid and base have entirely different chemical reactions when they exist in solutions. Successful control of chemical processing, therefore, necessitates that acid and base levels be carefully controlled to assure a desired outcome.

A more specific definition of pH refers to the number of ionized or free hydrogen ions (H^+) and hydroxyl ions (OH^-) in a solution. Acid has an abundance of H^+ ions, while base has large numbers of free OH^- ions. The pH value is, therefore, a measurement of the ratio of hydrogen and hydroxyl ions in a solution. When H^+ is predominant the solution is acid. When OH^- is predominant, the solution is base. If equal amounts of base and acid are present, the solution is a neutral salt.

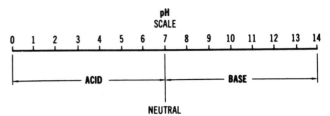

FIGURE 10-1 Standard numbering system for a pH scale.

The numbering system for a pH scale ranges from 0 to 14. The number 7 is at the center of this span and is considered to be an indication of a neutral solution. Acid levels occupy the position from 7 down to 0, with the smaller numbers indicating the highest acid levels. The numbers 7 to 14 represent the base scale, with the largest numbers indicating the highest base levels (Fig. 10-1).

pH MEASUREMENT

The pH level of a solution can be determined by direct measurement of the dc voltage developed between two electrodes immersed in the solution under test. A dc null-balancing potentiometer with a very high input impedance is commonly used to indicate the voltage developed by the measuring electrode probe. The electronics part of the instrument is then responsible for manipulating the voltage developed by the electrodes in such a way that it will develop an indication of pH. Hand-deflecting instruments, chart recorders, and digital readout displays are in common use today. A number of different display techniques are used in pH measurement. In this chapter (and the next), we will assume that the display device is a hand-deflection meter or a recorder. Recorders obviously provide a permanent record of pH levels for a variety of different time spans.

Figure 10-2 shows a typical pH indicator scale for a hand-deflection instrument. As you will note, there are 10 small graduations or divisions between each two numbers. Each division, therefore, represents 0.1 pH unit. If the indicating hand is deflected to the third small graduation to the right of the number 5, the pH level would be 5.3. This indicates an acid level of approximately 20%. A pH meter can also be

FIGURE 10-2 Typical pH indicator scale.

used to indicate positive and negative voltage values. In this particular indicator, full-scale deflection is +700 to −700 mV. In practice, the deflecting hand should come to rest at zero when measurements are not being taken.

The curved black strip of the pH meter scale of Fig. 10-2 represents a mirror finish. This part of the meter scale is designed to produce a reflection of the indicating hand. In practice, an operator looks at the scale in such a way that the hand and its reflection are exactly in line (i.e., the reflection cannot be seen). When this occurs, the indication should be quite accurate because the operator is not looking at the scale from an angle. Technically, this is described as reducing meter parallax.

pH PROBES

The probe or electrode part of a pH instrument is often thought of as a battery whose voltage varies with the pH level of the solution in which it is placed. It contains either two separate electrodes in a single probe housing or two distinct probes. In either case, one electrode, or part of the common probe, is sensitive to hydrogen. A special glass bulb or membrane material is used that has the ability to pass H^+ ions inside the sensitive bulb. When the electrode is placed in the solution, a voltage that is proportional to the hydrogen ion concentration is developed between an inner electrode and the outer electrode or glass bulb material. This pH-sensitive electrode is often called a *half-cell*.

A second discrete electrode, or the alternate part of the common probe, is used to develop a reference voltage. The reference electrode is primarily responsible for producing a stable voltage that is independent of solution properties. This electrode, or half-cell, develops a fixed voltage value when placed in the solution. When the reference half-cell and the pH glass-bulb half-cell are combined, they form a complete probe. Figure 10-3 shows a typical pH probe.

pH INSTRUMENTATION

Figure 10-4 shows a general diagram of the essential parts of a pH measuring instrument. In this case, pH is observed on a hand-deflection meter scale. Quantitative measurements of pH are produced by this type of instrument. A majority of the pH instruments used in industry today are of this type. They range from small portable units housed in convenient carrying cases to larger stationary units.

The essential parts of a pH instrument, regardless of its type or style, are the measuring half-cell electrode, the reference half-cell electrode, a high-imped-

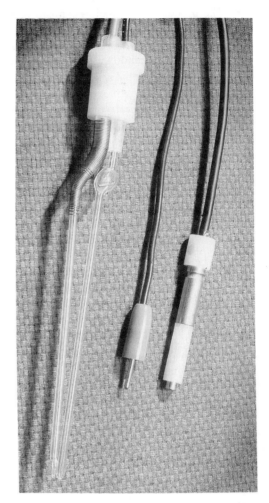

FIGURE 10-3 Typical pH probe. (Courtesy of Leeds & Northrup.)

ance amplifier, and an indicator. This type of instrument is normally classified as a direct-reading unit because it responds to voltage values produced directly from the solution under test. Nearly all industrial pH instruments are of this type as opposed to the indirect method that produces an indication due to color changes in a material sample.

The measurement of pH presents some interesting electronic circuits that have not been previously discussed Figure 10-5 is a block diagram of a typical pH indicator that shows some of these items. One of the new problems is the very high impedance of the measuring probe, which makes many different design features necessary. A negative-feedback measuring circuit, which is similar to a null-balancing system, is one of these necessary features. Careful shielding of the input circuits is also necessary. An emitter-follower amplifier is likewise needed to match the high impedance of the input circuit to the low input impedance of a high-gain solid-state voltage amplifier.

A great deal of care must also be taken to keep

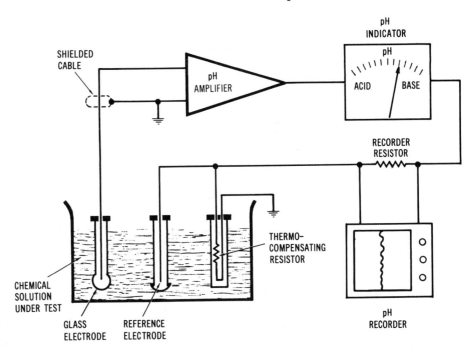

FIGURE 10-4 Simplified diagram of the essential parts of a pH measuring instrument.

the amplifier circuit from oscillating. Any ac output of the amplifier must be fed back so as to be 180° out of phase with the ac present in the amplifier. Of course, the simplest way to suppress oscillation is to keep the ac voltage in the feedback loop to a minimum. Some of the methods used to do this have already been mentioned. Shielding, for instance, helps keep spurious ac voltages out of the amplifier. Using all dc source voltages helps to reduce this problem. All ac voltages, unfortunately, cannot be kept out of the feedback loop. Oscillation can be prevented, however, if the ac signal is fed back out of phase, but circuits must be provided for this purpose.

The feedback measuring circuit has another purpose besides keeping the input impedance high. It also

helps to counteract fluctuations in amplifier gain due to circuit variations, line-voltage changes, etc., as long as the amplifier gain is high. From the block diagram, it can be seen that the electrode voltage is fed to the converter circuits. This dc voltage is changed to an ac voltage by an electronic converter. The ac signal is then fed to an emitter follower, which is the first stage of the voltage amplifier. As mentioned before, this circuit is for impedance-matching purposes. The output is amplified by a number of transistor high-gain voltage amplifiers and then fed into the phase detector. The dc output is then fed back to the amplifier input. Two bridge circuits are included in the feedback circuit; one is used to standardize the instrument, while the other provides temperature compensation.

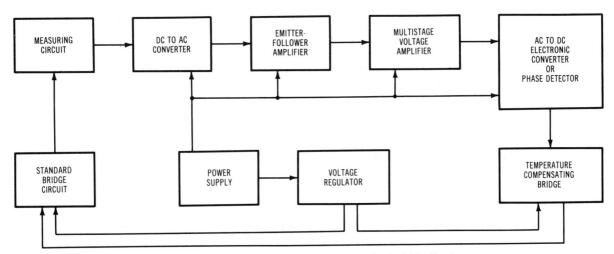

FIGURE 10-5 Block diagram of a typical pH indicator.

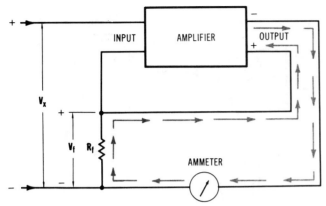

FIGURE 10-6 Feedback measuring circuit.

Figure 10-6 is a simplified amplifier diagram which illustrates the use of a negative feedback arrangement for making voltage measurements in a pH indicator. This diagram will help to show why variations in amplifier gain are minimized by using the negative feedback system. The amplifier output voltage (V_f) is fed back to the input in series with V_x, the unknown voltage that is being measured. Therefore, voltage V_f is designated as the feedback voltage, while resistance R_f is designated as the feedback resistance. This arrangement is called inverse or negative feedback because V_f is in opposition to V_x. It can be demonstrated that V_f is virtually equal to V_x, if the amplifier gain is high. Letting A_v represent the amplifier voltage gain, then

$$V_f = A_V(V_x - V_f)$$

from which we obtain

$$V_x = V_f\left(1 + \frac{1}{A_V}\right)$$

so that if A_V is sufficiently high, $1/A_V$ is very small and can be neglected. It then follows that

$$V_x \simeq V_f$$

This indicates that the operation of the circuit is not dependent upon the exact value of the amplifier gain, but only requires the gain to be high. Therefore, gain fluctuations caused by line voltage changes, circuit variations, and so on, have very little effect on measurement indications as long as the gain remains high.

POWER SUPPLY

The power supply of most pH instruments is similar in many respects to those used in other electronic instruments. Typically, the power supply corresponds to the type of components employed in the instrument. Discrete transistor circuits, for example, normally require a regulated low-voltage source that may range in value from 10 to 50 V. Older vacuum-tube instruments demand several hundred volts of dc with the addition of ac filament voltages. Integrated circuit instruments, by comparison, necessitate some type of a split low-voltage power supply. Digital circuits generally need +5 V for IC operation. Instruments often combine different components in their operation. It is not unusual for a power supply to deliver several different voltage values to its circuitry. Due to the general similarity of all instrument power supplies, we will investigate only those unique features that have not been discussed previously.

Figure 10-7 shows a simplified schematic diagram of the master power supply used in a laboratory pH analyzer. This particular power supply uses a bridge rectifier to achieve a split dc voltage of +15 V and −15 V with respect to the common or ground connection. The split 15-V outputs of this power supply

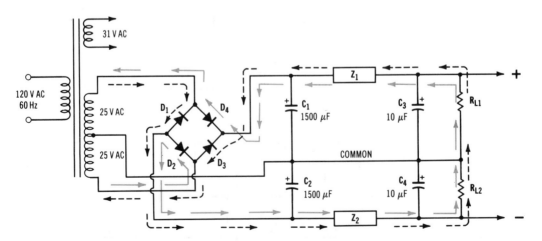

FIGURE 10-7 Simplified schematic of a split power supply.

are used to energize ICs and a number of discrete transistors in the composite circuit.

With 120 V 60-Hz ac applied to the primary winding of the transformer, 25 V ac appears across each half of the secondary winding. Assume now that the top of the secondary winding is positive and the bottom is negative during the first alternation. When this occurs, current conducts from the negative side of the secondary winding through D_2, Z_2, R_{L2}, R_{L1}, Z_1, D_4, and back to the positive side of the secondary winding. This alternation is indicated by solid arrows in the diagram.

The next alternation of the ac input causes the top of the secondary winding to be negative and the bottom to be positive. When this occurs, current conducts from the negative side of the winding through D_1, Z_2, R_{L2}, R_{L1}, Z_1, D_3, and back to the positive side of the secondary winding. This alternation is indicated by dashed arrows in the diagram.

As a result of the current conduction just described, full-wave rectification is achieved by the bridge circuit. After filtering, the dc output will appear as 30 V across R_{L1} and R_{L2}. If the common connection of these two resistors is connected to ground, the voltage is divided and becomes $+15$ V and -15 V with respect to ground. Resistors R_{L1} and R_{L2} are only used in this circuit to represent the load applied to the power supply. In practice, they are not specific components of the power supply. The positive half of the power supply is filtered by C_1, C_3, and Z_1. A duplicate filter is provided by C_2, C_4, and Z_2 for the negative half of the power supply.

TRANSISTOR VOLTAGE REGULATOR

Figure 10-8A shows a simplified schematic diagram of a series-connected transistor voltage regulator. Two of these regulators are used in the split power supply just discussed. In Fig. 10-7, Z_1 and Z_2 indicate the location of these regulators in the circuit. We will describe the operation of the regulator first, then show it placed in the split power supply. Through this approach, you should have a better idea of how transistor regulators are used to achieve improved power supply control.

Power supply voltage regulation can be greatly improved through the use of transistors. The series transistor regulator, in this case, behaves somewhat like a simple series-connected variable resistor whose resistance is determined by circuit operating conditions. Figure 10-8B is used to demonstrate the basic principle of transistor regulation.

Assume now that an unregulated dc input is applied to the variable resistor circuit. This voltage is labeled V_{in}. The variable resistor is R_T and the power supply load is represented by R_L. The voltage devel-

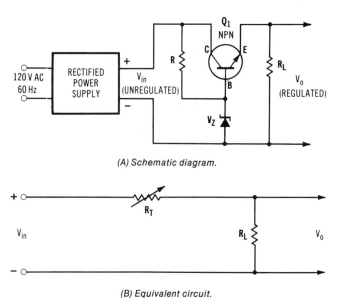

(A) Schematic diagram.

(B) Equivalent circuit.

FIGURE 10-8 Series-connected transistor voltage regulator.

oped by the load appears as V_o. For variations in R_L, if V_o is to remain at a constant value, the ratio of R_L to R_T must remain at a fixed level. This means that a change in load resistance, R_L, is compensated for by a similar change in the value of R_T.

Specifically, an increase in the resistance of R_L would cause a corresponding rise in V_o. To compensate for this, R_T should increase an equal amount. This, in turn, would lower V_o by causing an additional voltage drop across R_T. A decrease in the resistance of R_L would normally cause a corresponding decrease in the value of V_o. To compensate for this, R_T should be decreased an equal amount. This, in turn, would cause less voltage drop across R_T which would increase V_o accordingly.

The transistor regulator of Fig. 10-8A achieves control of V_o by changing its conduction capabilities according to voltage variations. Since the voltage across the zener diode is fixed, a decrease in V_o will result in a corresponding increase in the emitter–base voltage (V_{BE}) of Q_1. In this circuit, V_{BE} is determined by zener diode voltage V_Z minus V_o.

Assume now that a transistor regulator is placed in a circuit with an output voltage of 15 V. If V_Z is rated at a fixed value of 18 V, V_{BE} would be 3 V normally. Now assume that the resistance of the load increases in value, which causes V_o to rise to 17 V. As a result of this action, V_{BE} will decrease because 18 V (V_Z) $-$ 17 V (V_o) = 1 volt (V_{BE}). A decrease in V_{BE} will reduce the forward biasing of the transistor and cause it to increase in resistance. This, in turn, causes more voltage drop across the transistor, which returns V_o to its original 15 V.

In the same manner, a decrease in V_o is com-

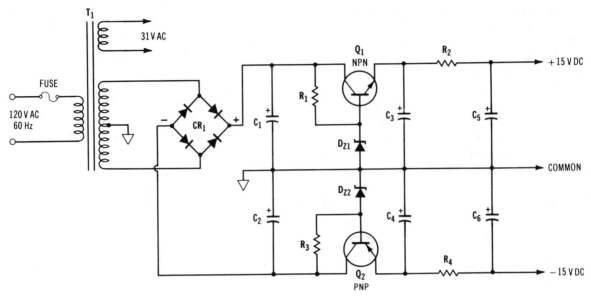

FIGURE 10-9 Split power supply with transistor voltage regulation.

monly caused by a reduction in the resistance of R_L. If V_o, for example, decreases to 13 V, it would cause a corresponding increase in V_{BE}. In this case, 18 V (V_Z) − 13 V (V_o) produces 5 V (V_{BE}). With this increase in V_{BE}, the transistor will conduct more and have a reduced internal resistance. This, in turn, will produce less voltage drop across the transistor and increase V_o to its original 15 V.

Figure 10-9 shows a ± 15-V split power supply with transistor voltage regulation. Note that a regulator is placed in each half of the supply. The transistors must be of the opposite polarity in order to develop the correct output polarity. Transistor Q_1 is an npn, while transistor Q_2 is a pnp. The operation of the pnp regulator is primarily the same as the npn circuit just described. Through series transistor regulator circuits, improved power supply regulation can be obtained. In some circuits today the entire transistor regulator is built on a single IC chip. This obviously simplifies the circuit and its physical construction.

CONVERTER CIRCUITS

Dc-to-ac conversion of the amplifier input is accomplished in much the same manner as was previously discussed. The converter circuit is shown in Fig. 10-10. Two low-impedance features of the previous converter circuits had to be eliminated, however. The first change involved the chopper itself. In previous amplifiers, the chopper has been normally closed. By this we mean that the contacts were closed more than 50%

of the time. The pH indicator chopper, on the other hand, is normally open. The switch contacts are actually closed only a very short portion of the operating cycle. This means that current occurs only for a very short percentage of each cycle. Average current, therefore, is very small. As a result, the converter acts as a high impedance.

The low-impedance effects of an input transformer also have to be eliminated. In place of the input transformer, three coupling networks were installed. The impedance was kept high by making R_2, R_3, and R_4 20-MΩ resistors. Capacitors C_2, C_3, and C_4 also have a fairly high impedance to the 60-Hz input. The effective impedance of the three coupling circuits is somewhat less than the effect of one RC circuit since they are essentially connected in parallel. Due to the large capacitive reactance of the coupling capacitors, some phase shift occurs in the amplifier input. This will be discussed in the section on phasing. Resistor R_1 is 1000 MΩ and also adds to the input impedance.

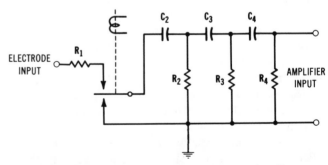

FIGURE 10-10 Dc-to-ac converter circuit.

EMITTER FOLLOWER

A schematic diagram of a typical emitter-follower amplifier is shown in Fig. 10-11. The collector of this circuit, as you will note, does not have a load resistor. If a collector resistor is used, however, a bypass capacitor returns the signal to ground. This capacitor effectively grounds the collector ac voltage. In some references, emitter-follower amplifiers are called common-collector or grounded-collector amplifiers because of this ground return connection.

Since there is essentially no collector load resistor in an emitter-follower amplifier, the only output load is the emitter resistor, R_E. The resistance value of R_E is largely responsible for the output impedance of the amplifier. In practice, R_E is kept low which means the output impedance is likewise low. With the input impedance of a typical amplifier high, a unique characteristic of the emitter-follower amplifier is high input impedance and low output impedance. Since the output voltage developed across the low-impedance emitter resistor is small, emitter followers are known to have a voltage gain of less than one. The power gain and current gain of this amplifier, however, may range as high as 100 in a representative circuit. Impedance matching and power amplification are the primary applications of an emitter-follower amplifier.

With no signal applied, the emitter-follower amplifier normally has only a few volts developed across the emitter resistor. This voltage, as indicated in Fig. 10-11 is positive with respect to ground. When the base signal goes positive, it causes a corresponding increase in emitter current through R_E. This, in turn, means an increase in positive emitter voltage. In a similar manner, when the base signal voltage swings negative, it causes a decrease in emitter current. This means a corresponding decrease in the emitter voltage. In a sense, the emitter voltage follows the value change in base input voltage. This is where the circuit derived the name emitter follower. This relationship between input and output voltage can also be described as being in phase.

For purposes of illustration, let us again look at the emitter-follower amplifier of Fig. 10-11 with no signal applied. In this case, note that the emitter voltage is $+5$ V and the base voltage is $+6$ V. The base is, therefore, 1 V positive with respect to the emitter. This adequately assures a suitable level of forward biasing. Assume now that the amplifier also has a voltage gain factor of 0.9.

When the positive 0.5-V half-cycle of the ac input signal is applied to the base, it will cause a corresponding increase in the emitter–collector current. With a gain of 0.9, a $+0.5$-V input signal will cause the emitter voltage to increase by 0.45 V, or 5.0 V $+$ 0.45 V $=$ 5.45 V. With the original no-signal base-emitter voltage of 1 V, this change in signal causes the emitter–base voltage to now become 6.5 V $-$ 5.45 V, or 1.05 V. This means that a 0.5-V input only causes a 0.05-V change in the emitter–base voltage.

In the same manner, when the negative 0.5–V half-cycle of the ac signal is applied to the base-emitter junction, it will cause a corresponding reduction in the emitter–collector current. The voltage gain will be -0.5 V $\times$ 0.9, or -0.45 V, which will produce an emitter voltage of 5.0 V $-$ 0.45 volt, or $+4.55$ V. The resulting base-emitter voltage will now be $+5.5$ V $-$ 4.55 V, or 0.95 V. This means that a 1.0-V peak-to-

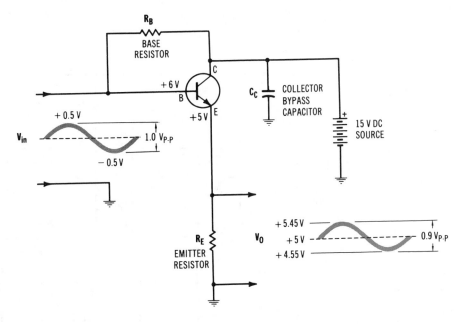

FIGURE 10-11 Emitter-follower amplifier circuit.

peak input signal only causes a 0.9-V peak-to-peak change (5.45 to 4.55 V) in the voltage across R_E. This effect is commonly called *degeneration*. Emitter-follower amplifiers have an inherent degeneration problem.

Due to the effect of degeneration in an emitter-follower amplifier, the base only draws a nominal amount of current. In transistor circuits of this type, degeneration has a rather significant influence on the total amount of signal gain achieved. Due to this characteristic, emitter-follower amplifiers are not used in applications that demand voltage gain.

VOLTAGE AMPLIFIERS

In pH instruments, voltage signals produced by the measuring circuit generally necessitate a rather high level of amplification in order to produce an output that will drive an indicator. In many instruments, this function is achieved by a combination of two or more discrete component *RC*-coupled transistor amplifiers. In this type of amplifier, the input signal is applied to the base of the first amplifier and removed from the collector. This signal, which is ac because of the converter, passes through a coupling capacitor to the base of the next transistor. The process continues, base to collector, through succeeding amplifiers.

Figure 10-12 shows a typical two-stage *RC*-coupled, transistor amplifier. In this circuit, each transistor uses emitter biasing. Transistor Q_1, for example, develops voltage across R_4 when the transistor is conducting. This voltage makes the emitter slightly negative with respect to ground. The base of the transistor is also made negative by a much larger value due to the voltage divider network composed of R_1 and R_2. Through this, the base is more negative than the emit-

ter, which results in forward biasing. Voltage variations across R_4 due to ac signal changes controlled by Q_1 are bypassed around R_4 by capacitor C_3. This reduces degeneration and maintains the emitter voltage at a constant level. Resistors R_5, R_6, and R_8, and capacitor C_4 achieve the same result for transistor Q_2.

Resistor R_3 connected to Q_1 is the collector load resistor. In practice, the resistance value of R_3 is quite large in order to achieve high levels of amplification. Capacitor C_2 is used to couple the ac output signal of Q_1 to the input of Q_2. Functionally, C_2 must pass the ac signal easily and block the passage of dc. Typical capacitor values are 50 microfarads. C_8 and R_7 of transistor Q_2 achieve the same function. Capacitors C_6 and C_7 are 0.002-μF RF noise filters. These capacitors provide a low capacitive reactance to high-frequency ac, and eliminate it by bypassing it to ground. The circuitry of Q_2 is identical to that of Q_1. After the signal has been satisfactorily amplified by the two transistors, the collector load resistor usually decreases in value and the emitter resistor increases. As a general rule, this type of design prevents distortion of the signal due to excessive amplification.

With the availability of high-gain operational amplifiers built on a single IC chip, multistage amplifiers are often replaced today with a single op amp. The primary principle of operation is very similar to that of the transistor circuit.

AC-TO-DC ELECTRONIC CONVERTER

In our representative pH instrument, we started with a dc signal from the sensor probe, converted it to ac, and then amplified it to a suitable level. The next step in this chain of events is ac-to-dc conversion. An electronic converter that achieves this operation is shown

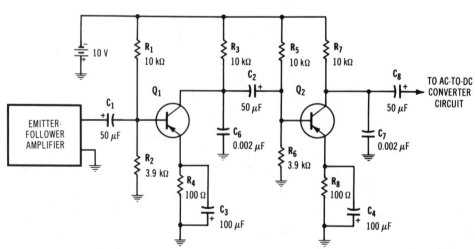

FIGURE 10-12   Two-stage *RC*-coupled voltage amplifier circuit.

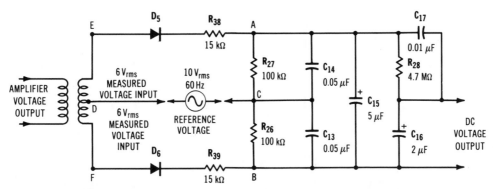

FIGURE 10-13   Ac-to-dc phase detector converter circuit.

in Fig. 10-13. This circuit is commonly called a phase detector and has a number of applications in electronic instrumentation.

A phase detector is essentially composed of two diodes that are subjected to two ac signals. One of these is the ac signal from the voltage amplifiers. This signal is the measured component to which the circuit is to respond. The second signal is a reference voltage that is placed in series with the diodes. The resulting output of the circuit is a dc voltage that is dependent on the phase relationship of the two input signals.

Three states of circuit operation will be discussed here. One of these occurs when no measured signal input is applied. The second two conditions occur when the measured signal is in phase or 180° out of phase with the reference signal. In practice, there are a number of phase variations between these two extremes.

A simplification of the phase detector converter is shown in Fig. 10-14 for the following explanation. Note that the transformer secondary winding is omitted in this circuit. In this case, we are assuming that no measuring signal is applied. Only the reference signal is applied to the circuit. This voltage is derived from a special power supply transformer winding. Typical reference voltages are 10 V rms, which produce a peak value of 14.14 V and a peak-to-peak value of 28.28 V.

With no measuring input signal applied to the phase detector, we can assume that only a reference voltage is applied to the diodes. Since the reference voltage is commonly fed between points D and C, both diodes will receive the same signal phase and voltage value at the same time. During the positive alternation, the diodes will be forward biased. Conduction occurs as indicated by the arrows and develops a voltage across R_{26} and R_{27}. No conduction occurs in either diode during the negative alternation. The direction of current in both circuits is determined by diode polarity. Current direction is, of course, from the cathode to the anode as indicated. This means that the current through R_{27} is from point C to point A in the circuit of diode D_5. The voltage at point A is, therefore, positive with respect to point C. Since D_5 offers very little resistance while conducting, the output will be the peak input voltage minus the diode voltage drop of 0.7 V, or 13.44 V. Diode D_6 produces an output voltage across R_{26}. The current through R_{26} is from point C to point B. Point B, therefore, is positive with respect to point C. The peak voltage of this output is the same as that across R_{27}. Referring back to Fig. 10-13 notice that capacitor C_{14} will charge to the peak voltage across R_{27}. Its discharge time is quite long, so for all practical purposes we will assume that the charge across C_{14} is 13.44 V. Capacitor C_{13} will charge to the

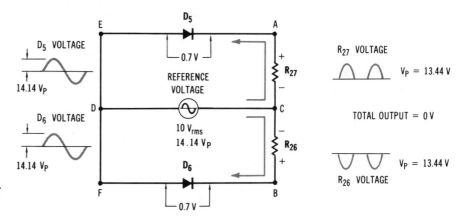

FIGURE 10-14   Phase detector with no input signal.

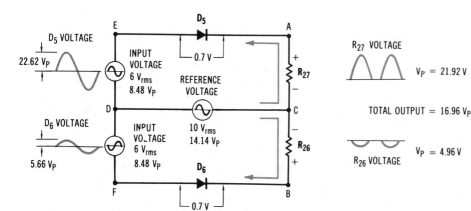

FIGURE 10-15 Phase detector with in-phase input to D_5.

voltage across R_{26}. Capacitor C_{15} will then charge to the sum of these two voltages. Since the difference of potential between these two points (A and B) is 0 V, C_{15} will have no charge.

The phase detector with a signal applied to the input is shown in Fig. 10-15. Notice that the transformer input between points E and D is in phase with the reference voltage. This, of course, means that the voltage from point D to point F is 180° out of phase with the reference voltage. An arbitrary voltage of 6 V rms, or 8.48 V peak, has been used for purposes of illustration. All voltages shown are peak values. The voltage applied to D_5 is between points E and C, which means the vector sum of the reference voltage and one-half the transformer input, since they are in series. These two voltages are in phase, so their vector sum is the direct sum of the peak voltages. In this case, it is 14.14 V plus 8.48 V, or 22.62 V peak. As mentioned previously, the diode output is developed across R_{27}. Its peak output voltage is 22.62 V minus the 0.7-V diode drop, or 21.92 V. This voltage will charge C_{14} of the original circuit.

The reference voltage of 14.14 V peak and the measured input voltage from points D to F of 8.48 V peak are applied to diode D_6. Since these voltages are

in series, they will add vectorially. In this part of the circuit, however, the voltages are 180° out of phase. The resulting voltage will, therefore, be the vector sum, which will actually be the difference of the two voltages. This means a diode voltage of 14.14 V peak plus -8.4 V peak, or 5.66 V peak. With a diode voltage drop of 0.7 V, the output across R_{26} will be 4.96 V peak. A rectified output with this amplitude is developed across R_{26}.

Regarding Fig. 10-13, you will recall that C_{13} will charge to the voltage developed across R_{26}. This voltage, in addition to that developed across C_{14}, is used to charge C_{15} to the total potential difference developed across points A and B. This would amount to 21.92 V across C_{14} plus a -4.96 V across C_{13}. Since these voltages are opposing, the total charge on C_{15} will be 21.92 $-$ 4.96 or 16.96 V peak. This essentially means that the ac voltage values applied to the phase detector will produce a corresponding value of dc voltage.

When the phase of the measuring input signal is reversed, the process changes, as indicated in Fig. 10-16. The voltage between points E and D is now 180° out of phase with the reference voltage. A peak voltage of 5.66 V is applied to diode D_5. The output across R_{27}

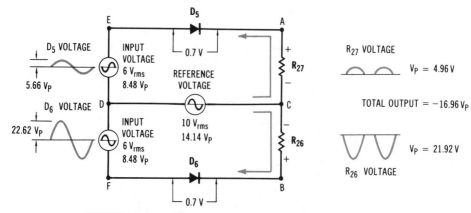

FIGURE 10-16 Phase detector with in-phase input to D_6.

is approximately 4.96 V. C_{14} will therefore charge to this value. At the same time, the reference voltage and the voltage from point D to point F are in phase. This causes a combined voltage of $8.48 + 14.14$, or 22.62 V peak to be applied to D_6. The rectified output across R_{26} has a peak value of 21.92 V. C_{15} will now charge to the combined peak voltages across C_{13} and C_{14}. The total charge appearing across C_{15} is 4.96 V due to C_{14} plus a -21.92 V due to C_{13}, or -16.96 V (point A with respect to point B). This essentially means that the second half of the measured input ac voltage produces a dc voltage value that is equivalent to the first alternation of the input, but of opposite polarity.

In normal operation, the value of the measured input voltage changes according to the pH level of the solution being tested. As a result of phase detector operation, a dc voltage is developed that can effectively be used to drive a recorder or that can be measured by a meter to indicate specific pH values.

FEEDBACK LOOP

It has been shown that a positive input produces a negative output from the amplifier. Figure 10-17 is a simplified schematic of the feedback loop. The negative output of the amplifier causes a current to conduct through R_{25}, the meter, R_{31}, R_{24}, S_1, R_{32}, and R_{44} in the direction shown. The amount of current depends on the output of the amplifier. This current is registered on the pH meter. The voltage drop across R_{31} also produces an output proportional to the feedback current for a recorder indication. S_1 is the adjustment for

manual temperature compensation. If automatic temperature compensation is used, S_1 and R_{32} are switched out of the circuit and a thermistor is switched in. This switching is accomplished by an auto-manual switch. The termistor automatically changes resistance with temperature variations.

A negative voltage is developed at the wiper arm of potentiometer S_1. This voltage determines the operating voltages of the standardization bridge. The battery indicates the voltage provided by the regulated power supply. S_2 is set with a standard buffer solution. This determines the operating curve of the amplifier. If the potentiometer were set at the center position, the voltage across the bridge would be zero. The voltage at wiper arm A would be the same as at wiper arm B. S_2 can be adjusted so that point A can be slightly more positive or more negative than point B. The negative voltage at point A produces a current through the reference half-cell of the electrode. This produces a voltage drop that causes the circuit common to go more positive, which subtracts from the input to the amplifier. This is, of course, negative feedback.

DIFFERENTIAL INPUT PREAMPLIFIER

In pH measurement, an unusual problem exists where any extraneous leakage currents to ground find a return path through the low-resistance reference electrode. With a single input amplifier, this extraneous voltage adds to the pH potential resulting in a significant error in the determined pH level. When the resistance of the reference electrode is relatively low,

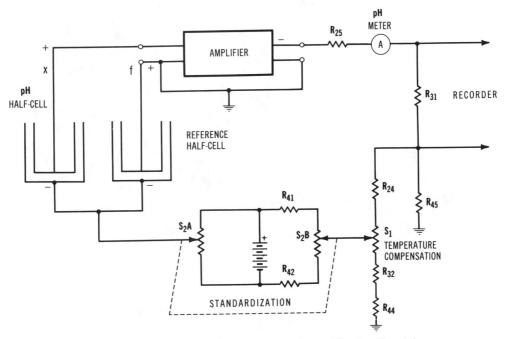

FIGURE 10-17   Simplified schematic of amplifier feedback loop.

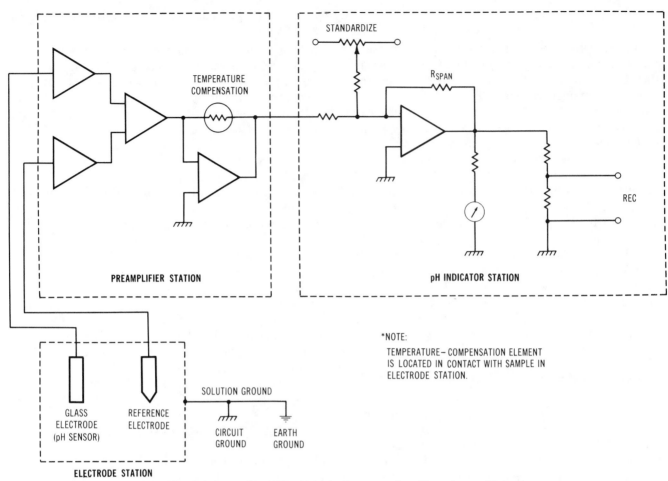

FIGURE 10-18 Simplified block diagram of a pH analyzer. (Courtesy of Beckman Instruments, Inc.)

FIGURE 10-19 Differential amplifier assembly of a pH analyzer. (Courtesy of Beckman Instruments, Inc.)

the amount of error is insignificant. Should there be an increase in electrode resistance due to coating or junction clogging, the voltage increases a great deal, which produces a significant error in the indicated pH level.

In pH instruments, extraneous leakage currents are reduced by connecting a preamplifier in both the pH sensor and reference electrode input. With this type of input a negligible amount of current conducts through the electrodes. The glass electrode only measures pH potential through the solution to ground while the reference electrode measures the voltage through the solution to ground.

The outputs of the preamplifiers are then applied to a differential amplifier circuit. The function of this amplifier is to cancel solution potentials and produce a signal that is equal to only the pH potential. The simplified block diagram of Fig. 10-18 shows the preamplifier/differential-amplifier unit. In practice, the entire circuit assembly is built on a printed circuit board and built into the top of the electrode assembly. Figure 10-19 shows a view of the assembly.

Signals from each electrode are applied to the noninverting input of a separate, high-gain op amp. With op-amp circuitry of this type, the input sees a high impedance. After a high-level gain has been achieved, each output is applied to the input of the differential amplifier. The pH signal is applied to the inverting input and the reference electrode signal is applied to the noninverting input. The output is thus representative of signal difference. The difference signal output is again amplified by a high-gain op amp. In this case, the amplifier is temperature compensated by a sampling signal from the solution under test. In practice, the temperature compensation signal is derived from an electrode that is in contact with the pH sample electrode.

A number of unique advantages in pH measurement are present in the circuit. These include elimination of drift and noise, reduced maintenance problems, remote pH instrument placement, and elimination of costly interconnecting coaxial cables.

DC AMPLIFICATION/CURRENT OUTPUT

Figure 10-20 is a schematic diagram showing the dc-amplifier/current-output circuitry of a pH analyzer. In this circuit, the dc output signal of the preamplifier/differential amplifier is applied to the input of the circuit at terminal 3 of terminal strip P_1. Dc amplification, a comparator amplifier, dc-to-dc inversion, and current output are all included in this part of the pH analyzer. The current output is ultimately used to indicate pH levels on a hand-deflecting instrument or to drive a chart recorder. The entire circuit is built on a printed circuit board and is called the current output module.

DC Amplification/Zero Adjust

Op amp AR_1 is a dc amplifier that has the dc signal applied to its noninverting input. Potentiometer R_6 is the zero adjust control used to compensate for circuit voltage variations that are generally due to input loading. Resistors R_1 and R_2 determine the total gain of the amplifier. Voltage amplification (A_V) is based on the ratio of R_1 to R_2 and is determined by the formula

$$A_V = \frac{R_1 + R_2}{R_1}$$

Comparator Amplifier

Op amp AR_2 is a comparator amplifier. It essentially compares the voltage levels of signals applied to its two inputs. The dc pH signal is applied to the noninverting input and a reference of feedback signal is applied to the inverting input. This op amp simply compares the measured pH signal with the reference and produces an output. If the pH measured input is larger than the reference, a positive output will appear at pin 6. If the input is smaller in value than the reference, a negative output will occur.

DC-to-DC Inverter

The term *inverter* is a common way of describing the primary function of components Q_1, Q_2, T_1, CR_1, and CR_2. In general terms, dc is first changed to ac by an oscillator composed of Q_1 and Q_2. The ac signal is then stepped up by transformer T_1. The two secondary winding voltages of T_1 are then rectified and filtered by CR_1–C_6 and CR_2–C_5, respectively. The resulting dc signals are either used to form a feedback signal, as in the case of CR_2-C_5, or to drive the output circuit, as achieved by ,CR_1-C_6.

The operation of a circuit similar to the dc-to-dc inverter was discussed in conjunction with the power supply of Chapter 9. The inverter of Fig. 10-20 is similar in operation to that of the previous circuit with the primary differences being input voltage values and application of the output circuitry. The circuit is called a dc-to-dc inverter here and was described as a dc-to-ac converter in Chapter 9.

Current Output

The current output section of the module is achieved by op amp AR_3, transistor Q_3, and the bridge rectifier CR_3. The circuit indicated has a 10- to 50-mA output capability. The current span is adjusted by R_{17}.

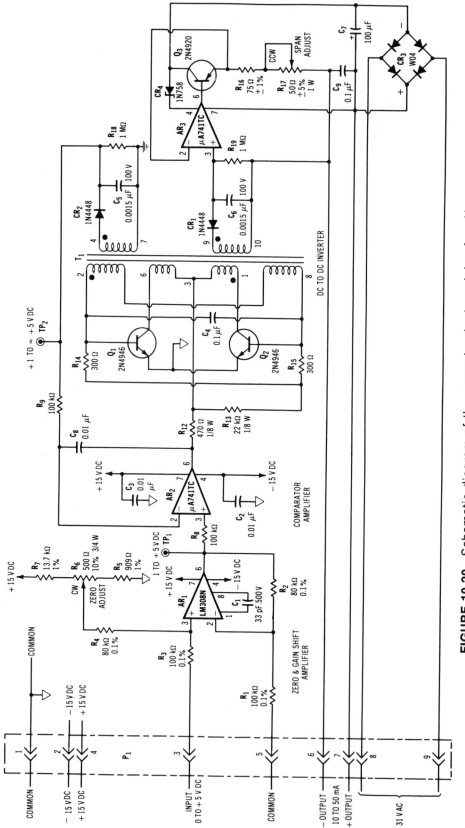

FIGURE 10-20 Schematic diagram of the current-output module of a pH analyzer. (Courtesy of Beckman Instruments, Inc.)

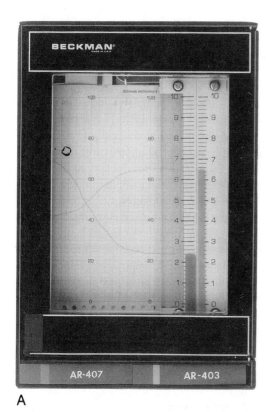

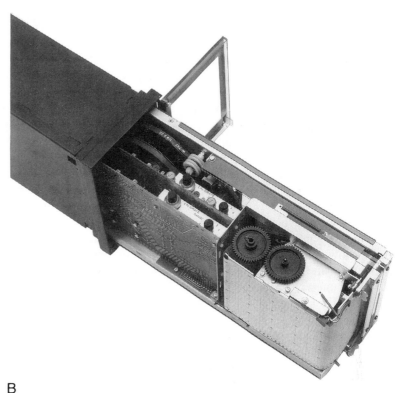

A   B

FIGURE 10-21   Strip-chart recorder pH analyzer. (Courtesy of Beckman Instrument, Inc.)

Negative dc voltages are developed by CR_1, filtered by C_6, and applied to the noninverting input of op amp AR_3. The op amp responds in this case as a noninverting amplifier. The negative dc output voltages appearing at pin 6 of AR_3 are connected directly to the case of pnp transistor Q_3. These voltages forward bias the base–emitter junction of Q_3. The collector–base junction of Q_3 is reverse biased by the dc output voltage of bridge rectifier CR_3. Higher negative base voltage values produce increased current output, while lower negative base voltages produce decreased current output. The span adjusting control, R_{17}, alters the emitter current conduction level of Q_3. When R_{17} is reduced in resistance, the current span is raised;

FIGURE 10-22   pH analyzer with hand-deflection indicating meter. (Courtesy of Beckman Instruments, Inc.)

when R_{17} is increased in resistance, the current span is lowered.

The bridge rectifier, CR_3, serves as a full-wave, high-current source for the output circuit. An ac voltage of 31 V from the main power transformer is supplied to the bridge through external connecting wires at terminals 8 and 9 of P_1. The bridge then rectifies the ac and filters it with C_7. Approximately 27 V of dc are applied to Q_3. With respect to the common ground, the dc output of CR_3 is approximately $+13.5$ V and -13.5 V. This voltage is also used to supply op amp AR_3. Zener diode CR_4 is used to regulate the -13.5-V side of the source.

The current output of the pH analyzer may be used to drive an indicating meter or a recorder. Figure 10-21 shows a strip-chart recorder pH analyzer, and Fig. 10-22 shows a pH analyzer with a hand-deflection indicating meter.

SUMMARY

The measurement of pH determines if a solution is either acid or alkaline (base). Specifically this refers to the number of positive hydrogen ions or negative hydroxyl ions in a solution.

The pH level of a solution is measured by the amount of dc voltage developed between two electrodes immersed in the solution under test. On a scale of 0 to 14, 7 is an indication of a neutral solution. The numbers from 7 down to 0 represent acid levels, while the numbers from 7 to 14 represent base levels.

The probe of a pH instrument is often called a half-cell or a cell. It develops a dc voltage when placed in a solution. A measuring half-cell and a reference half-cell may be combined in one probe or may be independent.

A pH instrument employs a measuring circuit, a standard circuit, dc-to-ac conversion, emitter-follower amplification, voltage amplifiers, phase-detector converters, and temperature compensation.

A split dc power supply is used to provide the operating voltages for the pH indicator. In this circuit, a bridge rectifier is used with series transistor regulators. The series transistor regulator responds as a variable resistor that changes value according to circuit operating conditions. An increase in the value of R_L is compensated by an increase in transistor resistance which maintains the output voltage at a constant value. A decrease in R_L would normally cause a drop in V_o. To compensate for this, conduction of the series transistor is increased to reduce its series resistance. As a result, V_o is increased. Through variations in transistor resistance the output voltage is maintained at a constant value.

In the differential input preamplifier, individual op amps are used in each of the two inputs. These high-impedance op-amp inputs are used to reduce extraneous leakage voltages to ground. The output is then applied to a differential amplifier. The pH electrode signal goes to the inverting input and the reference electrode signal goes to the noninverting input. The signal difference is then amplified to reduce drift, noise, and temperature variations.

An op-amp dc amplifier is connected to the output of the differential amplifier. The gain of this amplifier is determined by the formula $A_V = (R_1 + R_2)/R_1$.

The comparator amplifier is used to determine the levels of signals applied to its two inputs. The dc pH signal is applied to the noninverting input and a reference or feedback signal is applied to the inverting input. A comparison of the two signals is made so that when the pH signal is larger a positive output will occur. If the feedback/reference signal is greater a negative output will occur.

The pH instrument of this chapter employs a dc-to-dc inverter. Essentially this is similar to the dc-to-ac converter discussed previously.

The current output is achieved by an op amp and a power transistor. An output of 10 to 50 mA is typical for this instrument.

ACTIVITIES

pH INSTRUMENT CALIBRATION (One-Point)

1. Refer to the manufacturer's operational manual for specific calibration procedures that apply to the instrument used in this activity.

2. The temperature setting on the meter should correspond to the temperatue of the buffer solution used, or the instrument may use an automatic temperature compensator (ATC).

3. Turn the pH instrument to manual or ATC according to the temperature procedure being followed.

4. Place a clean electrode into a fresh, room-temperature, pH 7.00 buffer solution.

5. Adjust the pH reading to 7.00 using the zero-offset, standardize, or set control. This is considered to be a one-point calibration procedure.

6. Rinse the electrode in distilled water or clean deionized water.

7. As a rule, pH instruments should be calibrated at least once during an 8-hour shift.

pH INSTRUMENT CALIBRATION (Two-Point)

1. Follow the seven instruction steps for one-point calibration of the pH instrument.

2. Place the clean electrode into a second prepared buffer solution. It can be pH 4.00 or pH 10.00, depending on its intended measuring application. Some instruments may have a built-in buffer solution memory for automatic calibration.

3. Place the instrument in the CALIBRATE position.

4. Adjust the GAIN control to set the pH reading to the value of the buffer solution.

5. If the instrument is equipped with a SLOPE or DUAL SLOPE control, adjust it to display the correct value of the buffer solution.

6. Rinse the electrode in distilled water or clean deionized water.

7. Two-point calibration should be performed at least once per day.

MEASURING pH

1. Refer to the manufacturer's operational manual before attempting to measure a test solution.

2. Calibrate the instrument.

3. If the instrument is equipped with automatic temperature compensation, it will adjust the feedback and slope of its amplifiers according to the temperature of the test solution. If the instrument has manual temperature compensation, the solution temperature is measured and a manual control is adjusted to match the solution temperature. Set the temperature compensation to automatic or manual, and adjust the compensator control or note the automatic adjustment of the instrument.

4. Insert the pH electrode into the test solution. The probe should be positioned vertically in the solution or stand in an upright position.

5. Locate the indicator and read the pH value of the solution.

6. Clean the electrode with distilled or deionized water and test a different solution.

7. Test several samples and record the type of solution and its pH value.

8. Clean the electrode and turn off the instrument.

QUESTIONS

1. What does the number 7.00 indicate on a pH scale?

2. What part of the pH scale is used to denote a high-acid solution?

3. A moderate-base solution would be located on what part of the pH scale?

4. What does the reference electrode of a pH instrument measure?

5. What does the sensing or measuring electrode of a pH instrument determine?

6. What kind of signal is developed by the pH sensing probe?

7. What is the primary function of a phase detector in a pH instrument?

8. How is an acid solution determined by the pH electrode?

9. How does a base solution differ from an acid solution?

10. What is the difference between the reference electrode and the pH sensing electrode?

11. What is connected to the pH sensor and reference electrode inputs to reduce leakage currents?

12. What is the function of the differential amplifier of a pH instrument?

13. What is the function of the comparator amplifier of a pH instrument?

14. Why are the probes of a pH instrument often called half-cells?

CHAPTER 11

FLOWMETERS
AND CONVERTERS

OBJECTIVES

Upon completion of this chapter, you will be able to:

1. Explain the operating principle of a magnetic flowmeter.
2. Identify the functional blocks of the converter of a magnetic flowmeter.
3. Explain the function of selected blocks of the converter.
4. Define flow.
5. Explain the operating principle of a vortex flowmeter.
6. Identify different electronic flowmeters.

IMPORTANT TERMS

In this chapter we investigate the basic operating principles of flowmeters. Through this material you will have an opportunity to become familiar with instrument functions and operational theory. New words, such as *volumetric fluid flow*, *velocity flow*, *vortex*, *rate flow*, and *totalizing flow*, will begin to have some meaning for you. A few of these terms are singled out for further study. As a rule, the chapter will be more meaningful if these terms are reviewed before proceeding with the text.

Amplitude modulation (AM): A frequency mixing process in which the amplitude changes of a low-frequency signal are mixed or modulated with a higher-frequency signal.

Faraday's law: A law stating that voltage is induced into a conductor when it moves at right angles through a magnetic field.

Helical: A spiral form or coil-shaped structure.

Optical coupler: An electronic device that consists of a light source and a light detector. The light source is generally a light-emitting diode (LED). The light detector can be a phototransistor. This device is used to couple a signal between two circuits without a direct electrical connection.

Orifice: A hole, aperture, or opening in an object.

Piezoelectric element: A crystal sensing device that produces ac as a result of strain due to changes in pressure.

Quadrature voltage: An ac voltage that is 90° out of phase with a reference voltage.

Rate flow: A measuring procedure that determines the amount of fluid that moves past a given point at a particular instant.

Single-ended output: A stage of output amplification that employs one active device, such as a transistor or MOSFET, to control the output.

Strut: An obstruction bar placed in a flow path that produces swirls or vortices.

Swirl: A twisting, whirling mass or area such as that produced by a whirlpool.

Synchronous demodulator: An electronic circuit or device that picks out signal information or data from a composite signal when two or more signals are in step or synchronized.

Thermistor: A resistive device that changes its value with temperature.

Torsion: The twisting motion or movement of a body or material that is the result of an applied force at one end or location.

Totalizing flow: A measuring procedure that determines the total amount of material that flows through a system.

Transformer action: An operating condition where energy is transferred by an interaction of fields between two or more coils.

Turbine: A bladed assembly that produces rotary motion as a result of a flow stream passing through the blades.

Ultrasonic: An effect or device that vibrates or responds to a frequency that is 20 kHz or higher.

Velocity fluid flow: A condition that refers to the speed at which a stream passes through a closed pipe.

Volumetric fluid flow: A measure of the total volume of material passing through a pipe or tube.

Vortex: The center area of flowing fluid that twists or rotates at a high velocity.

Vortex shedding: A flow principle in which swirls are released from an obstruction placed in the flow path.

INTRODUCTION

Nearly all of the products manufactured by industry today are influenced in some way by the flow of materials. This type of processing ranges from the simple flow of fuel in a heating system to the control of large amounts of oil or gas passing through the pipeline of a regional area of the country. The measurement and control of flow is an essential function of industrial process control.

One way of classifying equipment associated with the flow process is to place it in two rather general divisions. Flow-rate instrumentation is one major division. This type of equipment is used to determine the amount of fluid that moves past a given point at a particular instant. Measurements range from liters per second to thousands of barrels per hour. Totalizing flow measurement is the other classification. This determines the total amount of flow that passes through a system. A gasoline pump indicates the total amount of gas that is removed from the storage tank. This is measured in gallons, liters, and barrels.

Totalizing and rate-flow instrumentation have a rather wide range of applications in industry. In general, the material that flows in the system is very important. Liquids, gases, and solid particles suspended in liquid are typical materials that respond to flow instrumentation. As a rule, some degree of instrument modification must take place in order for the instrument to respond to more than one type of material flow.

In this chapter we look at flow instruments that produce an electrical signal in response to a physical change in the amount of material passing through a pipe or tube. This includes the magnetic flowmeter and the vortex flowmeter.

MAGNETIC FLOWMETERS

A magnetic flowmeter is an instrument designed to measure the volume flow rate of electrically conductive fluids passing through a pipe or tube. This instrument is particularly applicable for the measurement of fluids which are somewhat difficult to handle. Corrosive acids, sewage, detergents, tomato pulp, crude oil, and paper pulp are common applications. An assembled magnetic flowmeter/converter is shown in Fig. 11-1.

A partial cutaway view of the magnetic flowmeter is shown in Fig. 11-2. Notice particularly the location of the electrode assembly, magnetic coils, and the nonmagnetic flow tube. This assembly is a volumetric fluid transducer that changes conductive fluid flow into an induced voltage when fluid flows through a magnetic field. The amplitude of the generated signal is directly proportional to the flow rate of the fluid.

Magnetic flowmeters are considered to be obstructionless metering instruments. An inherent advantage of this principle is that pressure losses are reduced to levels occurring in equivalent lengths of equal-diameter piping. This essentially reduces and conserves pressure source requirements compared with other metering methods.

Figure 11-3A shows a schematic representation of the magnetic flowmeter, while Fig. 11-3B is a pictorial representation of the basic operating principles involved. The flowmeter is constructed around a section of pipe that requires no orifices or obstructions within it. This means that the flowmeter itself has very little effect on the flow rate of the fluid through it. Around the metering section of pipe are wound two field coils: one above the piping, the other below it. When current passes through these coils, a magnetic field is produced in the direction shown in Fig. 11-3A. The flowmeter section of the pipe has two electrodes positioned so that the fluid passing between them is perpendicular to the magnetic field of the coils. Figure 11-4 shows these electrodes on each side of the pipe assembly. The electrodes are electrically insulated from the walls of the tubing.

We will first consider the operation of the flowmeter when a constant magnetic field is applied. Under these conditions, the operation of the flowmeter is based on Faraday's law of electromagnetic induction. Simply stated, the voltage induced across a conductor

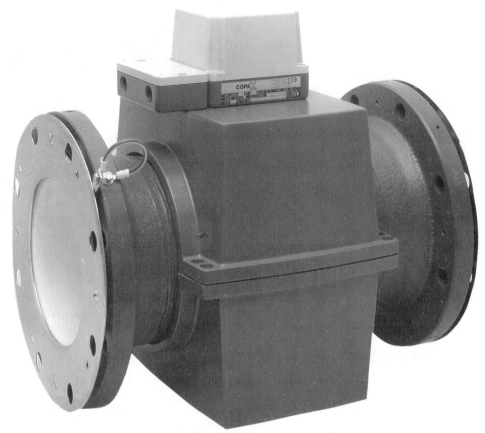

FIGURE 11-1   Assembled flowmeter. (Courtesy of Fischer & Porter Co.)

moving at right angles through a magnetic field is proportional to the velocity of that conductor, or

$$V = \frac{1}{c} Bdv$$

where c is a dimensional constant, B the flux density of the magnetic field, d the inside diameter of the pipe, and v the velocity of the conductor. The conductor in this case is that element of fluid that lies directly between the two electrodes. As fluid passes through the pipes, it is moving perpendicularly through the magnetic field B. The value of B depends on the amount of current through the windings of the field coil. For the present, this will be considered as a constant. The inside diameter of the pipe is, of course, also a constant. For most liquids with a reasonable conductivity, the constant c is considered to be 1. As a result, V is dependent only on the velocity of the conductor and is directly proportional to it. The induced voltage V is the flowmeter output, which is used as the amplifier input. This input is directly related to the velocity of the fluid through the flowmeter.

The magnetic field B, however, is not constant. An ac voltage is applied to the field coils. This means simply that the output voltage will be an ac voltage rather than a dc voltage. Its magnitude will still be proportional to the velocity of the fluid. The desired signal output of the flowmeter will also be in phase with the magnetic field current. That is, when B is maximum, maximum voltage is induced. B is maximum when the current through the field coils is maximum.

Since an ac voltage produces the magnetic field, some unwanted voltage will be produced when there is no flow. When there is fluid in the flowmeter and no flow, a stationary conductor is placed between the two electrodes. A voltage will be produced between these two electrodes, due to the changing field. In other words, this voltage is produced by transformer action. This voltage is maximum when a maximum change in the magnetic field occurs. Maximum change in the field is produced by maximum change in current, which occurs in an ac circuit when the current is passing through zero. Therefore, this unwanted signal is 90° out of phase with the desired signal. The feedback circuit will provide for elimination of this signal.

Another unwanted signal is also produced by transformer action. This voltage will be induced into the electrodes and their leads. These leads are formed into a twisted pair and shielded where possible to min-

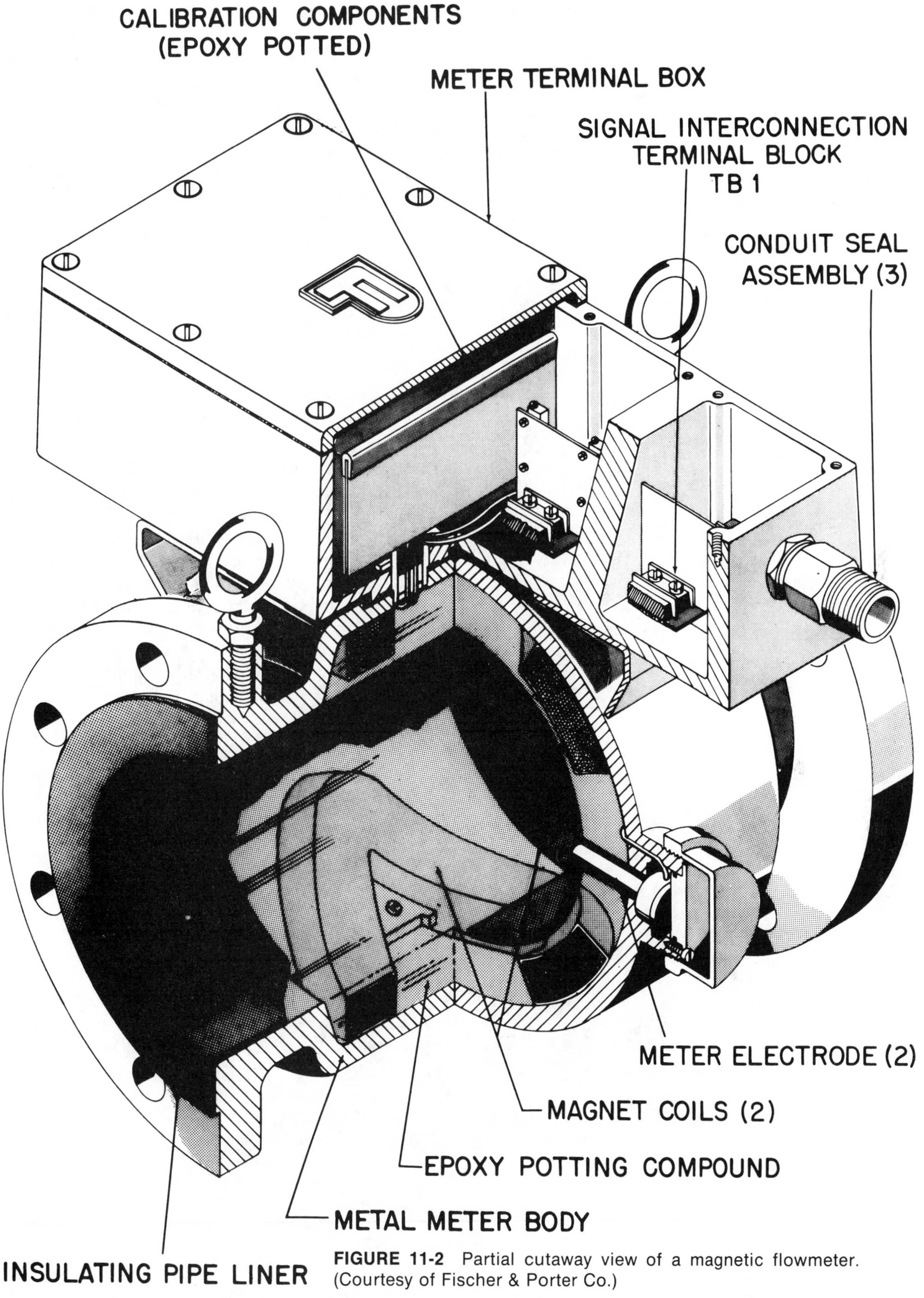

CALIBRATION COMPONENTS
(EPOXY POTTED)

METER TERMINAL BOX

SIGNAL INTERCONNECTION
TERMINAL BLOCK
TB 1

CONDUIT SEAL
ASSEMBLY (3)

METER ELECTRODE (2)

MAGNET COILS (2)

EPOXY POTTING COMPOUND

METAL METER BODY

INSULATING PIPE LINER

FIGURE 11-2 Partial cutaway view of a magnetic flowmeter.
(Courtesy of Fischer & Porter Co.)

193

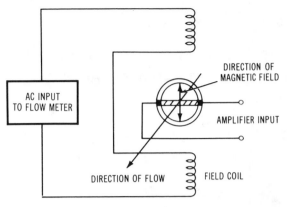

(A) Schematic representation.

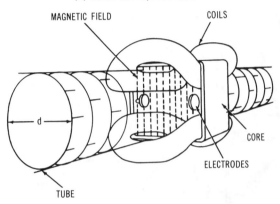

(B) Pictorial representation.

FIGURE 11-3   Magnetic flowmeter.

imize these unwanted signals. They must, however, be exposed near the electrodes and cannot be formed in a twisted pair around the pipe. This voltage must also be 90° out of phase with the desired signal. All signals 90° out of phase with the desired signal are called quadrature voltages. They will be canceled out in the feedback circuits and only the amplified signals will be proportional to the flow rate.

A zero-flow-rate signal is also developed in the feedback circuitry. This signal will compensate for any in-phase voltages fed into the amplifier at a zero flow rate. It will also compensate for any change in the magnetic field of the flowmeter due to changes in line voltages.

FLOWMETER SIGNAL CONVERSION

The signal converter of a magnetic flowmeter can be an integrally mounted assembly or a remote assembly housed in an independent enclosure. There is some difference in the circuitry of these two assemblies. Integrally mounted units are generally smaller and have optional features built on independent circuit boards. The remote units are more sophisticated and somewhat larger. Remote assemblies generally have more optional features and can be used with a number of different flowmeter units. The remote unit of Fig. 11-5 is a microprocessor-based signal converter. This unit is

FIGURE 11-4   Pipe assembly of a flowmeter. (Courtesy of Fischer & Porter Co.)

stitution is generally more economical than stocking a large variety of discrete components, ICs, transistors, and diodes. As a rule, troubleshooting is accomplished by observing signal waveforms and voltage measurements at key test points on the circuit boards. In working with this instrument, a technician must have some understanding of what should appear at the key test points. The following functional description of the circuitry is designed to assist the technician in understanding the general operation of a magnetic flowmeter.

FLOWMETER BODY

The flowmeter body is primarily responsible for signal development. The flowmeter body houses two signal electrodes and the electromagnetic flux producing coils as shown schematically in the block diagram of Fig. 11-6. The signal electrodes produce flow-rate information which is processed by the converter assembly. A reference signal, which is proportional to the coil exciting current, is also produced by the flowmeter body. The reference signal is derived from a resistance network that is connected in series with the magnet coils. This signal senses a change in coil drive voltage. It is used to compensate for variations in the magnetic coil voltage.

The electromagnetic coil assembly of the flowmeter body is energized with a low-frequency square-wave drive signal. The excitation rate of this signal is controlled by a gate circuit in the converter. Note the line frequency trigger and gate logic division blocks that are attached to the magnet driver. This part of the converter changes 60 Hz into a symmetrical square wave of 3.75 Hz. In this case, 60/16 = 3.75 Hz. The magnet coil drive signal can be observed at terminal M_1 of the output circuit board.

SIGNAL CONVERTER BOARD

The converter board of a flowmeter is primarily responsible for signal processing, logic gating, demodulation, frequency conversion, line-frequency triggering, feedback multiplication, and gain adjustments. A single circuit board accommodates all of this circuitry. It has selected test points which are identified by the letters TP and a number. An assembled board is equipped with a terminal block that permits it to be interconnected with other parts of the instrument. The block diagram of Fig. 11-6 will serve as a reference for the following functional description of the converter board.

FIGURE 11-5 Microprocessor signal converter. (Courtesy of Fischer & Porter Co.)

of the pulsed dc type. It supplies a pulsed constant current dc signal to the magnet coils of the flowmeter to establish the magnetic field. The frequency of the pulse signal is field selectable. A standard frequency is 7.5 Hz. A 15-Hz pulse signal is available as a noise reduction feature. This unit can have a display, be equipped to monitor flow in either direction, serve as a data link, and interface with a large data-base system.

A functional block diagram of an integrally mounted converter assembly of a magnetic flowmeter is shown in Fig. 11-6. This assembly has a converter board, an output/power supply board, and the flowmeter body. The converter circuit drives the magnet coils with 3.75 Hz from the 60-Hz line voltage. The entire circuit board assembly is energized by 120 V ac. It has an optional source capability that permits it to be energized with 12 to 24 V dc. An output signal developed by the unit can be either 4 or 20 mA of current or a frequency of 0 to 1 kHz. The entire electronics assembly is housed in a moisture-free enclosure.

The functional operation of the converter of Fig. 11-6 is accomplished with a number of integrated circuits. Some of the ICs are standard packaged items, while others are customized units. The circuit boards of this assembly can be easily replaced or exchanged in the field if the need arises. Servicing by board sub-

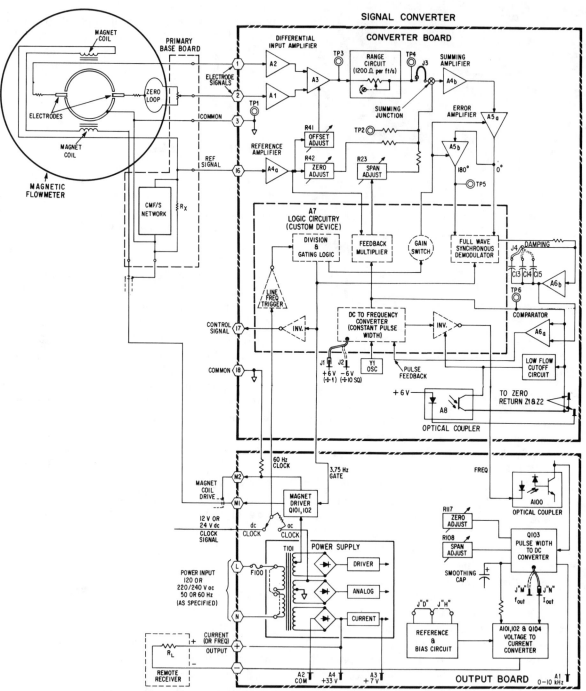

FIGURE 11-6 Functional block diagram of a magnetic flow-meter/converter. (Courtesy of Fischer & Porter Co.)

Differential Input Amplifier

When the coil assembly of the flowmeter body is energized, flow velocity will cause voltage to be induced into the two signal electrodes. Electrode signal voltage will appear at terminals 1 and 2 of the converter board. This signal is then applied to the noninverting inputs of operational amplifiers A_1 and A_2. The amplified output is then applied to the two inputs of differential amplifier A_3. This amplifier is used to convert the differential input signal to a single-ended output. The output signal appears at TP_3 of the converter board. The approximate gain of the input amplifier circuit is 10. This signal is directly coupled to the range control. The adjustable operational range of the circuit appears at TP_4.

Reference Amplifier

The reference amplifier of the converter is identified as A_{4a}. Its input is energized by a voltage reference signal that is developed from the magnet coil drive current. The reference signal appears at terminal 16 of the converter board. It is connected to the inverting input of operational amplifier A_{4a}. This op amp is a low-noise, low-power, wide-bandwidth amplifier that is equipped with a JFET input. The applied signal is amplified by a factor of 8. Its output is supplied to the feedback multiplier of A_7. The same output signal is applied to the offset adjust and zero adjust. The zero-adjust output is combined with the feedback multiplier output signal and connected to the summing junction. TP_2 shows the combined signals of this configuration.

Summing Circuit

The summing amplifier is responsible for measuring and amplifying the difference between the range adjust signal and the feedback multiplier signal. The summing amplifier is identified as A_{4b}. The inverting input of this amplifier serves as the summing junction. Its output is ac coupled to the noninverting input of the 0° error signal amplifier.

Error Amplifiers

Two error amplifiers appear in the circuit at this point. Amplifier A_{5a} is responsible for in phase or 0° amplification. The summing amplifier's output is connected to the noninverting input of the in-phase amplifier. This amplifier has a gain factor of 2000. Its gain is controlled by the gain switch of A_7. The high gain of A_{5a} and switching function of A_7 lets only the last half of each cycle appear as output. This means that only the steady-state part of the signal will produce an output. This is done to eliminate undesired noise and transient components. The output of the in-phase amplifier is applied to the inverting input of error amplifier A_{5b}. This amplifier serves as the out-of-phase or 180° error amplifier. The in-phase and out-of-phase error signal outputs of A_{5a} and A_{5b} are connected to the full-wave synchronous demodulator of integrated circuit A_7.

Comparator/Amplifier

The output signal of the synchronous demodulator of A_7 is processed by amplifier A_{6b}. This amplifier responds as an integrator. It removes the time-varying component of the signal and amplifies the dc component. The output of A_{6b} is coupled directly to the noninverting input of comparator A_{6a}. Simultaneously, a pulse signal is supplied to the inverting input of A_{6a} from the frequency converter of A_7. The output of comparator A_{6a} controls the logic level of a signal applied to the pulse feedback circuit of A_7. This signal controls the frequency converter. The duty cycle of the frequency is proportional to the input signal and pulse-width multiple of the crystal oscillator Y_1. The output of the converter board is developed by the dc-to-frequency converter of A_7.

OUTPUT/POWER SUPPLY BOARD

The output/power supply circuit board of a magnetic flowmeter has a dual responsibility. The converter board serves as an input signal source for the output circuit. The output can be current or frequency depending on a selected jumper connection. The power supply is responsible for providing dc and low-frequency ac to the flowmeter body and converter circuit. The power supply is energized by 120 or 240 V ac at 60 Hz. With some modification, the circuit can be energized by 12 or 24 V dc. The magnetic flowmeter block diagram of Fig. 11-6 will be used as a reference for the following functional description.

Power Supply

The power supply of a magnetic flowmeter is normally energized by the ac power line. The frequency of the power line can be either 60 or 50 Hz. This circuit also has a power transformer that is equipped with a dual primary winding. It can be adapted for operation with 120 or 220/240 V ac by selecting the appropriate jumper connections on the circuit board. The line voltage is stepped down to some rather small voltage values by the power transformer. Three independent secondary windings are used to develop the transformer's output. The top winding develops energy to drive the magnet coils. The 60-Hz secondary frequency is divided by 16 and appears as a 3.75-Hz drive signal. MOSFETs Q_{101} and Q_{102} of the magnet driver are switched on and off to control driver signal. This signal appears at terminals M_2 and M_1. The middle secondary winding is used to develop the $+6$ and -6 V dc for the active components of the converter and output circuits. This is identified as the analog supply of the block diagram. The bottom secondary winding of the transformer is used to develop current for the output circuit. The current output appears at the $+$ and $-$ terminals of the circuit board. This part of the power supply also has a voltage output that has the common connection at A_2, $+33$ V at A_4, and $+7$ V at A_3. All of the power supply functions, such as rectification, filtering, and regulation, are achieved on the circuit board. The assembled board is generally housed in the flowmeter body.

Output Circuitry

The output of a magnetic flowmeter has two operations that are accomplished through its circuitry. The 0- to 10-kHz output signal of the converter board is connected to optical coupler A_{100} on the output board. A_{100} is used to isolate the converter circuit from the output circuit. The converter signal is then connected to the pulse width to dc converter transistor Q_{103}. This circuit is responsible for changing the converter voltage signal into a frequency output signal. To perform this operation, jumper J from Q_{103} must be connected to the M connection. This feeds the converter signal voltage of Q_{103} directly into the gate of the MOSFET power amplifier Q_{104}. The output frequency then appears at the $+$ and $-$ terminals of the output board. This will develop an analog current signal that changes frequency from 0 to 1 kHz for retransmission to some remote instrument. For frequency measurement of the output, a resistor from 150 to 750 Ω must be connected between the $+$ and $-$ output terminals.

The alternate function of the output stage is converter signal conversion to current output. This is accomplished by changing jumper J from Q_{103} to the N position. This disconnects the pulse width-to-dc converter from the gate of Q_{104}. The signal now takes an alternate path through the voltage-to-current converter. Op amps A_{101} and A_{102} accomplish this operation. The output of A_{102} returns the signal to the gate of the MOSFET power amplifier Q_{104}. This MOSFET is now capable of delivering a constant-current signal into any load between 0 and 900 Ω. The current output signal is delivered to the $+$ to $-$ output terminals. These same two terminals also serve as the frequency output of the power amplifier. A representative current output signal can be measured by connecting a meter in series with a load resistor attached to the output terminals.

VELOCITY FLOWMETERS

Instruments that respond to the speed at which a stream passes through a closed pipe are classified as velocity flowmeters. This type of instrument is capable of high-performance measurements that are useful in precision blending operations and billing applications. It has linear operational characteristics with respect to the volume flow rate. A number of electronic instruments are placed in this general classification. The magnetic flowmeter is an instrument that uses the velocity flow principle in its operation. In addition to this, we have turbine instruments and vortex flowmeters.

Turbine Flowmeters

Turbine flowmeters are commonly used for accurate liquid measurement applications. The measuring element of this instrument consists of a multiple-bladed rotor assembly mounted inside a pipe. The imparting force of the flowing stream produces rotary motion on the turbine blade assembly. The rotational speed of the turbine is a direct indication of flow rate and can be sensed by a magnetic pickup or a gear-drive assembly. A simplification of the turbine flowmeter is shown in Fig. 11-7.

The resulting output signal of the simplified turbine flowmeter is developed by the signal pickoff coil. In the center of the pickoff coil is a small permanent magnet. The resulting field of the magnet is influenced by movement of the turbine rotor blades. When one of the blades passes through the magnetic field, it causes distortion by altering the path of the field. This change in reluctance causes the field to cut across the sensing coil, which generates a voltage by the induction process. Essentially, motion of the turbine blade generates a square-wave signal voltage into the pickoff

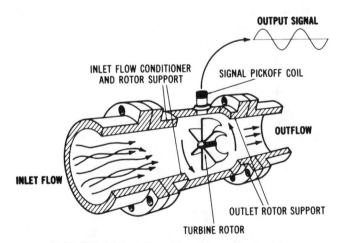

FIGURE 11-7 Simplification of a turbine flowmeter.

coil. The frequency of the generated voltage is used as an indication of flow rate. The pickoff coil assembly is commonly described as a variable-reluctance transducer. This type of signal sensor is used in a majority of the turbine flowmeters.

Vortex-Swirl Flowmeters

Swirl flowmeters are velocity instruments that respond to the vortex precession principle in their operation. The primary element of this flowmeter does not obstruct the flow path. Measurement occurs when a rotating body of fluid enters an enlarged area within the instrument. A simplification of the swirl flowmeter is shown in Fig. 11-8.

Fluid flow enters the swirlmeter on the left and is caused to rotate by passing through the stationary swirl-producing component. The center of the rotating fluid forms a twisting high-velocity area called the *vortex*. It tends to align itself axially in the center area of the instrument's meter body. Upon entering the enlarged area, the vortex leaves the center and forms a helical path around the inside diameter of the meter body. This action is called the *precession function*. The frequency of helical precession has proved to be proportional to the instrument's volumetric flow rate. A sensor placed in the enlarged area is then used to detect the frequency of the precession.

Just before exiting the flowmeter, all fluid is forced into the deswirl component. This part of the instrument restores the flow pattern to normal. The deswirl component is also used to isolate the measurement function from the downstream piping effects, which could impair the precession function.

The sensor or secondary element of a swirl flowmeter is usually a thermistor that detects velocity changes in the vortex precession. The frequency with which this velocity change occurs is then transposed into a changing voltage. The thermistor of Fig. 11-9 is placed in the precessional path so that it senses a

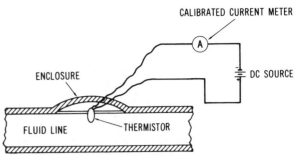

FIGURE 11-9 Thermistor flowmeter sensor circuit simplification.

change in temperature. Each resulting helical precession causes a corresponding cooling in the temperature due to the area of increased velocity. A constant-current source applied to the thermistor causes a change in voltage to be developed as a result of temperature variations. These voltage changes are subsequently amplified and filtered. A representative output signal is a $2\text{-V}_{p\text{-}p}$ square wave that varies in frequency with the volume of the flow rate.

The electrical output signal of a swirl flowmeter may be used to produce direct reading on a hand-deflection meter, an electronic recording instrument, or employ a digital readout. This element may be attached to the flowmeter body for direct-reading applications or placed in an independent housing for remote operation. Electrical energy must be supplied to both the integral mounted and remote assembly units of the instrument.

Strut Vortex Flowmeters

A strut vortex flowmeter is a common variation of the swirl flowmeter. The strut flowmeter responds to a principle known as the *vortex shedding phenomenon*. An obstruction bar or strut placed in the flow path produces a number of predictable swirls or vortices downstream. These vortices separate or are shed alternately from the strut and travel downstream in a predictable pattern. The number of vortices that appear downstream in a given period is directly proportional to the velocity of the flow. The strut of this instrument is a vertical bar located near the center of the flow tube body. Flow striking this strut gives off small vortices that alternately change direction as shown in Fig. 11-10.

An exploded view of a vortex flowmeter is shown in Fig. 11-11. In operation, the incoming flow stream enters the instrument at the left. Upon reaching the vortex shedding bar it alternately divides into swirls or vortices. The rotation of each swirl changes direction as it is detached from the shedding bar. The fre-

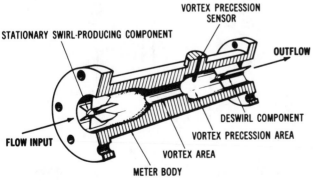

FIGURE 11-8 Simplification of the swirl flowmeter.

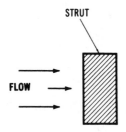

(A) Flow approaching strut.

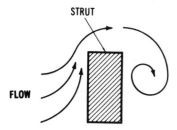

(B) Vortex generated clockwise.

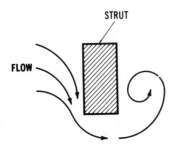

(C) Vortex generated counterclockwise.

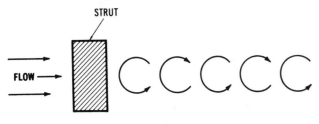

(D) Generated vortices.

FIGURE 11-10 Vortex-shedding principle.

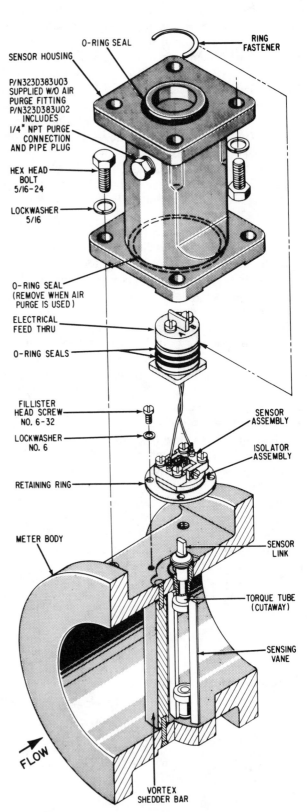

FIGURE 11-11 Exploded view of a vortex flowmeter. (Courtesy of Fischer & Porter Co.)

quency of the vortex pattern is directly proportional to the flow velocity.

After the vortices have been developed they move past the shedding bar. They then encounter the sensing vane. The twisting action of each swirl has torque. This torque causes the sensing vane to pivot according to the direction of each swirl. The developed torque occurs alternately in the clockwise and counterclockwise directions. This action causes the torsionally supported sensing vane to oscillate at a frequency that is proportional to the flow rate of the fluid. The minute deflection of the sensing vane is coupled to a piezoelectric sensing element.

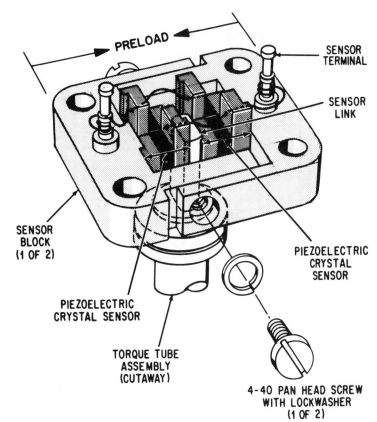

FIGURE 11-12 Detail drawing of a piezoelectric element. (Courtesy of Fischer & Porter Co.)

A detail drawing of the piezoelectric sensor of a vortex flowmeter is shown in Fig. 11-12. This assembly consists of two pairs of balanced polarized piezoelectric elements that are mounted between two preloading blocks. The sensor link has a flat face on each side of the two piezoelectric pairs. Movement of the sensing link causes a force to be applied to the piezoelectric elements. This produces ac at a frequency that depends on the clockwise and counterclockwise rate change of the vortices passing across the sensing vane. The resulting ac signal must be conditioned by electronic cir-

cuitry to produce a voltage, current, or frequency output that is suitable for driving an instrument. In general, the electronic circuitry of this converter is similar to that of the magnetic flowmeter.

Ultrasonic Vortex Flowmeters

A recent development in vortex flowmeters utilizes ultrasonic signals as a method of detecting flow rate. This method of sensing, shown in Fig. 11-13, combines both the vortex shedding principle and ultrasonic detection

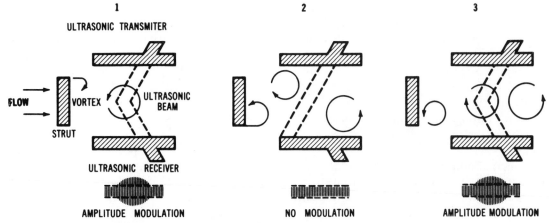

FIGURE 11-13 Operation of an ultrasonic vortex sensor. (Courtesy of Brooks Instruments Div., Emerson Electric.)

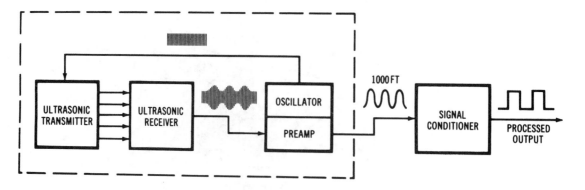

(A) Block diagram.

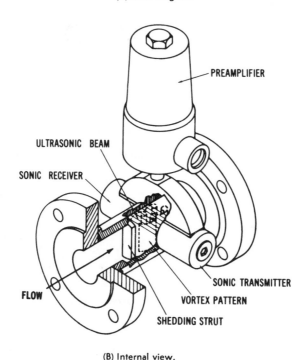

(B) Internal view.

FIGURE 11-14 Ultrasonic vortex flowmeter.

in its operation. An ultrasonic beam is transmitted across the vortex pattern as it travels downstream from the strut. As each vortex passed through the ultrasonic beam it causes a corresponding change in signal amplitude. This produces a form of amplitude modulation (AM) that is similar to that of a radio communication system. Each amplitude change in the signal represents a vortex passing through the beam. The number of amplitude changes that occur can then be used to indicate liquid flow rate.

The modulated output signal of the primary element is then detected and processed through electronic signal conditioning equipment. A block diagram of a representative unit is shown in Fig. 11-14A. The entire assembly is attached to the flow tube body. To energize the instrument, 120/240-V, 60-Hz electrical energy is needed. Remote instruments up to 1000 ft away may be actuated by the output of the instrument. An internal view of the instrument is shown in Fig. 11-14B. This type of instrument is commonly used as a high-performance flowmeter specifically designed for long-term reliability and a wide range of operating characteristics.

SUMMARY

Nearly all industrial products are influenced in some way by the flow of materials. In this regard, there is rate flow and totalizing flow. Measurement is achieved by employing the magnetic and velocity principles in the operation of flowmeters.

Magnetic flowmeters are obstructionless instruments that measure volume flow rate. This instrument has a flowmeter body, a signal converter, and an output/power supply. The flowmeter body is responsible for signal development. It contains two electromagnetic flux–producing coils and two signal-sensing electrodes. The converter is responsible for signal processing, logic gating, demodulation, frequency conversion, line-frequency triggering, feedback multiplication, and gain adjustments. The output/power supply develops frequency or current output and provides dc and low-frequency ac to the flowmeter body and converter circuits.

Velocity flowmeters respond to the speed at which a stream passes through a closed pipe or tube. A turbine flowmeter consists of a multiple-bladed rotor assembly mounted inside a pipe. Flow causes motion of a turbine blade, which generates a square-wave signal voltage. Swirl flowmeters are velocity instruments that respond to the vortex precession principle. Flow entering this instrument passes through a swirl-producing element. The twisting effect created by this element is called a vortex. The vortex forms a helical path around the inside diameter of the meter body. The frequency of the helical precession is proportional to the instrument's volumetric flow rate. Strut vortex flowmeters have an obstruction bar placed in the flow path. Flow striking this strut causes the shedding of vortices. A sensing vane detects the twisting action of the vortices. This effect causes force to be applied to a piezoelectric sensing element that generates an ac that changes in frequency. The ultrasonic vortex flowmeter generates a high-frequency beam across the flow path. Each vortex passing through the beam causes a change in signal amplitude. The number of amplitude changes that occur in a given unit of time can be used to indicate liquid flow rate.

ACTIVITIES

FLOWMETER FAMILIARIZATION

1. Refer to the manufacturer's operational manual for the flowmeter to be used in this activity.
2. Identify the type of flowmeter being used.
3. Turn off or disconnect the electrical power from the flowmeter.
4. Remove the exterior housing or covering from the flowmeter.
5. Describe the construction of the flowmeter body.
6. Describe the placement of the circuit boards.
7. Identify the key test point locations on the circuit boards.
8. Replace the instrument's cover or housing.

FLOWMETER CIRCUIT EVALUATION

1. Refer to the manufacturer's operation manual of the instrument being used in this activity.
2. Remove the cover or housing from the instrument.
3. With a VOM measure the power supply voltage values supplied to the instrument. Make a voltage chart and record the measured values.
4. With an oscilloscope, observe the waveforms at the converter input and output. Make a drawing of the observed waveforms.
5. If a magnetic flowmeter is being used, identify the voltage source of the flowmeter body.
6. Connect an oscilloscope to the flowmeter body source and observe the waveform. Determine the frequency of the observed waveform and make a sketch of the display.
7. Identify the converter circuit board of the instrument.
8. With the aid of the operational manual, connect an oscilloscope to each of the key test points. Make a sketch of the observed waveforms found at each test point. Note the frequency and amplitude value of each waveform.
9. Replace the cover or housing of the instrument.

QUESTIONS

1. What is meant by the term *flow*?
2. What is meant by the term *rate flow*?
3. Explain the meaning of the term *totalized flow*.
4. What are the basic parts of a magnetic flowmeter body?
5. Why is a magnetic flowmeter considered to be an obstructionless instrument?
6. What is the difference between an integral mounted and a remote converter assembly?
7. Explain how the magnetic flowmeter body produces a signal.
8. What is the function of the converter of a magnetic flowmeter?
9. What is the function of the power supply of a magnetic flowmeter?
10. What is meant by the term *velocity flowmeter*?
11. What are some common types of velocity flowmeters?
12. What is a vortex?
13. Explain the operation of the vortex-swirl flowmeter.
14. What is meant by the term *vortex shedding*?
15. How is flow velocity sensed by the vortex shedding principle?

CHAPTER 12

RECORDERS AND INDICATORS

OBJECTIVES

Upon completion of this chapter, you will be able to:
1. Discuss the operation of a servomotor type of recorder/indicator.
2. Identify the functional parts of a recorder/indicator from a block diagram.
3. Explain the role of the servomotor in the operation of a recorder/indicator.
4. Explain how an electronic indicator produces a display without any moving parts.

IMPORTANT TERMS

In the study of recorders and indicators one frequently encounters a number of new and somewhat unusual terms. These terms play an important role in the presentation of this material. New words, such as *flux bridge, torque motor, servomotor,* and *electronic indicator,* will begin to have some meaning for you. A few of these terms have been singled out for further study. As a rule, the chapter will be more meaningful if these terms are reviewed before proceeding with the text.

Astable multivibrator: A free-running generator that develops a continuous square-wave output.
Bar graph: An electronic display device that shows information by illuminating vertical or horizontal segments that represent specific signal values.
Comparator: An electronic device or circuit that compares an input voltage value with a predetermined reference value and develops a corresponding output.
Damping: An action that extracts energy from a value-changing system and suppresses the changing action.
Electronic indicator: An instrument that displays or measures the value of an applied signal or data application.

Flux bridge: A component network that responds to invisible lines of magnetic force and generates a voltage signal.
Flux gate: A mechanism that responds to invisible lines of magnetic force due to a change in its position.
Gas-discharge display: A device that illuminates vertical segments of a display due to the ionization of gas when electrodes are energized.
Multivibrator: An electronic generator circuit that changes back and forth between conduction and nonconduction while producing an output signal.
Passive damping: An electronic circuit function that suppresses signal changes by passing the signal through resistors, capacitors, or inductors.
Servomotor: A rotating machine that produces some form of physical displacement of an armature mechanism when it is in operation.
Symmetrical: A condition of balance where parts, shapes, and sizes of two or more items are the same.
Torque motor: A servomotor or machine that produces some form of twisting or turning motion in its operation.

INTRODUCTION

In this chapter we investigate electronic recorders and indicators that have a wide range of applications in the

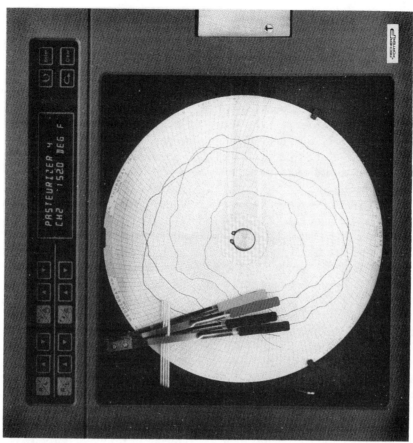

FIGURE 12-1 Circular chart recorder. (Courtesy of Fischer & Porter Co.)

industrial process control instrumentation field. Numerous companies make recorders and indicators that will drive a pen or deflect a hand to produce a reading on a scale. Figure 12-1 shows a representative circular chart recorder. This instrument is a microprocessor-based recorder that delivers high accuracy and responsiveness in recording, indication, and controlling applications for up to four process variables.

The first part of the chapter is devoted to a servomotor-controlled recorder. In this part of the chapter we investigate a block diagram of the recorder and see how different functions of the instrument are achieved. We then discuss the role of the servomotor in the operation of the instrument. This will permit you to see how the instrument achieves balance when the input changes. A feedback signal is developed and compared with the value of the input signal to control the servomechanism. The final part of the chapter describes a method of instrument indication that has no moving parts. Indications are made on an electronic bar graph that is viewed from the front of the indicator. This type of indicator is widely used in electronic instruments.

ELECTROMECHANICAL INDICATOR/ RECORDERS

An electromechanical indicator/recorder is an instrument that displays the response of a process variable through the operation of an electromagnetic mechanism that is controlled electronically. The electromagnetic mechanism of this instrument is a servomotor. The servomotor is responsible for moving the pen mechanism or deflecting hand of the instrument. The instrument contains an electronic converter circuit, a servomotor mechanism, a recording pen assembly, and a recording chart mechanism.

In the operation of a recorder/indicator a process variable signal is applied to its input. This signal is an analog voltage or current value. The input signal is applied to one input of a differential amplifier. The output of this amplifier is immediately applied to the servomotor driver amplifier. This amplifier controls the current applied to the two windings of the servomotor field. As a result, the armature of the servomotor positions the pen assembly on the chart. The position is sensed and returned to the other input of the differential amplifier. The differential amplifier compares the value of the feedback signal with the process variable signal. If the pen position compares with the process variable input value that originally produced it, no change or correction will be made. The pen will remain at this position. This is considered to be its balanced position. A change in the process variable input will cause the instrument to repeat the sequence just described. In a sense, this instrument moves the pen or hand mechanism, then stabilizes its position, and alters its position according to the value of the applied process variable input.

A functional block diagram of a representative electromechanical recorder is shown in Fig. 12-2. This diagram shows an analog process variable applied to its input. Its value can be 0 to 200 mV, 1 to 5 V dc, or 4 to 20 mA. If a current signal is supplied to the input, it must be converted to a proportional voltage value before being received by the converter circuit. This is accomplished by having current flow through a precision shunt resistor connected across the input. Typically, the input of a recorder or indicator is designed to accept analog voltage signals.

In the operation of an electromechanical recorder the analog process variable signal is first applied to a voltage divider and a passive damping network. This section of the assembly is responsible for developing an adjustable servo-response signal. The signal is then

applied to the noninverting input of the differential amplifier. The differential amplifier controls the clockwise (down) or the counterclockwise (up) signal applied to the output driver amplifiers. A driver amplifier controls the duty cycle of the signal applied to the respective upscale or downscale windings of the servomotor.

The servomotor or torque motor shaft of a recorder is mechanically linked to the recorder pen assembly and a contactless flux bridge. The flux bridge is used to develop the feedback signal. The feedback signal is demodulated and applied to the inverting input of the differential amplifier. The feedback signal is used to balance the servomotor when it reaches a specific operating position.

The reference voltage generator of the instrument is responsible for developing a 10-kHz, controlled-amplitude, symmetrical square-wave signal. The oscillator uses an operational amplifier as an astable multivibrator. The output of the multivibrator is a square wave. It is then shaped and limited before being applied to the primary winding of a coupling transformer. The secondary winding of the transformer applies the signal to the flux bridge assembly. The armature of the flux bridge is mechanically linked to the torque motor. As the permanent magnet of the flux bridge armature turns in step with the servomotor, it develops feedback voltage in the velocity coil. This feedback voltage is

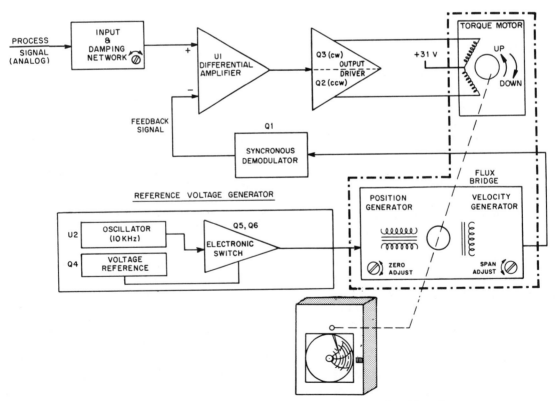

FIGURE 12-2 Functional block diagram of a recorder. (Courtesy of Fischer & Porter Co.)

demodulated and applied to the inverting input of the differential amplifier. The differential amplifier then develops an output signal that controls the position of the torque motor assembly. As a result, this instrument moves the pen mechanism, detects its position, reaches a position of balance, and readies itself for the next change in the process variable input.

SERVOMOTOR MECHANISMS

The servomotor of a recorder/indicator is primarily responsible for producing movement of the pen mechanism or deflection of an indicating hand. This mechanism is described by a variety of names. One manufacturer calls this a Torq-er mechanism, while another manufacturer calls it a servomotor assembly. Others refer to this device as a pen-error servo and a torque-detector motor. Functionally, this mechanism is a rotating machine that changes the mechanical position of the pen assembly. It may be mechanically linked to other components to combine different functions. The torque motor, flux bridge, and velocity generator assembly are combined in a single unit called a Torq-er receiver by Fischer & Porter Company. This assembly will serve as the basis of our discussion of the servomotor.

A simplification of Fischer & Porter Company servomotor assembly is shown in Fig. 12-3. The torque motor serves as the prime mover for the combined assembly. It has a permanent-magnet armature that is actuated by two independent direction windings. The duty cycle of a square-wave signal applied to the directional coils determines the amount of angular displacement produced by the armature. A 50% duty cycle is considered to be a holding current value. It causes an equal amount of current to flow into each directional coil at the same time. This causes the armature's position to be stabilized or balanced. An increase or decrease in the duty cycle of the square wave will cause one directional coil to be more conductive than the other. This permits the coil with the largest current flow to be the dominant factor in the rotational procedure. The position of the armature can change only 35°. This represents the maximum angular displacement of the pen mechanism.

The flux bridge of the assembly is mechanically linked to the servomotor armature. The flux bridge is a contactless position feedback device. The position of the flux gate of this assembly determines the amount of feedback developed by the bridge assembly. The primary coils of the bridge are energized by a 10-kHz signal. The voltage developed by the secondary coil is proportional to the position of the flux gate assembly. The secondary voltage is a feedback signal that rep-

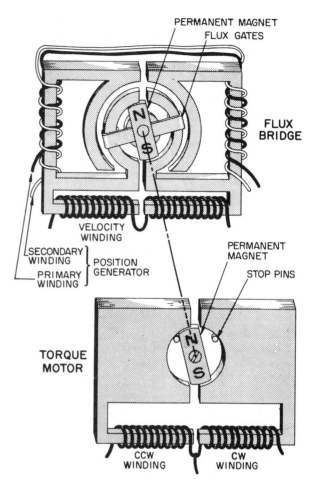

FIGURE 12-3 Torque motor. (Courtesy of Fischer & Porter Co.)

resents a positional change in the torque motor's armature shaft.

Another function of the flux bridge is the generation of a rate feedback signal that is proportional to the speed changes of the torque motor's armature. A permanent magnet attached to the flux gate mechanism induces voltage into the velocity winding of the unit. The resulting output is considered to be a degenerative feedback signal. This signal is used to reduce the deflection speed of the pen mechanism. It is compared with the process variable input signal and determines if the servomotor is in balance or needs to produce a corrective change. The flux bridge and velocity winding are mechanically linked to the torque motor to form a composite electromagnetic recorder mechanism.

RECORDER/INDICATOR CIRCUITRY

The circuitry of a recorder/indicator is similar in many cases to that of other electronic instruments. A block diagram of a basic recorder shows that it contains a power supply, input and damping network, differential amplifier, output driver, reference voltage generator, and a synchronous demodulator/feedback assembly. The servomotor of this assembly is actuated by the output driver. The torque of this motor alters pen mechanism position and generates a feedback signal that is compared with the input signal. A recorder/indicator is an instrument that employs automatic feedback to control its operation. Some of the circuitry of this instrument has been discussed previously.

Power Supply

The power supply of a recorder/indicator is primarily responsible for developing the ac and dc voltage values that are needed to make the instrument functional. Recorders and indicators are generally energized by the ac power line, with 120-V 60-Hz ac being the most common primary source. This voltage is stepped down by a transformer, rectified, filtered, and regulated to some specific values. One manufacturer has a power supply circuit that develops $+31$ V dc to drive the servomotor assembly. Additional dc voltages are $+15$, -15, and $+7.5$ V. Other manufacturers use similar voltage values, with the largest voltage used to supply the motor assembly.

The power supply of a recorder/indicator is frequently equipped to supply low-voltage dc for a two-wire transmitter source. This power source is generally an optional item. Typical supply voltages are 24 to 26 V dc. The two-wire supply must be equipped with short-circuit protection. Normal operating current values are in the range 4 to 20 mA. Series current-limiting resistors are used for short-circuit protection. The two-wire transmitter source generally has a termination point near to the power supply circuit board. This voltage is readily accessible for measurement and circuit connections.

Input Divider and Damping Network

The input/damping network of a recorder is responsible for current conversion to voltage and filtering unwanted noise from the process variable input signal. Current-to-voltage conversion is accomplished by attaching a precision shunt resistor across the input terminals. The input signal voltage is then applied to an adjustable damping network that feeds a high-impedance voltage divider. The output-to-input signal ratio is in the range 20:1. This represents a 200-mV output for a 4-V input. This signal is then applied to the noninverting input of the differential amplifier.

Differential Amplifier

A representative differential amplifier for a recorder/indicator is shown in Fig. 12-4. This part of the circuit is responsible for comparing the magnitude of the process variable and feedback signals applied to its inputs. The process variable signal is applied to the noninverting input and the feedback signal is applied to the inverting input. When these two input signals are of equal magnitude, the output will be a symmetrical square wave. This wave will have a 50% duty cycle. Any signal difference between the two inputs will cause a change in the duty cycle of the output. An increase in the process variable signal will cause an increase in the on time of the duty cycle. This will cause a corresponding increase in the upscale or clockwise rotation of the servomotor. Rotation will continue until the feedback signal equals the applied process variable value. This indicates that the balance or holding condition has been met. This is sensed by the flux bridge assembly, which is mechanically linked to the servomotor. The differential amplifier is essentially used to distinguish between the voltage value of signals applied to its input.

Reference Voltage Generator

The reference voltage generator of a representative recorder is responsible for producing a square-wave signal of 10 kHz. This signal serves as an ac voltage source for the synchronous demodulator and the position generator of the flux bridge. Figure 12-5 shows a schematic of the reference voltage generator.

The operational amplifier of the circuit is used as an astable multivibrator. This type of multivibrator is free-running and self-starting. It is used to generate timing frequencies. The shape of the waveform and its frequency are determined by an *RC* network attached to its input. The multivibrator of this circuit uses positive or regenerative feedback from the output to the input. This causes the amplifier to be alternately driven between saturation and cutoff. The output is a square or rectangular wave that has a constant frequency.

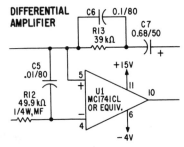

FIGURE 12-4 Differential amplifier. (Courtesy of Fischer & Porter Co.)

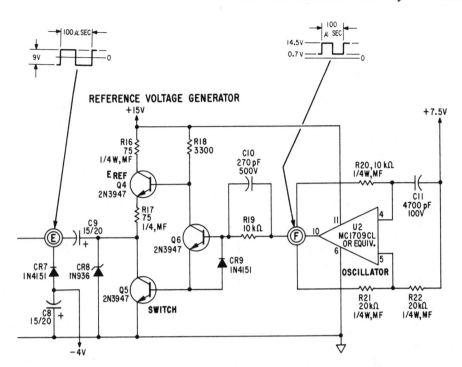

FIGURE 12-5 Reference voltage generator. (Courtesy of Fischer & Porter Co.)

The output of the oscillator is used to drive two transistors that are used as an electronic switch. When the output goes positive it forward biases Q_6 and Q_5. As a result, conduction of Q_5 shorts out the zener diode and Q_6 removes the bias voltage of Q_4. When the output of the oscillator swings low during the next half of the alternation, Q_5 and Q_6 are driven to cutoff. This causes Q_4 to be conductive. Capacitor C_9 charges to 9 V, which is the zener voltage rating of CR_8. The output that appears at test point E is a square wave

that has a symmetrical shape and is of a constant frequency and amplitude.

Synchronous Demodulator and Feedback Assembly

The demodulator/feedback circuitry of a representative recorder is shown in Fig. 12-6. The reference voltage signal is supplied through two windings of a coupling transformer. The center-tapped coil of the trans-

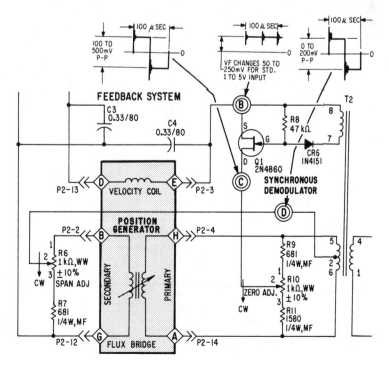

FIGURE 12-6 Demodulator/feedback circuit. (Courtesy of Fischer & Porter Co.)

former supplies reference voltage to the flux bridge assembly. The other transformer winding supplies reference voltage to the synchronous demodulator. The reference voltage has a frequency of 10 kHz that has its corners slightly rounded.

The synchronous demodulator is a junction field-effect transistor that is used to rectify the reference voltage signal. Functionally, the gate bias voltage for the JFET is developed by resistor R_8 and diode CR_6. This voltage is essentially a high-to-low impedance gating signal that causes rectification of the reference feedback signal applied to the drain of the JFET. Capacitors C_4 and C_3 filter the feedback signal and bypass any remaining 10-kHz signal to ground. Ultimately, the filtered dc signal appearing at test point B is coupled to the inverting input of the differential amplifier through the position coil of the flux bridge. The resulting signal is then compared with the process variable signal being applied to the noninverting input.

Servomotor Driver

The servomotor driver of a recorder mechanism is responsible for controlling the current flow into the CW and CCW windings of the motor assembly. A schematic of a representative driver circuit is shown in Fig. 12-7. Note that this part of the recorder employs two bipolar transistors as driver amplifiers. These two transistors are designed to alternately control current to the two windings of the servomotor. Transistor Q_2 is driven by the output of the differential amplifier. When the output of the differential amplifier rises above $+0.7$ V it causes saturation of Q_2 and cutoff of Q_3. This actuates the clockwise (CW) winding of the servomotor and turns off the counterclockwise (CCW) winding. Conduction continues for the duration of the positive alternation. When the differential amplifier output drops below $+0.7$ V it turns off Q_2 and turns on Q_3. This causes the CCW winding to be actuated and the CW winding to be off. This condition will continue for

the remainder of the negative alternation. Current flow through the two windings depends on the duty cycle of the applied input signal. A 50% duty cycle signal has equal conduction time of the two windings during any single cycle of operation. This represents a condition of balance. An increased or decreased duty cycle will cause one winding to be more conductive than the other. This will cause the mechanism to move until balance is restored. Operation of the assembly is a process of changing from a balanced condition to unbalanced and returning to the balanced condition in a suitable period of time.

SUMMARY

Electronic recorders and indicators have a wide range of applications in the industrial process control instrumentation field. An electromechanical instrument produces a display through the response of a servomotor mechanism. In its operation a process variable signal is applied to its input. This signal is filtered and applied to one input of a differential amplifier. The output of this amplifier is applied to the servomotor driver. This circuit controls current flowing to the two windings of the servomotor. The armature of the servomotor positions the pen assembly on the chart. Its position is sensed and returned to the other input of the differential amplifier. The differential amplifier compares the value of the feedback signal with the process variable signal. If the pen position compares with the value of the process variable input, the pen has reached its balanced position. A change in the process variable input will cause the process to be repeated until balance is reached for the new pen position.

The servomotor of a recorder/indicator is responsible for producing movement of the pen mechanism. Some manufacturers attach other mechanisms to the servomotor assembly. A flux bridge and rate movement sensor are often linked to the servomotor.

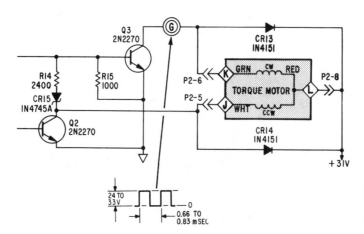

FIGURE 12-7 Servomotor driver stage. (Courtesy of Fischer & Porter Co.)

These added mechanisms are used to generate and control the feedback signal.

A recorder/indictor has an input/damping network, differential amplifier, power supply, output driver, reference voltage generator, demodulator/feedback assembly, and a servomotor. The input/damping network is responsible for current conversion to voltage and filtering unwanted noise from the process variable input. The power supply is responsible for developing the ac and dc voltage values that are needed to make the instrument operational. The differential amplifier compares the magnitude of the input and feedback signals. The reference voltage generator produces a square-wave signal. The servomotor driver is responsible for controlling current flow into the CW and CCW windings of the motor assembly. The armature of the servomotor moves according to the current flow into its field windings. This movement drives the pen mechanism. The position of the pen is sensed and a feedback signal is generated and mixed with the reference generator signal. After demodulation a signal is returned to the differential amplifier for comparison with the process variable. The entire assembly responds by changing from a balanced condition to unbalance and returning to the balanced condition on a continuous basis.

Electronic indicators that have no moving parts display information on a bar graph. Each bar contains 200 gaseous display lamps with separate cathodes and a common anode. When the lamps are scanned with ac they will be illuminated until one segment is turned off by a comparator circuit.

ACTIVITIES

RECORDER CIRCUIT INVESTIGATION

1. Refer to the manufacturer's operational manual of the electronic recorder that is to be used in this activity.
2. The electrical power should be turned off or disconnected from the instrument.
3. Open the front door or remove the exterior housing from the instrument.
4. Locate the circuit boards of the instrument.
5. Prepare an oscilloscope for operation.
6. Apply power to the instrument.
7. Locate the test points on the circuit board. Connect the vertical probe of the oscilloscope to the reference voltage generator, synchronous demodulator output, differential amplifier inputs, differential amplifier output, and servomotor winding test points.
8. Make a sketch of the observed waveforms appearing at each test point. Indicate the amplitude value of each observed signal.
9. If possible, change the value of the process variable input while observing the response of the instrument. Describe how this alters the operation of the instrument.
10. Turn off the power to the instrument and return its housing.

QUESTIONS

1. What is the primary function of a recorder?
2. What is the primary function of an indicator?
3. What is a servomotor?
4. What is the function of a servomotor in the operation of a recorder?
5. Describe the operation of an electronic recorder.
6. What is the primary function of the input/damping network of a recorder?
7. What is the primary function of the differential amplifier of a recorder?
8. What does the reference voltage generator produce?
9. What is the function of the synchronous demodulator of a recorder?
10. What does the duty cycle of a signal applied to a servomotor do to its operation?
11. What causes the position of the pen mechanism to be balanced?
12. What causes the pen mechanism of a recorder to change from a balanced condition to an unbalanced condition?
13. What is a multivibrator?
14. Where is a multivibrator used in the circuitry of a recorder?
15. Why are the servomotor and flux bridge mechanically linked?

CHAPTER 13

ELECTRONIC CONTROLLERS

OBJECTIVES

Upon completion of this chapter, you will be able to:

1. Describe the primary function of an electronic controller.
2. Explain the difference between pure control and combinational control.
3. Identify the mode of operation achieved by a given op-amp circuit.
4. Analyze the operation of a proportional controller.
5. Analyze the operation of a PID controller.
6. Define commonly used controller terms, such as *error amplifier, integral action, derivative, proportional,* and *modes of operation.*

IMPORTANT TERMS

In the study of electronic controllers one frequently encounters a number of new and somewhat unusual terms. These terms play an important role in the presentation of this material. A few of these terms are singled out for study before proceeding with the chapter. A review of these terms will make the reading of the text more meaningful.

Anticipation function: The ability of a circuit to determine or visualize a future action.

Automation: The technique of making an apparatus, a process, or a system operate automatically.

Controlled variable: A process that is being regulated by a controller.

Derivative control: A mode of control that provides an output that is related to the time rate of change that occurs between a process variable and its set-point.

Differential amplifier: An amplifier designed so that its output is proportional to only the difference between signals applied to its two inputs.

Discrete components: Individual electronic devices that are capable of only one function, are packaged separately, and are used individually.

Error signal: The algebraic difference between the set-point value and the process variable value applied to the input of a controller.

Final control element: A device connected to the output of a controller that alters the process being manipulated by an instrument.

Integral control: A mode of control which develops an output that is the time integral of the difference between the set-point and the value of the process variable being measured.

Mode of control: A method or type of control operation.

Offset: A measurable output signal from a controller or device when the input is zero.

On–off control: A mode of control that alters its output by switching it on and off for periods of time.

Process: The dynamic variables of a system that are used in manufacturing and production operations.

Process variable: Any process parameter, such as temperature, flow, liquid level, or pressure, that changes its value during the operation of a system.

Proportional band: A band or range of values that a process variable operates within when a controller is active in the proportional mode of control.

Proportional control: A basic mode of control in which the output is a linearly related function to the difference between a set-point value and the process variable being controlled by a system.

Proportional plus derivative (PD) control: A mode of control which provides an output that is a linear combination of proportional and derivative control modes in a single instrument.

Proportional plus integral (PI) control: A mode of control that has proportional and integral operations combined in a single instrument.

Proportional plus integral plus derivative (PID) control: A controller that has proportional, integral, and derivative modes of control combined in a single instrument.

Rate action: Another name for the derivative mode of control.

Reset action: Another name for the integral mode of control.

Set-point: A desired value setting of a process variable.

Steady-state error: The difference between a controller's output and input after the input has been applied for a prolonged period of time.

Summing amplifier: An operational amplifier circuit that adds two or more of the inputs applied to its summing input point.

Time constant: A number referring to the time required for the output or a device, a circuit, or a system to reach approximately 63% of the final value following a step change of its input.

INTRODUCTION

An electronic controller is the instrument that is primarily responsible for industrial automation (the automatic control of process variables). Applications range from simple on/off control operations to completely automated systems that respond to signals initiated by direct access to a computer-based system. Control may apply to only one variable such as temperature, pressure, electrical conductivity, or fluid flow, or it may respond to literally hundreds of process variables simultaneously.

In this chapter we look at the electronic circuitry of some representative industrial controllers. Electronic controllers have been used in industry for a number of years and have gone through several significant transitions. Electronic controllers were first introduced in the 1940s, when they were constructed with vacuum tubes. These units were rather massive assemblies housed in large metal cabinets. With the development of solid-state devices in the 1950s, the size and outward appearance of the controller changed

FIGURE 13-1 Typical solid-state controller. (Courtesy of Leeds & Northrup.)

rather significantly. The transition to all solid-state controllers, however, was rather slow. Most manufacturers were somewhat reluctant to change their popular selling vacuum-tube controllers. There was a period when controllers employed both vacuum tubes and solid-state devices. These units necessitated both

FIGURE 13-2 Indicating controller partially removed from its housing showing plug-in modules. (Courtesy of Moore Products Co.)

high and low voltage power supplies to energize the devices. Many of these "hybrid" units are still in operation today.

All major controller manufacturers today produce a wide variety of solid-state instruments. These units, in general, are small in size and employ literally thousands of discrete components. Figure 13-1 shows a typical solid-state controller of this type. It has unusually precise control capabilities with exceptional stability.

The development of the solid-state controller has brought some innovative design features that have had a decided impact on controller maintenance. Components, for example, are mounted on removable printed circuit cards or boards for easy replacement. Figure 13-2 shows an example of an indicating controller that has been partially removed from its metal housing. The printed circuit modules of this controller can be easily removed by pulling the board from an edge-connector socket. This controller also has a great deal of versatility through construction of this type. Figure 13-3 shows a functional diagram of the potential location of alternate modules that can be utilized in this unit.

The next trend in controller technology found large numbers of discrete solid-state components re-placed by integrated circuits. These controllers are significantly smaller than their discrete component solid-state counterparts. Maintenance, in this case, is based on faulty IC determination and PC board replacement.

Computer-based controllers are now an integral part of the industrial instrumentation field. These controllers are classified as smart instruments. They may be controlled by a microprocessor in a dedicated instrument or be part of a network of units controlled by a central computer system. Microprocessors are ICs that house computer functions on a single chip. This type of system is classified as a universal digital controller. The data is shown on a digital display and control can be programmed according to the demands of the system. This unit has memory and responds to all signals through digital signal changes. Digital controllers are compact self-contained units that may be attached directly to the instrument or machine being controlled. Figure 13-4 shows an example of a universal digital controller.

With the wide range of diversity that exists today in controller technology, it is difficult to single out a particular controller that is representative of the field. In this regard, we first discuss some common solid-state controller circuitry, then show some typical IC

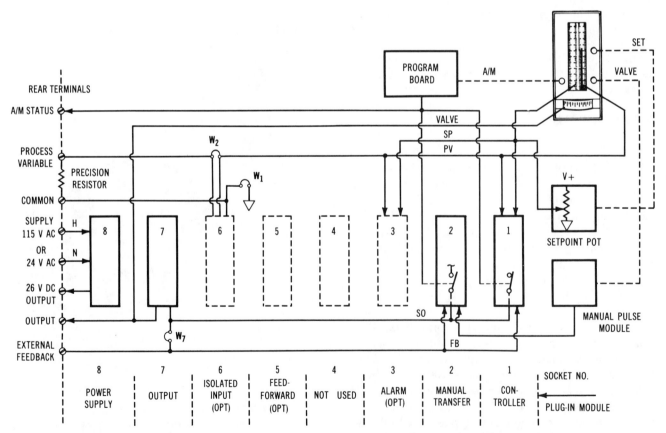

FIGURE 13-3 Functional diagram of alternate module locations for the controller of Fig. 13-2. (Courtesy of Moore Products Co.)

FIGURE 13-4 Universal digital controller. (Courtesy of Honeywell, Inc.)

diagram of a controller is shown in Fig. 13-5. Operation is achieved by processing the applied error signal. The error signal is parallel connected to the derivative, proportional, and integral circuit components. The derivative circuit serves as one input for the differential amplifier. The proportional and integral circuit outputs are combined with the feedback signal and applied to the other differential amplifier input. The differential amplifier drives the output amplifier, which ultimately controls the load or final control element. Feedback is developed across the load and returned to the differential amplifier for internal correction. This composite circuit permits the load or final control element to be controlled automatically through a self-correcting operation.

MODES OF CONTROL

The operational response of a controller is often described as its mode of control. Several different types of control are available today. In some cases only a single mode of control is needed to accomplish an operation. This is described as a pure control operation. On–off, proportional, integral, and derivative are examples of pure control. More sophisticated control is achieved by combining two or more pure modes of operation. This is described as a composite mode. Proportional plus integral, proportional plus derivative, and proportional plus integral plus derative are examples of composite control modes. The specific mode of operation utilized by an instrument is determined by the control procedure of the application.

On–Off Operation

An on–off or two-state controller is the simplest of all process control operations. The output or load of this instrument is automatically switched on or off without any intermediate level of operation. Control of this type is very popular and rather inexpensive to accomplish.

The operation of an on–off controller is determined by the position adjustment of its set-point value. As a rule, the output continually changes or oscillates above or below the set-point value. Figure 13-6 shows the response of an on–off temperature controller. Note the location of the set-point and how the temperature of the system rises and falls above this value. The final control element of this system is a heater. The electrical energy source supplying the heater is turned on and off with respect to time. When electrical energy is applied to the heating element it produces heat. Turning off the energy source stops the process. In systems where on–off cycling occurs rapidly, there can be some damage to the final control element. On–off

applications using operational amplifiers. Through this approach you will become familiar with the characteristic differences in controller circuitry and be able to select the information that is particularly applicable to your controller needs.

CONTROLLER FUNCTIONS

A controller is an instrument that employs pneumatic, electronic, and/or mechanical energy to perform a system control operation. This instrument is designed to maintain a process variable at some predetermined value by comparing an existing value to that of a desired system value (the set-point) and altering its output in response to this difference. A functional block

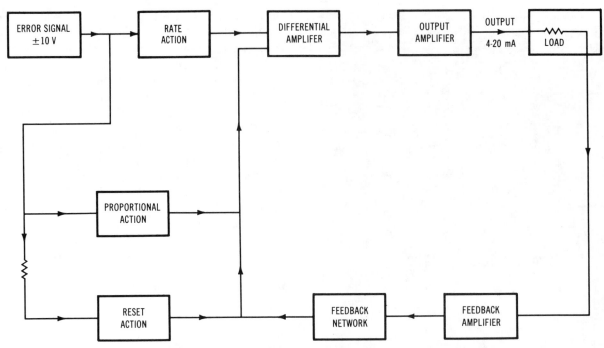

FIGURE 13-5   Block diagram of a typical controller.

control is generally used in systems where precise control is not necessary. To be effective, these systems must have a rather large capacity that changes slowly. A common application of an on–off controller is the heating/cooling system of a building.

Proportional-Only Control

Proportional-only control determines the ratio between the controller output signal and the input signal. If the proportional-band control is set at 100%, the

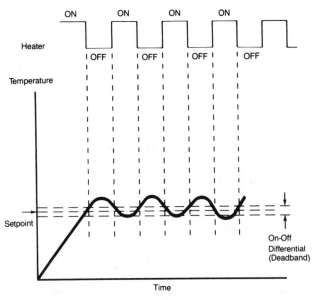

FIGURE 13-6   On–off temperature control action.

controller output will result in a change that is directly proportional to the error signal. For insensitive transducer inputs, the proportional-band control will be set for less than 100%. The error signal will, therefore, produce an output that is proportionally greater. This is called *narrowband control*. A narrow range in error signal can produce a full range of controller outputs. For settings of the proportional-band control greater than 100%, the error signal range will be greater than the controller output. This is called *wideband operation*.

In conventional operation proportional-only controllers cannot reduce a steady-state error signal to zero. This type of controller introduces a permanent offset in the output signal. Control applications are limited primarily to processes where the gain can be made large enough to reduce the steady-state offset to acceptable limits.

Integral Control

Integral control, which is sometimes called *reset action,* has an output whose rate of change is proportional to the error signal applied to its input. When there is a large error signal, the output changes rapidly to correct the error. As the error signal gets smaller, the output changes more slowly. This action is purposely done to minimize for overcorrection. As long as there is an error, the output of this control operation will continue to change. Once the error is driven to zero, the output change also goes to zero. This essentially means that the controller has inertia that tends to hold the output that was necessary to eliminate the error signal applied to its input. Pure integral control has a very poor transient response. Should the error signal produce a step input, it will respond by beginning to ramp its output. As a rule, pure integral control is not used alone because it acts only on long-term or steady-state errors.

Derivative Control

Derivative control responds to the rate of change of the applied error signal. At one time derivative control was called rate action. This was due to its reaction to rate changes in the control variable or error signal. Its effect on controller output is twofold: If the controller output rate were dependent on proportional-band control only, it is possible for the error signal to become so great that the controller could not possibly zero itself. On the other hand, if the rate action were too fast, the controller would oscillate or hunt. The input error signal of this controller is differentiated so that its rate of change can be detected and produce a proper output signal. Derivative control is not used alone in controller operation. It only reacts to measurements that are changing and does not change with steady-state errors.

Proportional Plus Integral (PI) Control

Proportional and integral modes of control are frequently combined to provide automatic reset action, which eliminates proportional offset of the output. This combination is called proportional plus integral (PI) control. The integral mode produces a reset action by constantly changing the controller output until the error signal is reduced to zero. The proportional mode causes a change in output that is proportional to the error signal.

PI control is used on processes that have large load changes and frequent set-point adjustments. Proportional control alone is not capable of altering this type of process by reducing the offset to an acceptable level. When the reset function of integral control is combined with the proportional mode, it automatically eliminates the offset problem. Combining these two operational modes permits control of large loads with frequent set-point adjustments quite effectively. This permits the process to be controlled without prolonged oscillations, no permanent offset, and have a quick recovery after a disturbance.

Proportional Plus Derivative (PD) Control

Proportional and derivative modes of control are often combined to reduce the tendency for oscillations and to allow for a higher gain setting. This combination is called proportional plus derivative (PD) control. The proportional mode of this controller provides a change in output that is a proportion of the error signal. Derivative control, on the other hand, provides an additional change in output that is due to the rate of change of the error signal. Derivative control anticipates the future value of the error signal and changes the output accordingly. This anticipating action makes the derivative mode of control very useful in controlling processes that have a sudden change in the load. PD control is frequently used in motor servo systems and in systems that have small but quick process parameter changes.

Proportional Plus Integral Plus Derivative (PID) Control

When proportional, integral, and derivative control operations are combined in a single instrument it is considered to be a three-mode or PID controller. This type of instrument has a proportional response to signal error changes, with automatic reset to reduce output offset, and an anticipating action that reduces errors caused by sudden load changes. The PID controller is a complex instrument that is designed to control difficult industrial processes.

The PID controller is a very unique instrument that is widely used to control a number of difficult processes with a great deal of precision. This type of controller is generally more expensive than other units, and it is more difficult to prepare for operation. In some instruments each mode of operation can be selected for independent use by programming in the desired operation. Each mode of control must be individually adjusted or tuned to make it functional. PI and PD control can also be accomplished by instrument programming. PID controllers are generally not used for all controller applications today. Controller selection is determined by such things as the amount of precision control needed, difficulty of the process being controlled, initial setup and tuning procedure, characteristics of the process being controlled, and the initial cost of the controller.

CONTROLLER INPUT CIRCUITRY

The input circuitry of a controller shows how the different modes of operation are connected to produce a control function. Figure 13-7 is a simplified schematic diagram of the input circuitry of a controller. This circuit shows the discrete components of a PID instrument. The blocks of this diagram represent different amplifier functions. Amplification can be achieved with bipolar transistors, MOSFETS, or operational amplifiers.

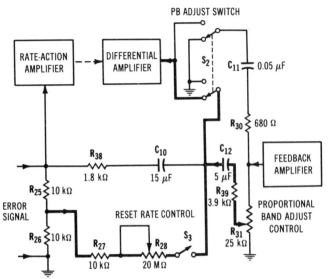

FIGURE 13-7 Simplified schematic diagram of controller input circuits.

The controller circuit simplification of Fig. 13-7 shows that the error signal is applied to a voltage divider made up of R_{25} and R_{26}. This signal is a dc voltage with a magnitude of 10 V or less. A plug attached to the controller can be positioned in either direct or reverse position. In the direct position, a positive input produces an increase in controller current. In the reverse position, the positive input produces a decrease in controller current. This input is fed directly to the derivative amplifier. Its output serves as one of the inputs to the differential amplifier.

The first input to the differential amplifier is a signal that combines integral and proportional control operations. One-half of the error signal is developed across R_{26}. This voltage is fed to the integral control circuit. The integral circuit is made up primarily of R_{26} and C_{12}. R_{27}, R_{39}, and a portion of R_{31} are also included in the circuit. R_{28} determines the number of repeat operations per minute. Its value is much larger than the other resistors all combined together. Its value is largely responsible for the time constant of the circuit. This time constant determines how often the proportional response is repeated. In other words, C_{12} charges to the change in the error-signal voltage; R_{28} determines the length of time it takes C_{12} to charge. The voltage across C_{12} is the second input to the differential amplifier. The amplifier is stabilized by a feedback voltage that is equal and opposite to the input. How this signal is produced will be discussed as feedback in the solid-state controller amplifier section.

Proportional control is determined by the percentage of feedback amplifier output that is applied to the differential amplifier input. The percentage of feedback is determined by the setting of the proportional-band control, R_{31}. The voltage at the wiper arm of R_{31}

determines proportional control as well as a reference voltage for C_{12}. The integral control voltage or the charge across C_{12} adds to this voltage. The sum of these two voltage values makes up the input to the differential amplifier.

The proportional-band adjust switch (S_2) plays an important role in the operation of the solid-state controller. This switch is incorporated in the proportional-band adjust (R_{31}) procedure by being a push-and-turn type of adjustment. When the switch is pushed, the normal differential amplifier input is grounded. The other pole of S_2 feeds an input into the amplifier that represents the charge voltage across C_{11}. During normal operation C_{11} maintains a charge that is equal to the voltage across R_{31}. When S_2 is thrown, the ground is removed from one side of C_{11} and applied to the input of the differential amplifier. Since this voltage is the same as that maintained previously, no change in output is assumed until after the adjustment has been made. The result is a smooth transfer from one proportional-band setting to another.

SOLID-STATE CONTROLLERS

Nearly all controllers placed into operation in recent years are of the solid-state type. Discrete component solid-state controllers were first introduced around 1970. These controllers have become an industrial standard for controlling process variables. Figure 13-8 shows a representative schematic diagram of a discrete component solid-state controller.

Power Supply

The power supply of our representative discrete component controller has two dc outputs. They are +36 and +46 V, respectively. Each power supply uses silicon diodes to provide full-wave rectification. See the power supply in the lower-left corner of the schematic diagram. Transformer T_3 supplies ac for the two power supply circuits.

Filtering of the power supply is not quite as obvious in the controller circuit as is generally displayed in other schematics. The +36-V source, for example, is filtered by a pi-section filter near the center of the diagram. See printed-circuit locating points 32, 33, and 10. The +46-V supply, by comparison, employs an LCR filter composed of L_2, C_{12}, and R_{31}. These components are located between printed-circuit locating points 26 and 27.

Amplifier Circuitry

The amplifier of our representative solid-state controller can be divided into three basic sections. These are the impedance bridge and internal feedback loop,

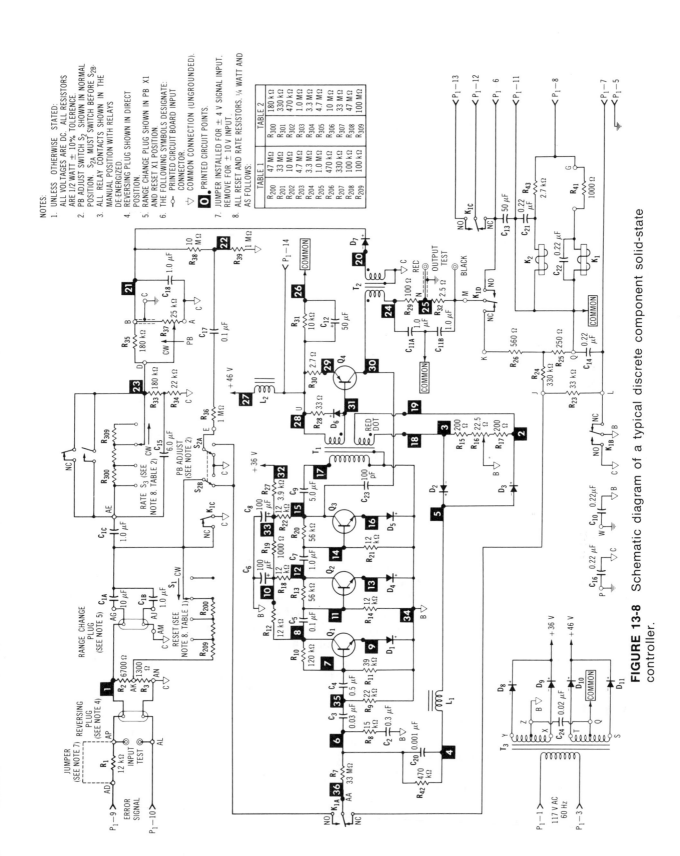

FIGURE 13-8 Schematic diagram of a typical discrete component solid-state controller.

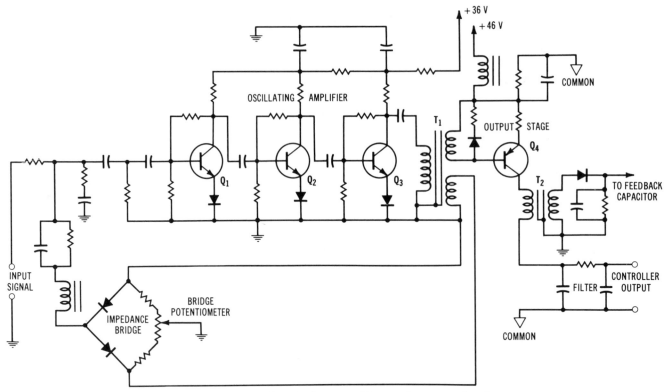

FIGURE 13-9   Oscillating amplifier and output circuit.

the three-stage oscillating amplifier, and the output stage. A simplified schematic for this part of the controller is shown in Fig. 13-9.

INTEGRATED CIRCUIT CONTROLLERS

Integrated circuit controllers are solid-state instruments that employ operational amplifiers to accomplish controller functions. This type of circuit construction has in general replaced the discrete component solid-state controller. Through this construction technique the IC has simplified the circuit, enhanced the control capabilities, reduced the size of the instrument, and simplified the explanation of its operation.

Error Amplifiers

The input or error signal amplifier of an IC controller receives signals from the set-point or desired value and the process variable or actual circuit value. The resulting output or error signal is the difference between these two input values. The error signal is either negative or positive, depending on whether the process variable signal is above or below the set-point value.

An operational amplifier can be used to perform the error amplifier function of a controller. Figure 13-10 shows an operational amplifier circuit that responds as an error amplifier. The process variable input is la-

beled V_{pv} and the set-point input is identified as V_{sp}. The resulting output signal is the difference between the two inputs. This operational amplifier responds as a difference amplifier. The developed error signal (V_{error}) equals the set-point (V_{sp}) minus the process variable (V_{pv}) or $V_{error} = V_{sp} - V_{pv}$.

Proportional Control

Operational amplifiers can be used very effectively to accomplish proportional control. The equation for this operation is the sum of two linear input quantities. One input is the developed error signal voltage (V_{error}) of the error amplifier. The other input is the offset voltage (V_{os}) adjustment. These inputs are brought together and form a summing point connected to the inverting

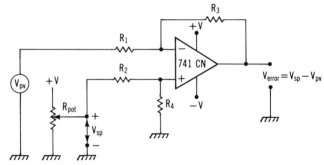

FIGURE 13-10   Op-amp error amplifier.

input. The op-amp is used as a summing amplifier that establishes the proportional band of operation. The gain factor of the amplifier is determined by the resistance values of R_f/R_i. If the resistance ration is $1:1$, the proportional band is 100%. Under these conditions, a full-scale input change produces a full-scale output change. A change in the resistance ratio changes the proportional band. A range of proportional bands of $1:100\%$ or $10:1000\%$ can be established by the resistance ratio. In addition to a variation in range of the proportional band, a change within the range can be provided by using a potentiometer for R_f. This potentiometer (R_1) then becomes the proportional-band adjust control of Fig. 13-11. By adjusting this control, the amount of output that is used in the feedback circuit can be varied. With a decreasing feedback signal, a larger output is developed. An increased output for a given input is merely a narrowing of the proportional band. A full range of output can therefore be produced by small input signals. The range of input in percent that produces a full range of the output signal is considered to be the proportional band of the circuit.

Integral Control

An operational amplifier integral control circuit is shown in Fig. 13-12. This circuit has a voltage divider formed across the input and a variable resistor placed in parallel with capacitor C_{in} for reset action. The voltage divider and variable resistor provide a path for current whenever an error signal is applied to the input. This current provides a continuous amplifier input whenever the process variable signal differs from the set-point value. Feedback current continues to charge or discharge the feedback capacitor as long as there is an input. The amplifier then produces a continuous change in output as long as an input signal exists. This signal is the integral function or reset action of the

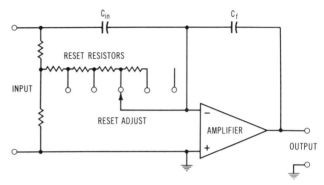

FIGURE 13-12 Integral controller.

circuit. In discussing the proportional circuit, it was evident that there was current only when there was a change in the input signal. In the integral circuit, the only time that the input current stops is when the error voltage is zero. By decreasing the value of the reset resistors, input current is increased for a given input signal. The output current must increase at a faster rate. Increasing the size of the reset resistor results in a slower reset rate. To produce a very low reset rate, a voltage divider has been incorporated in the input circuit. The voltage divider causes the reset resistors and the amplifier to see a smaller portion of the input signal, resulting in a smaller charging current. The input capacitor still has the full input voltage applied to it. The value of the reset resistor determines the number of times that control operation is repeated per minute.

Derivative Control

The derivative mode of control can be achieved with an operational amplifier. Figure 13-13 shows an op-amp derivative control circuit and the mathematical equation of its components. In this diagram, notice that

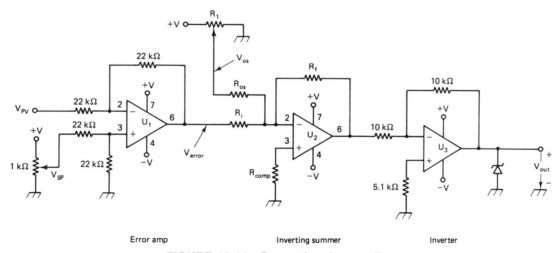

FIGURE 13-11 Proportional controller.

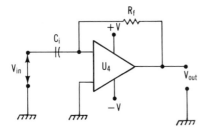

FIGURE 13-13   Derivative controller.

the process variable or input signal is applied to capacitor (C_i) connected to the inverting input. Feedback for the circuit is achieved by resistor (R_f). Derivative gain is the product of these two values. Proper selection of R_f and C_i permits the derivative control circuit to anticipate changes in the process variable signal by quickly reacting to input changes. This makes the circuit respond faster, which ultimately permits the set-point to be followed more closely.

In a derivative-mode op-amp circuit, the gain is set by adjusting the time constant of R_f and C_i. In some circuits this can be made a variable adjustment. The

value of R_f can be determined by a potentiometer. The larger the value of the R_f, the longer the time constant and the greater the effect it will have on the response of the output. Conversely, a shorter time constant has less effect on the response of the output. The time constant of this circuit is often called its *rate action*. The rate action of a derivative control circuit will not respond very effectively to an input signal with a large constant error. A derivative circuit is nearly always used in combination with another mode of control.

Proportional-Integral-Derivative Control

Proportional plus integral plus derivative control can be achieved very effectively by combining op-amps in a special circuit configuration. Figure 13-14 shows a simplified schematic diagram of a parallel-connected PID operational amplifier circuit. Op amp U_1 is the error amplifier. V_{pv} is the process variable input voltage and V_{sp} is the set-point value. The output of the error amplifier (V_{error}) is applied to U_2, U_3, and U_4. U_2 achieves the proportional control function. U_3 is a junction field-effect transistor op amp that achieves the

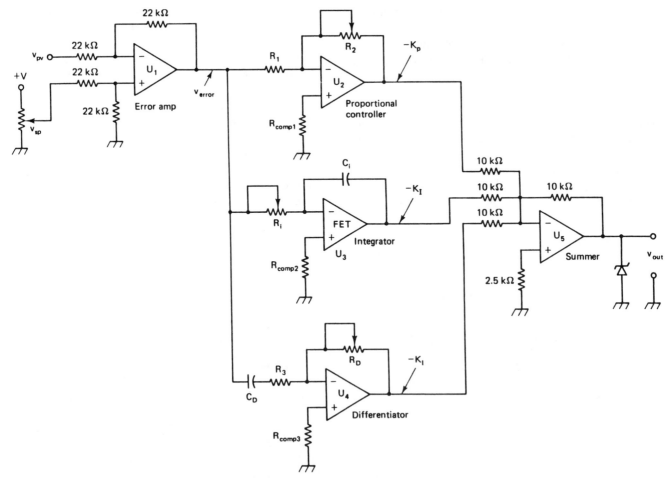

FIGURE 13-14   Three-mode controller.

integral function. U_4 achieves the derivative control operation. The outputs of U_2, U_3, and U_4 are added together at the inverting input of U_5. Operational amplifier U_5 is a summing amplifier that adds equally weighted input values and inverts the sum at its output. An op-amp PID controller can be used in practically all process control applications simply by tuning the time constant and gain of the respective control circuits.

AUTOSELECTOR CONTROL STATIONS

The autoselector permits the use of several controllers to operate a single output load such as a valve. Each of these controllers monitors related variables. In the situation chosen for demonstration of the autoselector, two controllers are feeding the autoselector. We want the one with the highest output to operate the valve, as shown in Fig. 13-15. Other modes of operation are available with the use of the autoselector; the situation that we illustrate here merely demonstrates its operation.

Controller 1 in Fig. 13-15 is assumed to be deliv-

ering a 40-mA signal. The current develops a voltage drop across dummy load R_1. We will assume that the output of controller 2 is 10 mA. This current develops a voltage across R_2. R_1 and R_2 have equal resistances. The voltage at point A will then be higher than the voltage at point B. The voltages thus produced are such that diode D_1 will conduct and D_2 will not. The signal fed to the autoselector is then coming from controller number 1. If the output of controller number 2 increases, the voltage drop across R_2 increases. When the voltage drop across R_2 exceeds the voltage drop across R_1, D_2 will conduct and D_1 will not. The autoselector current then comes from the second controller.

RATIO-SET STATION

It is often desirable to use two controllers whose outputs maintain a constant ratio. For example, controller 1 will have a variable output and the second controller output should be twice its value. To perform such a function a ratio-set station, represented schematically in Fig. 13-16 is used. The output of controller number

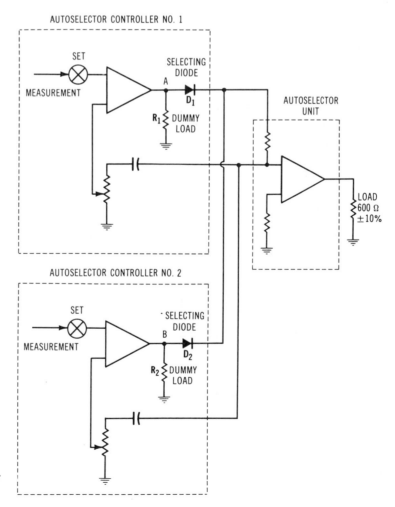

FIGURE 13-15 Autoselector controller.

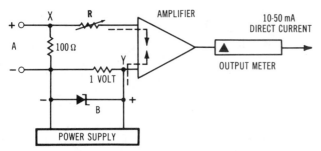

FIGURE 13-16 Ratio-set station.

1 is represented by the letter A. This current develops a voltage drop across the 100-Ω resistor. At 0% signal (10 mA dc) a 1-V signal is developed. A zener diode–regulated voltage source develops a 1-V signal at point B. Since the polarities of these two voltages oppose each other, the potential between points X and Y is 0 V. The magnetic amplifier is so biased as to produce a 10-mA output. As the signal at point A increases, a current passes through variable resistor R and the control winding of the magnetic amplifier. The setting of R determines the current through the control winding; R can be set so that the amplifier output is the desired ratio of the input at A.

SET-POINT STATION

The set-point station is a computer link with a process controller. The computer output to the set-point station is either an upscale or a downscale pulse. These pulses operate two relays. An upscale pulse will close the relay, which provides a voltage to a reversible motor in such a direction as to drive the set-point control upscale. When the desired set-point is reached, the relay will be deenergized. The set-point control so operated is much the same as the manual set-point mentioned previously. A downscale pulse closes the second relay which provides a voltage to reverse the motor and drive the set-point down scale. The motor also drives a dial indicator and a feedback potentiometer. The voltage of the feedback potentiometer feeds a signal to the computer that corresponds to the input of the controller.

ALARMS

Foxboro alarm units are designed to open or close contacts when a measurement signal exceeds a preset limit. Although units are available for duplex alarms, only a single alarm will be discussed here. A simple wiring change will cause the alarm to function on a high signal rather than a low signal as demonstrated here.

Figure 13-17 shows a simplified schematic diagram of a single alarm unit. Three voltage sources provide the input to a two-stage transistor amplifier. When a pulsating current passes through the primary of the coupling transformer T_x, it induces a voltage in the transformer secondary. The amplifier strengthens this ac signal to a level sufficient to trigger the alarm relay.

Two of the three inputs to the alarm amplifier are dc voltages and the other is a pulsating or ac voltage. The first of these voltages (V_i) is developed across resistor R_i by transmitter or controller output. This current is in the range 10 to 50 mA. The second dc voltage, V_s, is developed across resistor R_s by a zener-regulated power supply. The amount of this voltage appearing in the alarm unit input is determined by the limit control, potentiometer R_s. The polarity of V_s opposes that of V_i. The voltage across diode D_1 is the difference between these two dc voltages. The diode will conduct only when the amplitude of V_s is greater than that of V_i.

The alternating voltage source, V_{ac}, is rectified by diode D_2. This half-wave rectified voltage is developed across resistor R_p. Since the ac voltage is rectified it will always have the same polarity. The polarity of this voltage is in such a direction as to add to V_i; therefore, it will further subtract from V_s. If diode D_1 is reverse biased (nonconducting) by V_i and V_s, V_{ac} will simply strengthen the reverse bias during each pulse. If, however, D_1 is forward biased (V_s greater than V_i), the V_{ac} will decrease the amount of forward bias during each half-cycle. This will produce a changing current through T_x, and an amplifier input will trigger the alarm.

The only wiring change necessary for a high-limit alarm is to change the polarity of diodes D_1 and D_2. Changing the polarity of D_1 means that it will conduct when V_i exceeds V_s. Changing D_2 will change the polarity of the half-wave pulses across R_p. Therefore, when D_1 is forward biased (V_i greater than V_s), V_{ac} will subtract from this forward bias. The resulting

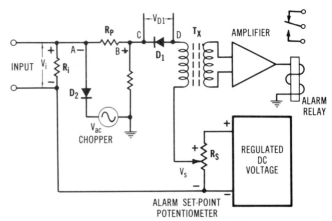

FIGURE 13-17 Alarm unit.

changes in current through T_x will again sound the alarm.

SUMMARY

An electronic controller is an instrument that is responsible for the automatic control of process variables. A process variable can be temperature, flow, liquid level, pressure, or any physical parameter that changes its value during a manufacturing operation.

The operational modes of a controller include on–off, proportional, integral, derivative, proportional plus integral, proportional plus derivative, and proportional plus integral plus derivative.

On–off or two-state control is the simplest of all control operations. This mode of control switches the load or output on or off without any intermediate level of operation. The output changes or oscillates above or below the set-point value during this control operation.

Proportional-only control determines the difference between a set-point value and the process variable being controlled by a system. In conventional operation this controller has a difficult time reducing a steady-state error signal to zero.

The integral mode of control develops an output that is the time integral of the difference between the set-point and the value of the process variable being measured. Pure integral control has a very poor transient response and is not used alone because it acts only on long-term or steady-state errors.

The derivative mode of control develops an output that is related to the time rate of change that occurs between a process variable and its set-point. The input of this controller is differentiated so that its rate of change can be detected and produce a proper output signal. Derivative control is not used alone because it only reacts to measurements that are changing and does not change with steady-state errors.

When proportional and integral modes are combined they provide automatic reset action, which eliminates offset of the output. PI operation is used to control large load changes that have frequent set-point adjustments. This permits a process to be controlled without prolonged oscillations, no offset, and have a quick recovery after a disturbance.

Proportional and derivative modes of control are combined to reduce oscillations and to allow for a higher gain setting. PD control is used in systems that have small but quick process paramater changes.

Proportional, integral, and derivative control modes are combined in the operation of a PID controller. This combination of control has proportional response to signal error changes, with automatic reset to reduce output offset, and an anticipating action that reduces errors caused by sudden load changes. A PID controller is widely used to control difficult processes with a great deal of precision.

Integrated circuit operational amplifiers can be used today to accomplish all controller modes of operation. Through the use of op amps the circuit is simplified, has enhanced control capabilities, reduced size, and simplified the explanation of its operation.

ACTIVITIES

CONTROLLER FAMILIARIZATION

1. Refer to the manufacturer's operational manual of an electronic controller.
2. Describe the type of control achieved by the instrument used in this activity.
3. The following activity steps should be performed with the power turned off or disconnected from the instrument.
4. Locate the wiring connections for the controller. Draw a diagram of the wiring layout. Describe the location of the wiring connections.
5. Locate the controls and indicators of the instrument.
6. Identify the type of indicator used in the controller.
7. Locate the set-point control. Describe how the set-point is adjusted.
8. Describe the different adjustments that can be made on the controller.
9. Remove the housing or covering from the controller.
10. Identify the location of the power supply and primary circuit boards.
11. Notice how the circuit boards are attached to the instrument. Are they hard-wired or do they plug-in to connectors?
12. Describe the general construction of controller.
13. Does the instrument have any internal controls? Identify the function of each control.
14. Reassemble the controller.

CONTROLLER OPERATION

1. Refer to the manufacturer's operational manual of the controller to be used in this activity.
2. Identify the controller type, its model number, and any of its distinguishing features.
3. Connect an input and output to the controller. For example, use a thermocouple for the input and a heater for the load.

4. Turn on the controller or apply power to the instrument.

5. Adjust the set-point of the controller to a temperature value slightly above the ambient temperature. Place the thermocoupole near the heating surface of the heater load.

6. The controller should respond by altering the heating element temperature. Describe the operation of the controller's output with respect to the set-point value.

7. Program the controller for PID operation.

8. Describe the programming procedure.

9. Adjust the set-point to a higher temperature value.

10. Describe the operation of the controller.

11. Describe the response of the proportional time, proportional band, derivative time rate, and integral time of reset.

12. Does this controller have a tuning procedure? If so, review it in the operational manual and locate the controls that are adjusted.

13. Turn off the controller and disconnect the input and output.

QUESTIONS

1. In your own words, what is a controller?

2. Define the term *process variable*.

3. What is an on–off controller?

4. What is a limitation of on–off control?

5. Describe the operation of a proportional-only controller.

6. Define the term *offset*.

7. Which mode of control has an offset problem?

8. Describe the term *integral control*.

9. Why is integral control not used alone?

10. What is another term for integral control?

11. Describe what is achieved by derivative control?

12. Why is derivative control not used alone?

13. What is meant by the term *combinational control?*

14. What are the advantages of using operational amplifiers to achieve controller functions?

15. What is meant by the term *set-point?*

CHAPTER 14

PROGRAMMABLE LOGIC CONTROLLERS

OBJECTIVES

Upon completion of this chapter, you will be able to:

1. Define a programmable controller.
2. Explain the operation of a relay ladder diagram.
3. Identify the basic components of a programmable controller system.
4. Identify the parts of a programmable controller block diagram.
5. Explain how each block of a programmable logic controller diagram functions.
6. Explain the operation of a programmable logic controller input module.
7. Explain the operation of a programmable logic controller output module.
8. Describe the operation of a motor control circuit actuated by a programmable logic controller.

IMPORTANT TERMS

In this chapter we are going to investigate some of the basic principles of the programmable logic controller. In this study you will have a chance to become familiar with a number of programmable logic controller functions. New words, such as *ladder diagrams, sinking, scanning,* and *isolation,* will begin to have some meaning for you. A few of these terms are singled out for further study. As a rule, the chapter will be more meaningful if these terms are reviewed before proceeding with the text.

Actuating: The process of moving or controlling something indirectly or by a functioning machine or instrument.

Assembly language: A symbolic representation of a binary code that is very efficient.

Cathode-ray tube (CRT) terminal: A digital system video display and keyboard assembly.

Executive program: A firmware package of instructions that tells a microprocessor what sequence to follow in performing its operations.

Input/output (I/O) port: The hardware or circuitry of a microprocessor that deals with the transmission or reception of logic information between peripheral devices.

Ladder diagram: A circuit representation that resembles the structure of a ladder. The vertical lines or rails represent the power source, while the horizontal lines or rungs are used to connect components to the source.

Menu: A list of program options that a user can select to be processed by a computer system.

Message storage: A memory allotment that stores information pertaining to operating conditions of a microprocessor.

Microcomputer: An assembly of components that has a central processing unit, memory, input/output interface, and bus network housed on IC chips that will accomplish computer functions.

Mini-PCs: Microprocessor based programmable controllers that are capable of accomplishing a few relay functions, timing, and counting operations.

Programmer: A hardware unit that is used to program a programmable logic controller.

Rail: The vertical part of a ladder diagram that represents the power source of the circuit.

Read-only memory (ROM):   A memory device or circuit in which memory cells contain fixed information imparted to them during the manufacturing process.

Relay language: A programming language that uses the relay contact as its basic construction element.

Rung: The horizontal part of a ladder diagram that represents a number of components that are used to perform a sequential step in the operation of a circuit.

Scanning: A cycling process of automatically checking the condition or status of information placed in memory locations due to programming or input data.

Scanning rate: The time that it takes to cycle through data placed in the memory locations of a programmable controller.

Simulator module: A plug-in component that enables the system operator to reproduce a condition that is likely to occur in the actual performance of a programmable controller.

Sinking: The ability of a device or circuit to dissipate, give off heat, or pass current to the common terminal of the electrical power source when in its operational state.

INTRODUCTION

For a number of years industrial control had been achieved by electromechanical devices such as relays, solenoid valves, motors, linear actuators, and timers. These devices were used to control manufacturing operations, industrial processes, and heavy equipment where only switching operations were necessary. Most control was of the two-state type which simply called for a machine to be turned on or off. The control circuitry was hardwired to the machine and considered to be a permanent installation. Modification of the system was rather difficult to accomplish and somewhat expensive. In industries where production changes were frequent, this type of control was rather costly. It was, however, the best way and in many cases, the only way that control could be effectively achieved.

In the late 1960s, solid-state devices and digital electronics began to appear in industrial controllers. Circuitry that utilized this type of control was designed to replace the older electromechanical devices. The transition to solid-state control has, however, been more significant than expected. Solid-state devices, digital electronics, integrated circuit technology and computer-based systems have lead to the development of programmable controllers or PCs. These devices have capabilities that far exceed the older electromechanical controllers. Programmable controllers permit flexible circuit construction techniques, have reduced downtime when making changeovers, operate with improved efficiency, and can be housed in a very small space.

The first programmable controllers could only perform a limited number of functions. Two-state control, AND, OR, and some limited timing functions were the extent of the control capabilities. Today, this type of controller can perform all logic functions, do arithmetic operations, and can sense analog changes in a manufacturing operation. It can accept a millivolt signal from a thermocouple, multiply it by a constant, and display the results in degrees Celsius. The resulting control operation can be stored in memory for future use, displayed on a cathode-ray tube, or used to energize an alarm. The unique features of a modern PC are flexibility, operational efficiency, and versatility.

PROGRAMMABLE CONTROLLER SYSTEMS

A programmable controller should be viewed as an operational system. It has a source of electrical energy, a path for it to follow, control, and load devices to do the intended work function. The PC is primarily responsible for system control. It achieves this operation in a rather unusual way when compared with other electronic controllers. A PC is applied to system applications where a moderate to low-cost electrical assemblage is desired along with less than 100,000 logical operations per second as a requirement. Generally, a PC can economically replace an electromechanical assemblage of 10 or more relays with the added advantage of significantly more operational versatility.

A block diagram of a simplified programmable controller is shown in Fig. 14-1. Note that the control function of this system is achieved entirely by the programmable controller. PCs in general are more complex than indicated by a single block of the diagram.

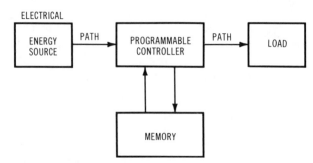

FIGURE 14-1 Programmable controller system.

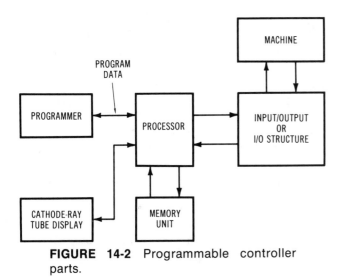

FIGURE 14-2 Programmable controller parts.

A programmable controller, for example, has a number of unique parts or functions in its physical makeup. Specifically, these are described as the input/output or I/O, the processor, memory, the display, and the programmer. The programming panel of this system is not considered to be an essential part of an operating PC. It can be disconnected from the system once the programming function has been achieved. The system will operate satisfactorily without the programmer connected. Figure 14-2 shows an expansion of the PC indicating these functions.

The block diagram of a programmable controller is very similar to that of a microcomputer system. In fact, most programmable controllers are classified as computer-based systems. This type of system is de-

signed to perform a number of control functions in the operation of an instrument, machine, or manufacturing process. The degree of sophistication or "power" of a PC depends on its internal construction. Some units are equipped with a large number of I/O connection ports. This type of unit can accommodate a number of loads and input devices by simply making the appropriate connections to a module. The power capability of a PC is determined by its I/O capacity. This is achieved by plug-in modules. Figure 14-3 shows a representative PC with I/O modules plugged into a track assembly. Connection points to a respective module are made at the top and bottom of the track assembly. The processor, memory, power supply, and power distribution function are all housed in the control box of the track unit. The programmer is completely independent of the system. A data access and display module or DAD/M is also included with the unit. Expansion of this system is achieved by simply adding a number of I/O modules to the track assembly. Several track units can be interconnected to one controller assembly. A larger track assembly can also be used to accommodate more modules if the need arises.

The programmable controller of Fig. 14-3 is classified as a mini-PC. This type of unit is designed to control a small number of machine operations and a variety of manufacturing processes. Mini-PCs are classified as systems that can economically replace as few as four relays in a control application. They are capable of providing timer and counter functions, as well as relay logic, and are small enough to fit into a standard 19-in. rack assembly. Most systems of this type can accommodate up to 32 I/O ports or modules. Larger

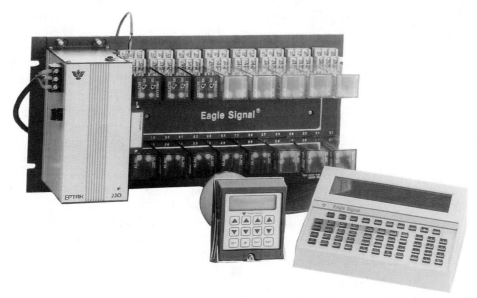

FIGURE 14-3 Mini-programmable controller. (Courtesy of Eagle Signal Co.)

units can accommodate up to 400 I/O ports or devices. Mini-PCs, in general, can achieve control similar to that of larger units, but on a smaller basis. These units are less expensive, easier to use, smaller, and more efficient than the larger PC units. In the future, most relay applications of industry will be accomplished by mini-PCs.

PC OPERATION

A circuit controlled by a PC is uniquely different from that of an equivalent hard-wired electrical circuit. Figure 14-4 shows an electrical circuit connected in a ladder diagram format. This circuit diagram is presented so that it resembles a ladder. The rails or vertical part of the diagram represent the power source or electrical distribution lines. Power flows from the left rail to the right rail by horizontal lines. The horizontal lines represent the rungs of the ladder. Devices, contacts, or connections on the left side of the rung are used to achieve control. Electromagnetic devices, loads, and outputs are connected on the right side of the rung. The operational sequence of the diagram starts at the top rung and is followed by the succeeding rungs. Most PCs are programmed and displayed in a ladder diagram format.

The hard-wired circuit of Fig. 14-4 is connected in a conventional manner. In this case, PB_1 controls lamp L_1 and PB_2 controls lamp L_2. The switch in both circuits is connected directly to the source and controls each lamp. The normally open (NO) push button PB_1 turns on lamp L_1 when *it* is pressed, while the normally closed (NC) push button PB_2 will turn off lamp L_2. Operation is achieved by direct connection of the control device (PB_1 and PB_2) to load device (L_1 and L_2) and the power source.

Figure 14-5 shows an equivalent PC system that achieves control of two loads. This circuit has several differences when compared with a hard-wired circuit. The switches, for example, and not connected directly

to the lamps. PB_1 and PB_2 are connected to input modules. Each input module performs a switching function that is applied to the processor. The lamps are connected to the processor through output modules. Control of each output module is directed by the processor. Essentially, switches are used to control the input function which is monitored by the processor. The processor then directs the output module to conform with programmed information that controls power flow to the load. Respective loads are controlled by the output modules. In a strict sense, the processor is isolated from the switching function by the input module, and from the power source feeding the load through the output module.

Input Module Circuitry

The input module of a PC is extremely important in the operation of a circuit. It is responsible for connecting an external input source to the PC so that it modifies the operation of the processor. The input source usually necessitates some degree of electrical isolation to protect the delicate input of the processor. As a rule, input module circuitry has a prescribed operating voltage and current rating. This varies a great deal among different manufacturers. Typically, the input voltage has an operational range of several volts. Some representative values are 20 to 28 V ac/dc, 10 to 55 V dc, 105 to 130 V ac/dc. The current needed to actuate the input is generally of a nominal value. Operational ranges of 10 to 50 mA are very common. Different modules are also made to accommodate a number of inputs, ranging from one to eight in current PCs.

Figure 14-6 shows a wiring diagram for a representative 24-V ac/dc input module. This particular module will accommodate eight input circuits. Typical input devices include push buttons, limit switches, selector switches, and relay contacts. The input will interface either 24 V ac or dc from an external source to the processor. The actuating voltage source is derived from circuitry outside of the input module. When the module accommodates a number of inputs, each one has a reference location or number designation. This number is the same for each module. It will vary however, according the working position of the module in the system. In practice, the number has three or more digits. The least significant place value refers to the module circuit number. The next most significant place value usually denotes the input module location in the system. The most significant place value generally denotes the function of the module, such as input or output.

The circuitry of one input of the previous module is shown in Fig. 14-7. This circuit responds to either an ac or dc energy source. The circuit actuates an LED

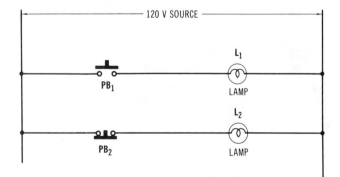

FIGURE 14-4 Hard-wired electrical system control.

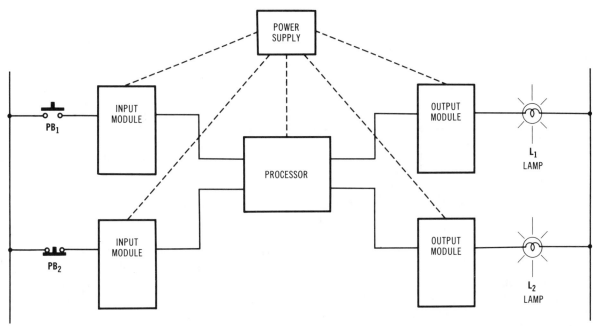

FIGURE 14-5 PC system.

in its output. With ac applied, the bridge will be energized and produce dc to energize the LED of the optical coupler. When the optical coupler is energized, output energy is transferred to the processor through a phototransistor. The coupler isolates the processor from the input source voltage. A similar response will be produced by either polarity of dc applied to the input. The LED on the left or line side of the circuit serves as an indicator to show when the input is being energized.

One circuit of the input module of Fig. 14-7 uses 90 mW of power from its source when actuated. A representative circuit might respond to 9 volts dc at 10

mA. Each of the eight input circuits will use the same amount of power when actuated. All eight circuits in operation at the same time will use 8 × 90 or 720 mW of power from the external energy source. Essentially, the operating energy of the input module does not detract from energy being supplied to the processor.

Output Module Circuitry

The output module of a PC is responsible for connecting the processor to a load device being energized by an outside source. These modules vary a great deal in their design and operation. Some units may house only one output circuit per module, while others may incorporate several individual output circuits in a single module. The circuitry of a module will vary a great deal among different manufacturers. The components of a module must be capable of sinking the energy source supplied to the load device. *Sinking* refers to the ability of a device to dissipate or give off heat. An output module usually contains a power control device such as a transistor, silicon-controlled rectifier (SCR), or triac. The power dissipation rating of this device determines its sinking capability. This function of the output module is dependent on ambient temperature. Most output modules have rated operating temperatures that must be followed in order to assure that the output device will not be damaged. Some modules fuse each circuit to protect the output device from damage. Figure 14-8 shows a wiring diagram of a typical output module. This particular module will accommodate four outputs. Note that these are numbered as 070, 071, 072,

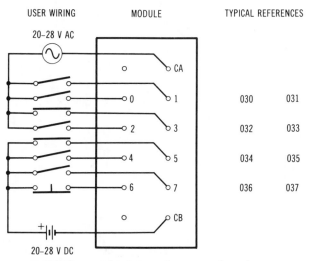

FIGURE 14-6 Wiring diagram of an input module. (Courtesy of General Electric—Programmable Control Dept.)

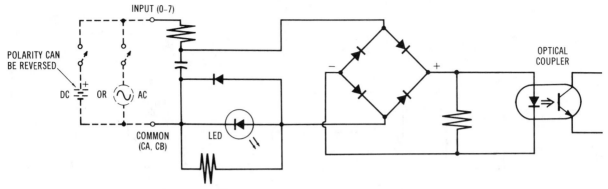

FIGURE 14-7 Circuitry of an input module. (Courtesy of General Electric—Programmable Control Dept.)

and 073. The loads are energized by a dc source of 5 to 24 V dc.

The circuitry of one section of the output module is shown in Fig. 14-9. This particular circuit has an optical coupler connecting the output of the processor to the module. This is used to isolate the processor from the power source of the load device. The output device of the circuit is an n-channel MOSFET of the depletion type. A string of zener diodes connected across the source–drain of the FET is used to regulate the voltage to a value that will not destroy the device. These diodes will conduct if the voltage rises above a prescribed value. The output of this circuit is also protected by a fuse. If the source–drain current exceeds a prescribed value, the output device will be protected from damage. An external source is used to energize the output transistor. A signal from the processor actuates the gate of the FET, causing it to conduct. The signal, in a sense, is the connecting link between the processor and load device operation. When an appropriate signal is applied to the input of the module, the output is energized. This action ultimately controls the load device. The circuitry of an output module varies a great deal among different manufacturers. The circuit shown here is only representative of that used by one manufacturer. The external source voltage and sinking capability of the module generally dictate the circuitry and the type of output device used. Mini-PCs use low-level output sinking circuits, while larger PCs may use heavy-duty modules with high-level sinking capabilities. As a rule, the output circuit develops only two states or conditions of operation—on and off. The load device changes according to these conditions.

The Processor

The processor is the heart and brains of a programmable controller. The processor is essentially a powerful computer housed on an IC chip. This chip is mounted on the main printed circuit board and may have several other auxiliary chips connected to it to form the composite processor. The combined board

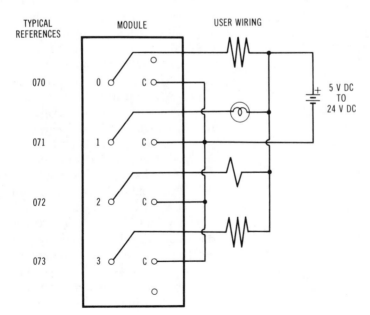

FIGURE 14-8 Wiring diagram of an output module. (Courtesy of General Electric—Programmable Control Dept.)

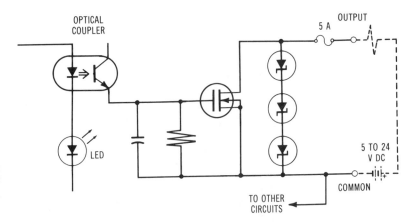

FIGURE 14-9 Circuitry of an output module. (Courtesy of General Electric—Programmable Control Dept.)

and chip assembly performs the functions needed to make the PC operational.

Figure 14-10 shows a PC with the processor chassis mounted in the lower right-hand corner of the assembled unit. This unit has input and output modules mounted on the same rack assembly. The processor chip and circuit board layout cannot be observed from this view. Industrial equipment of this type generally has the processor and its associated components housed in an environment protected enclosure.

A view of a disassembled processor board is shown in Fig. 14-11. This unit has a power supply,

memory chips, signal conditioning units and the processor on a single board assembly. This type of board construction permits the chips to be interconnected without using external wires. Digital signals applied to the assembly move along the foil lines of the printed circuit board. Off-board components can be connected to the board by an interconnecting cable. Sixteen I/O modules of this system are attached to an off-board assembly by a ribbon connector. This type of construction permits a number of external components to be connected to the processor board using a minimum of conductors.

FIGURE 14-10 Programmable controller. (Courtesy of Industrial Solid State Controls, Inc.)

FIGURE 14-11   View of a disassembled processor board. (Courtesy of Analog Devices.)

Processor Operation

Operation of the processor of a PC is somewhat complex when compared to other instruments. A person does not need to know a great deal about the overall operation of the internal workings of the processor in order to use it effectively. In general, only a few of the basic operational facts of the PC are needed to make it functional.

The processor of a PC has a master program permanently stored in its memory. This program is needed to make the processor operational when it is turned on initially. In a sense, this program material is similar to the firmware of a computer-based system. Without this program material in the processor, the system would not function when it is energized. The master program or firmware does several things in the normal operational sequence. It must energize the system, tell the processor to send a startup menu to the CRT terminal or display, and open the operational channels to energize the keyboard. This permits the user to see an operational menu on the display unit and make the necessary selections to start programming of the keyboard. The firmware of a PC is generally called the executive, or ROM, program. This program also monitors the master switch of the PC panel. The switch is used by the programmer or PC user to determine whether the processor uses a program keyed into memory, or if it operates from a program stored in memory. Some PCs have the option of selecting a test program from memory without energizing the outputs.

Allen Bradley calls the positions of their master select switch "run program," "run," "test program," and "program." Modicom and other manufacturers use specific switch selections instead of a master key switch to set up the operation of their system. Once the selection is made, the processor begins running through the executive program automatically.

If the user selects the program switch position, the processor recognizes signals from the keyboard and begins the program development procedure. If the run or test switch position is selected, the processor begins to execute the user's program while looking for input signals, and develops appropriate output signals. To stop the program sequence but leave the PC powered up, the select switch is placed in the stop processor position. The operator of the PC has control over the switch selection procedure, and the processor responds according to its directions.

Memory

The memory of each processor is generally somewhat different because of the chip used and the operations it must perform in a PC. The differences are usually in the size of the storage space for programs, counters, timers, messages, and registers. Figure 14-12 shows a layout of the memory allocation of a typical PC. Notice that this shows the memory divided into distinct parts. These parts are made of binary bits. Sixteen bits are grouped together to form words. Memory allotment is made in total words. The distribution of memory is

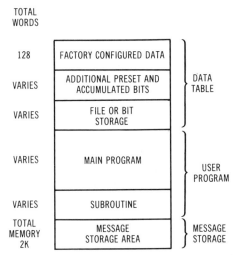

FIGURE 14-12 Memory layout of a PC. (Courtesy of Allen Bradley Co.)

divided into three groups called the data table, user program, and message storage. The data table uses 128 words for factory configured data and a variable amount of data space for preset data and file/bit storage. The user program section houses the main program data and some of the subroutines of operator-controlled programs. The third group deals with message storage. This space allotment is variable and stores messages pertaining to operating conditions. If more memory is needed for a particular operation or control function the system can be expanded to accommodate this.

The memory size of a PC can be expanded to meet the needs of the system where it is being used. A typical system is generally purchased with a rather small memory to control one machine with a limited number of I/O ports. If the need arises, the memory can be expanded to meet the demands of a larger operation. Most PC manufacturers feel that this is the best solution for memory selection. The system can be designed to fit any application by expanding memory according to the operations being performed.

The memory of a PC varies a great deal among different manufacturers. Presently, memory is specified in bytes. A group of 8 bits is called a *byte*. Computer-based systems are usually described according to the word size of an instruction. A *word* is also 8 bits in length. The terms word and byte can be used interchangeably. A 16-bit computer-based system is said to have a word size of two bytes. Microprocessors used in PCs can have an 8-bit word size or one-byte instructions. The letter K as used with computers is a symbol equivalent to the numeral 1024. Memory is generally specified in 1000 bytes or 1K increments. Typical memory sizes of a PC are 1K, 2K, 4K, 8K, 16K, 32K, 64K, 128K, and 256K. As a rule, the more

memory that a PC has, the more operations it can perform.

Program scanning

When a processor checks the memory of a PC and executes the program stored there, it must follow a procedure that tells what is stored at each memory location. This procedure is called program scanning. Some PCs scan the memory by vertical columns. A vertical scan usually starts at the top left corner and moves to the bottom of the first column, then up through the next column, down to the bottom of the next column, and continues through the remaining columns in the same manner. Horizontal scanning is similar to eye movement when reading a printed page. The memory elements are scanned across vertical columns in a horizontal line or one rung of the ladder diagram at a time. The scanning procedure used by a particular system is determined by the manufacturer.

Most PCs may have 10 or 11 vertical columns in their format. The number of columns refers to the number of element functions of a rung in a ladder diagram. The Modicon 484, for example, can support 10 elements plus one output in each of its rungs. This element number is largely determined by the processor being used and the programming plan of the system. Obviously, the number of elements in a specific rung cannot exceed the number of columns established by the system.

The scanning procedure of a PC deals with the program material placed in memory. Essentially the entire program is scanned many times per second. Scanning rates vary from 4 ms to several hundred milliseconds, depending on the memory size. A millisecond is one thousandth of a second. This means that a great deal of program scanning can occur in a very short period of time. Scanning is done to update the inputs, outputs, timers, counters, and math registers. In mini-PCs the scanning time is not of any major concern. In large PC systems scanning time may be an important issue. Data registers, for example, may not be updated rapidly enough to keep the data accurate in a long program. In some systems, parts of the program can be skipped or subroutines may be developed that will make critical functions more accurate.

PC PROGRAMMING

Instructions for the operation of a programmable controller are given through push buttons, a keyboard programmer, or a magnetic disk drive assembly. Each PC has a special set of instructions and procedures to make it functional. How the unit performs is based on its programming procedure. In general, PCs can be programmed by ladder diagrams, logic diagrams, or Boolean equations. These procedures can be expressed as language words or as symbolic expressions on a CRT. One manufacturer describes these methods of programming as assembly language and relay language. Assembly language is used by the microprocessor of the system. Relay language is a symbolic logic system that employs the relay ladder diagram as a method of programming. This method relies on relay symbols instead of words and letter designations. We use the relay ladder method of programming in this presentation.

Relay Logic

The processor of a PC dictates the language and programming procedure to be followed by the system. Essentially, it is capable of doing arithmetic and logic functions. It can also store and handle data and continuously monitor the status of its input and output signals. The resulting output being controlled is based on the response of the signal information being handled by the system. The processor is generally programmed by a keyboard, program panel or CRT terminal. Figure 14-13 shows the layout of a relay ladder programmer. This particular panel uses a liquid crystal display. The

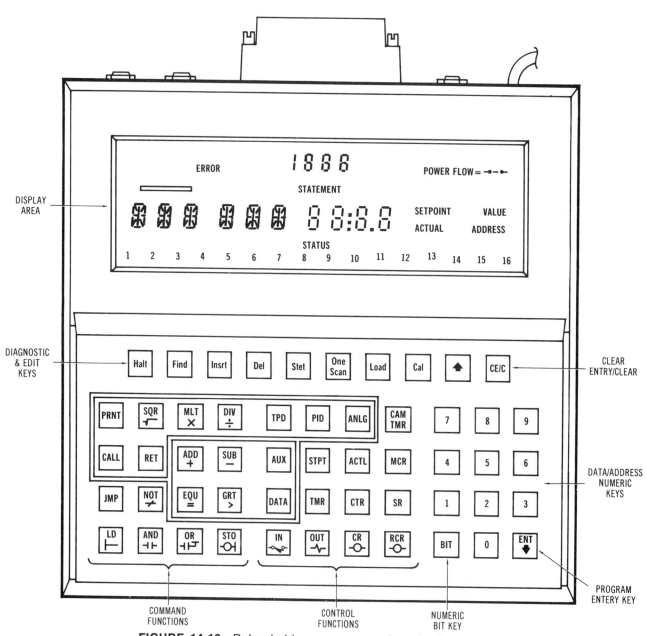

FIGURE 14-13 Relay ladder programmer layout. (Courtesy of Eagle Signal Co.)

display area is divided into three fields or areas. The top field shows the program statement number, error indicator, and power flow information. The middle field shows the alphanumeric details of the contents of a statement number or the dynamic status of data registers, timers, or counters. The bottom field displays the dynamic status of the input/output (I/O), control relay (CR), retentive control relay (RCR), shift registers (SR), and cam timers (CAM TMR).

In a relay language system, the basic element of programming is the relay contact. This contact may be normally open (NO) or normally closed (NC). Figure 14-14 shows the symbolic expression of these contacts. Note that the normally open contacts are on the left and the normally closed contacts are on the right. The line on the right side of the two lower contacts is for connection to branch circuits. This is generally an optional circuit possibility. Below each contact is a four-digit reference number. This number is used to identify specific contacts being used in the system. The contact is then connected in either series or parallel to form a horizontal rung of the relay ladder diagram.

Once the relay program has been entered into the PC, it can be monitored and modified if the need arises. For systems with a CRT, modification is accomplished on the display. This is achieved by simply moving the cursor of the CRT to the component being altered and making the change with a key stroke. For systems without a CRT, modification is made by reviewing the program one step at a time. Changes are made by altering the program statement so that it conforms with the desired procedure. Most systems of this type may have simulator modules that can be placed at strategic locations to monitor program operation. The program can be stepped through to see if the sequence is correct.

All the control components of a PC are identified by a numbering system. As a rule, each manufacturer has a unique set of component numbers for its system. One manufacturer has a four-digit numbering system for referencing components. The numbers are divided into discrete component references and register references. A discrete component could be used to achieve on and off control operations. Limit switches, pushbuttons, relay contacts, motor starters, relay coils, solenoid valves, and solid-state devices are examples of discrete component references. Registers are used to store some form of numerical data or information. Timing counts, number counts, and arithmetic data may be stored in register devices. All component references and register references are identified by the numbering system. Each manufacturer has some distinct way of identifying system components.

Each I/O module of a programmable controller has a distinct reference number or address to identify its location in the system. The number in general identifies the location of the module in the system. If a track system is used, the reference number refers to a specific track location. Some systems may identify an input module with a four digit number beginning with a one (1). The output module is then identified by a four digit number beginning with a zero (0). The prefix number cannot be altered in the programming procedure.

Assume now that a relay diagram has been placed into the PC by selection of proper number data entries and symbol selections. The PC must then examine this network and solve the interconnected logic elements in the proper sequence. In doing this, the first rung or network of the ladder must be solved. Then networks 2, 3, and 4 must be solved in order. The solving of each sequence is achieved by a series of scanning pulses. These pulse scans occur at a rather high rate. Each pulse passes through the network in a specific sequence. Scanning occurs in a PC when power is first applied and continues as long as the system is energized. This permits each network rung of the ladder to be solved from the left rail to the right and from the top to the bottom in an appropriate sequence. This assures that each network is solved according to its numerical step and not by the value assigned to a specific coil or contact.

Programming Basics

Programmable controllers are provided with the capability to program or simulate the function of relays, timers, and counters. Programming is achieved on a format of up to 10 elements in each horizontal row or rung of a relay diagram, and up to seven of these rungs connected to form a complete network (Fig. 14-15). A network can be as simple as a single rung or a combination of several rungs as long as there is some interconnection between the elements of each rung. The left rail of the ladder can be the common connecting element. Each network can have up to seven coils connected in any order to the right rail of the ladder. These

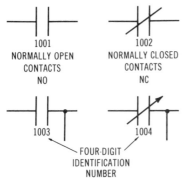

FIGURE 14-14 Programming format of relay contacts.

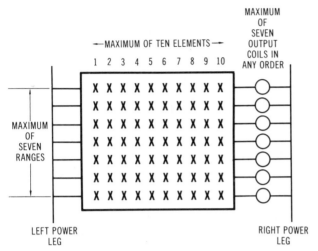

FIGURE 14-15 Relay ladder programming board.

coil numbers can only be used once in the operational sequence. The quantity of discrete devices and registers available for use depends on the power or capacity of the system.

When programming a relay ladder diagram into a PC, the discrete devices and registers are placed in the component format of Fig. 14-14. Each component in this case is assigned a four-digit identification number. The specific reference number depends on the memory size of the system. In a low-capacity system, number assignments could be 0001 to 0064 for output coils and 0258 to 0320 for internal coils. A system with a larger capacity might use number assignments of 0001 to 0256 for output coils and 0258 to 0512 for internal coils. Any coil output or internal coil can only be used once in the system. References to contacts controlled by a specific coil can be used as many times as needed to complete the control operation. Output coils that are not used to drive a specific load can be used internally in the programming procedure.

When programming the response of a particular input module, it may be identified as a relay contact. In this regard, the symbol may be a normally closed contact or a normally open contact. The coil or actuating member of the contact takes on the same numbering assignment. The coil, however, is identified as a circle on the diagram and the contacts are identified by the standard contact symbol. Figure 14-15 shows some examples of the symbol identification procedure. The number designation is used to identify specific devices and contacts. The contacts can be programmed to achieve either the NO or NC condition, according to its intended function.

Any external input that is considered to be normally closed, such as a safety switch, overload switch, or stop pushbutton, must be treated differently. An external NC pushbutton, for example, would not be entered on a CRT as closed contacts. It would produce the opposite effect internally from that of a NO contact. Inverting the external contact function, as well as its signal, constitutes a double inversion operation. It is for this reason that all normally closed external contacts or switches are programmed as normally open on the CRT.

Assume now that the simple start–stop motor controller of Fig. 14-16 is to be connected by a programmable controller. The start and stop buttons are located externally. Pushing the start button energizes the relay coil. This action latches the relay coil by closing contacts CR_1 across the start button. Contacts CR_2 close at the same time, completing the energy path to the motor, thus causing it to run. The motor continues to run as long as energy is supplied from the source. Pushing the stop button turns off the motor and removes the latch from the start button.

A programmable controller equivalent of the motor starter of Fig. 14-16 is shown in Fig. 14-17. The number assignments refer to the specific components of the PC. Input devices are numbered 1001 and 1002. This includes the input module and its resulting switching operation. The output device is numbered 0049. The start and stop buttons are externally connected

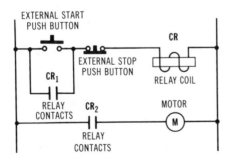

FIGURE 14-16 Simple start–stop motor controller circuit.

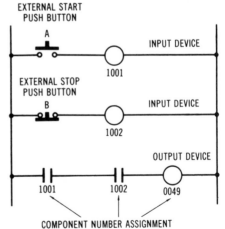

FIGURE 14-17 PC equivalent of motor controller circuit.

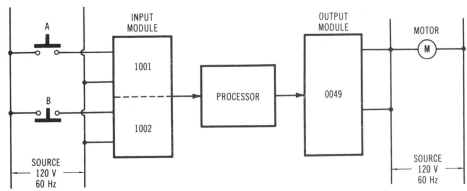

FIGURE 14-18 Actual PC circuit of a motor controller.

and do not have a module number assignment. Operation of the PC equivalent circuit will achieve the same control procedure as the original relay ladder diagram.

The actual PC circuit for a motor start-stop control operation is shown in Fig. 14-18. This diagram shows how the I/O modules are interfaced with the processor. Note that the input modules and output modules are treated as independent parts of the system that are controlled by the processor. This circuit would be displayed on a CRT type of indicating system and could be modified with a few simple keystrokes. The PC equivalent is somewhat more complex than its ladder diagram equivalent. It is, however, more versatile and can be modified very quickly by a program change. Programming is simply a process of entering the appropriate component number assignment and then designating the function to be achieved by each component. The procedure can then be placed in memory and retained for future use, or used immediately according to the needs of the system.

SUMMARY

For years the control circuitry of industrial machinery had been achieved by hardwired devices connected in permanent installations. Modifications were difficult to accomplish, rather expensive, and caused a rather large amount of down time. The development of programmable controllers reduced these problems.

A programmable controller of PC should be viewed as an integral part of a functioning electrical system. The PC is actually responsible for system control. The PC has a number of functional parts in its physical makeup. These are described as the input/output, processor, memory, the display, and the programmer. A block diagram of a PC is very similar to that of a microcomputer system.

A PC and the circuitry that it controls are generally viewed in a ladder diagram configuration. The

rails or vertical members of the ladder represent the power source. Power flows from the left rail to the right side through rungs. The rungs represent discrete lines of a circuit. The operational sequence is from the top rung of the ladder to the bottom rung.

The input module is extremely important in the operation of a PC. It connects the external input source to the processor so that it modifies circuit operation. This module isolates the input source from the delicate input of the processor.

The output module of a PC is responsible for connecting the processor to a load device being energized by an outside source. The components of the output module must be capable of sinking the energy source supplied to the load device. The module also isolates the load source from the output of the processor.

The processor is the heart and brains of a programmable controller. It is essentially a computer housed on a printed circuit board. As a rule, several other auxiliary chips are included on a board to form the composite processor.

The processor has a master program or firmware instructions permanently stored in its memory. This program tells the processor what to do when it starts and sends operational instructions to the keyboard.

The memory of a processor varies a great deal among different units. It is distributed between the data table, user program, and message program. Memory is measured in thousands of bytes.

PC operation is achieved by scanning the vertical columns and rows of data placed in memory. This material is scanned several times per second to update the operation of the system. Scanning occurs on a continuous basis when the PC is operational.

Instructions for the operation of a programmable controller are given through push buttons, a keyboard programmer, or a magnetic disk drive assembly. Each PC has a specific set of instructions that it must follow in order for it to be functional. In general, programming is done by a ladder diagram format. In this procedure, the relay contact is the basic element of the programming structure. Once the program has been entered, it can be modified or changed as the need arises. Each component of the system is identified by a number. When components are placed in the system and referenced with a number, operation is achieved by examining the networks line by line in a scanning sequence.

Programming of a PC is accomplished by using a format of 10 elements in each horizontal row of the ladder and up to seven rungs in a network. The network can be as simple as a single rung, or any combination of up to seven rungs. Programming is simply a process of entering the appropriate component number assignment and then designating the function to be achieved by each component. All of this is placed in memory and used immediately or retained for future use.

ACTIVITIES

PC FAMILIARIZATION

1. Refer to the manufacturer's operational manual for the programmable controller to be used in this activity.
2. Attach the programming unit to the PC and connect a load device to a selected output. Connect an input device to selected inputs.
3. Note the location of the inputs and outputs being used by the controller. These will be used in the programming procedure.
4. Prepare a program that will accomplish a simple control operation that will turn on an output when an input is energized.
5. Program the unit to achieve this control operation.
6. Test the operation of the program to see if it achieves the desired level of control.

PC PROGRAMMING

1. With the aid of the manufacturer's operational manual, explain the component addressing procedure for the inputs and outputs of the PC being used.

2. Using the programming procedure outlined for your system, develop a program that will achieve the motor control circuit of the Fig. 14-16.
3. Energize the power source for the I/O of the system.
4. Test the operation of the program to see if it achieves the desired level of control.
5. Erase the program and clear the display.

PROGRAMMING

1. Using the programming procedure for your system, develop diagrams that will permit the inclusion of the following items into a functional system: (a) AND gate, (b) OR gate, (c) timer, and (d) event sequencer.
2. Each item should be programmed into the system independently.
3. Make a sketch of each diagram employed noting the assignment of the components used.
4. Explain the general operating procedure of the circuit.

QUESTIONS

1. What are the fundamental parts of a programmable controller?
2. Define the term *programmable controller*.
3. A block diagram of a programmable controller is similar to that of what type of system?
4. What is a ladder diagram?
5. What is the primary function of an input module?
6. What is the primary function of an output module?
7. Explain what is meant by the term *isolation*.
8. What is the function of the processor in the operation of a programmable controller?

9. What does the memory of a programmable controller do?
10. In general, the programming done on a PC is in what type of format?
11. The memory of a programmable controller is divided into three groups. What are they?
12. What is meant by the term *scanning?*
13. How are instructions delivered to a programmable controller?
14. Of what significance is memory to a programmable controller?

CHAPTER 15

ELECTRONIC CONVERTERS

OBJECTIVES

Upon completion of this chapter, you will be able to:

1. Identify the major parts of a block diagram of an electronic converter.
2. Explain the operation of the measuring circuit of an electronic converter.
3. Given a schematic diagram of an mV/I converter, trace the signal path from the input to the output.
4. Explain the functional role of an electronic converter.
5. Define commonly used electronic converter terms, such as *demodulator, chopper,* and *push-pull output.*

IMPORTANT TERMS

In the study of electronic converters one frequently encounters a number of new and somewhat unusual terms. These terms play an important role in the presentation of this material. A few of these terms are singled out for study before proceeding with the chapter. A review of these terms will make the reading of the text more meaningful.

Chopper: An electronic device or circuit that interrupts the flow of current or a change in voltage at a regular rate.

Converter: An electronic circuit that changes a measured parameter applied to its input into a different parameter of an equivalent value at its output.

Demodulator: An electronic circuit or system function that picks out or detects the changing component of a modulated electronic signal.

Diffusion: A process in the formation of solid-state materials that occurs when current carriers move from one material to the other during its fabrication.

Filter: A selective network made of active or passive components that permits the passage of a band of frequencies while rejecting the passage of other frequencies.

Full-wave rectifier: A diode circuit that permits both positive and negative alternations of an ac waveform to pass through its output or load. Through this circuit configuration, an ac voltage is changed to a single directional dc voltage.

Modulation: The process by which a characteristic of one signal is varied in proportion to the information contained in another signal.

Negative feedback: An output to input signal path where the output signal is 180° out of phase with the input, which causes a reduction in circuit gain and distortion.

Parameter: An arbitrary constant factor whose value changes define the operating characteristics of a system.

Pneumatics: A branch of mechanics that deals with the mechanical properties of air or gas.

Pressure regulator: A device that varies or prevents variation of pressure in a system.

Push–pull amplifier: A balanced amplifier circuit that uses two similar transistors that work in phase opposition.

INTRODUCTION

An electronic converter is responsible for changing a measured parameter applied to its input into another type of parameter of an equivalent value at its output. Electronic converters are of the voltage input to current output (V/I), millivolt input to current output (mV/I), current input to voltage output (I/V), current input to pressure output (I/P), voltage input to frequency output (V/f), analog input to digital output (A/D), and digital input to analog output (D/A). A converter is often described as the "electronics" of a process control system. The system input employs a transducer or sensor that changes the form of energy supplied to its input. Temperature to millivolts, pressure to resistance, light level to resistance, and liquid flow to voltage are typical transducer functions. A transducer's output is then applied to the input of the electronic converter. The output of the converter is designed to meet the demands of the system to which it is connected. Typically, the output is a current value of 4 to 20 mA. This is used to drive the final control element, a controller, a transmitter, or a recording instrument. Essentially, an electronic controller interfaces the transducer to the load or work function of the system.

The mV/I converter of Fig. 15-1 and the mV/P converter are discussed in this chapter. Both circuits produce an electrical current output with a range of 4 to 20 mA. The mV/P converter, however, incorporates a current-to-pressure transducer in its output. Pressure is used to drive the final control element. The electronics of both converters is identical. They will be discussed simultaneously.

The discrete component converter described in this chapter is divided into three sections. The first section involves the measuring circuit and its associated power supply. This section of the instrument incorporates a zener diode that provides a regulated dc suppression voltage. The measuring circuit is designed to handle a standard millivolt input from a thermo-

couple. When the thermocouple input is used, cold junction compensation is provided. From the measuring circuit the signal passes through a filter and a chopper circuit. In this part of the circuit any stray ac signals or transients are filtered out of the input signal. The filtered dc input is then changed to a 60-Hz voltage by the chopper circuit. Negative feedback from the amplifier is fed into the circuit to produce an error signal.

The third section of the converter involves an amplifier and a demodulator circuit. The amplifier has a very high gain. Its input is an error signal that is only 0.5% of the measured signal value. The advantages of using a high-gain amplifier with negative feedback have been discussed previously. You may recall that it minimizes variables within the amplifier and reduces the loading effect of the measuring circuit. The converter output is a current value of 4 to 20 mA that is proportional to the millivolt input.

MEASURING CIRCUIT AND POWER SUPPLY

Figure 15-2 is a simplified schematic of the input or measuring circuit. You will notice from the diagram that two voltages are added algebraically to the input millivolt signal. The first of these voltages is added to the input by a bridge circuit in series with the negative input terminal. The reference or suppression voltage for this bridge is provided by a zener diode regulated power supply that makes the use of a standard cell or constant standardization unnecessary. The values of R_1, R_2, R_3, and R_6 vary with the range of the input. P_1 is a zero-adjust potentiometer. With a millivolt input, the value of RN_1 is zero. Under these conditions, the bridge adds a signal to the millivolt input so that the amplifier input range is within the capabilities of the instrument.

With a thermocouple input, RN_1 provides for cold-junction compensation. This resistor is wound

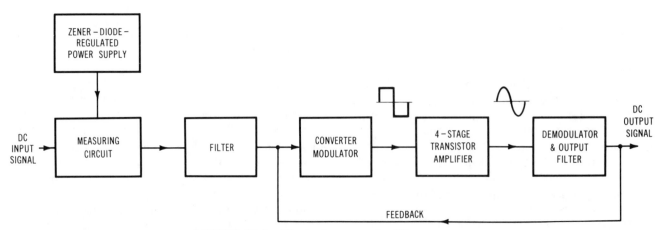

FIGURE 15-1   Block diagram of mV/I converter.

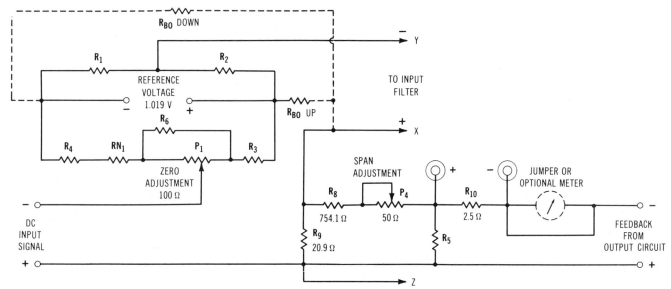

FIGURE 15-2 Simplified schematic diagram of measuring circuit.

around the cold-junction compensating block. A voltage drop across this resistor subtracts a voltage corresponding to the ambient temperature. R_{BO} provides protection in the case of a thermocouple failure. Figure 15-2 shows the position of this resistor for either upscale or downscale deflection. The bridge circuit adds to the negative input potential a voltage that compensates for ambient temperature and differences in range.

Feedback voltage is added algebraically to the potential at the positive input terminal. Output current develops the negative feedback voltage across R_5. This voltage is also developed across the parallel voltage divider made up of P_4, R_8, and R_9. The voltage across R_9 subtracts from the potential present at the positive input terminal. The portion of the voltage developed across R_9 is controlled by the span-adjust potentiometer, P_4.

The zener diode regulated power supply is shown in Fig. 15-3. This supply provides the 1.019 V necessary for the operation of the measuring circuits. The power supply input is taken from a 90-V center-tapped

winding of the power transformer. Full-wave rectification is accomplished by diodes D_6 and D_7. The rectifier output is filtered by C_9. This dc output is applied to a voltage divider made up of R_{19} and two zener diodes. A zener diode always maintains a constant voltage across it. The zeners used in this power supply produce a regulated 7 V. Two of these in series produce 14 V across the voltage divider made up of R_{12} and R_{13}. The potential between R_{12} and R_{13} is the negative input to the bridge. R_{CU} provides for temperature compensation. P_2 adjusts the standardizing potential. This adjustment needs to be checked only biannually. The positive potential for the bridge circuit comes from a second stage of regulation. The 14-V output of the power supply is developed across R_{11} and another zener diode, across which the potential is again 7 V. This is the potential at the positive side of the bridge circuit.

A zener voltage regulator is produced by connecting a reverse bias across a pn junction. If this voltage is high enough, a breakdown of the diodes occurs

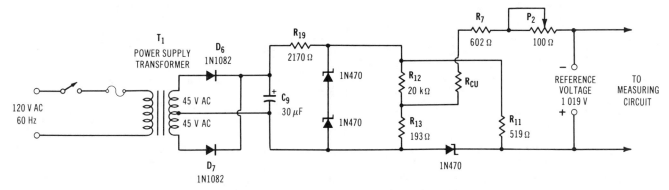

FIGURE 15-3 Zener diode regulated reference voltage power supply.

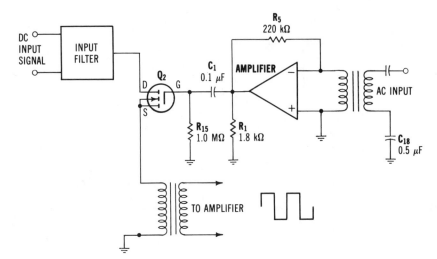

FIGURE 15-4 Electronic dc to ac chopper.

and current flows. The voltage drop so produced will be constant and is called the *zener voltage*. The value of this voltage is determined by how the pn junction is formed. The zener voltage can be raised by diffusing the pn junction; that is, the higher the zener voltage, the wider the area in which there is a gradual transition from a p-type to an n-type crystal.

FILTER AND INPUT CONVERTER

A discrete component electronic converter employs a signal filter and a dc to ac chopper in its circuitry. Figure 15-4 shows a representative example of this cir-

cuit. The filter is used primarily to eliminate any stray ac pickup that may have occurred in the measuring circuit. The circuit must also stabilize any input voltage supplied to the MOSFET that is developed by the feedback loop. Unstable conditions that occur in the chopper circuit are due to the changes of impedance that take place between conduction and cutoff of the MOSFET. The chopper operates the same as those used in the instruments previously discussed. The MOSFET of this circuit is synchronized with 60 Hz supplied by the ac power source. As a result, the applied dc input will produce a representative square wave that is supplied to the primary winding of the

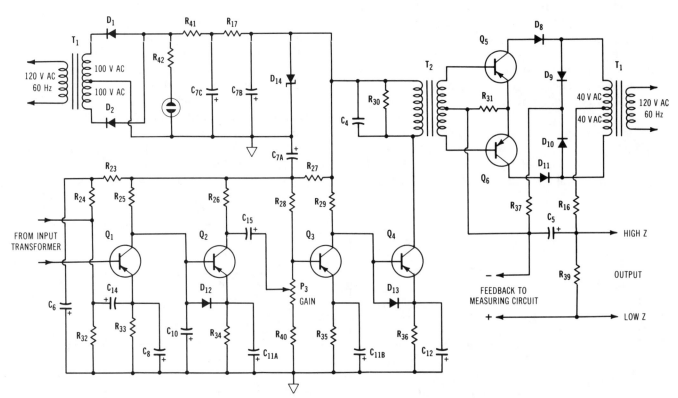

FIGURE 15-5 Amplifier and demodulator circuits.

input transformer. This signal contains an ac component riding on a dc voltage value. This is often described as a fluctuating dc signal. When this signal is applied to the primary winding of the input transformer the fluctuating component produces an ac voltage across the transformer's secondary winding. This signal voltage is then applied to the input amplifier. Its amplitude and phase depend on the value of the dc input voltage from the measuring circuit. This circuit is primarily responsible for filtering, changing a dc signal to ac, which is often called *modulation,* and applying the ac component of the modulated signal to an amplifier for further processing.

AMPLIFIER AND OUTPUT CIRCUITS

Figure 15-5 is a schematic diagram of the amplifier and output (demodulator) circuits. Four stages of voltage amplification are produced in this circuit. A very high gain is produced in this amplifier. All four stages of

FIGURE 15-6 Electric-to-pneumatic converter. (Courtesy of Leeds & Northrup.)

amplification use common-emitter, pnp transistors. The first and second stages are dc coupled, as are the third and fourth stages. The second and third stages are capacitance coupled. Transistor Q_3 incorporates a gain control in its base circuit. The first three transistors develop their outputs across a resistive load in the collector circuit. Q_4 uses the primary of an output transformer as its collector load; collector current through this primary produces a voltage across the secondary. This voltage is the input to the output or demodulator stage.

The amplifier power supply produces about -32 V dc. The power supply input comes from a 200-V ac center-tapped secondary of the power transformer. Full-wave rectification is provided by diodes D_1 and D_2; filtering is accomplished by capacitors C_{7C}, C_{7B}, and C_{7A}, and resistors R_{41}, R_{17}, and R_{27}.

A push-pull output stage is composed of transistors Q_5 and Q_6. The amplifier output is applied to the bases of these transistors by the center-tapped secondary of the output transformer. The collector voltage of these transistors is provided by an 80-V center-tapped power transformer secondary. Half-wave rectification of this voltage is provided by D_8 and D_{11}. As long as the base voltage and collector voltage are in phase, the collector will conduct current. The amount of this current, of course, depends on the amplitude of the base voltage. Transistor Q_5 will conduct during the other half-cycle. This pulsating current is filtered by capacitor C_5. As a result, the dc output current is the average collector current of Q_5 and Q_6.

MILLIVOLT-TO-PRESSURE CONVERTERS

Millivolt-to-pressure (mV/P) converters serve as a link between an electrically energized controller and a pressure-actuated system. Figure 15-6 shows a representative electric-to-pneumatic type of converter. This converter, or transducer, is essentially a motor-operated pressure regulator which converts varying dc signals into 3 to 15 lb/in^2 (20.7 to 103.4 kPa) air-pressure signals. Design features are specifically aimed at eliminating abrupt process upsets. This includes such things as self-locking valve position settings with power failure, and smooth valve adjustment without continuous balance adjustment. Figure 15-7 shows the converter with the moisture-resistant cover removed.

SUMMARY

Converters are devices that form a link between a process transducer input and the controlled output. Typical converters have low-voltage inputs (millivolts) and outputs that are a value of current or pressure.

A millivolt-to-current (mV/I) converter employs

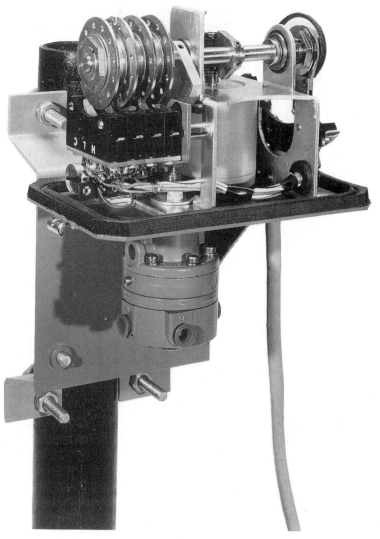

FIGURE 15-7 Converter of Fig. 15-6 with cover removed. (Courtesy of Leeds & Northrup.)

a zener diode–regulated reference voltage, measuring circuit, filter, dc-to-ac converter, amplifier, demodulator, and output circuitry.

The measuring circuit algebraically adds two voltages to the input millivolt signal. The first one is added to the input by the bridge circuit. Feedback voltage is then added to the positive input terminal. The resulting output is filtered and applied to the converter where it is changed from dc into ac.

The ac signal is then amplified by several common-emitter transistor stages. The output is a push-pull stage composed of two transistors. These transistors amplify the signal on alternate half-cycles which

would normally develop a sine-wave output. By employing diodes in the output, the ac is rectified. As a result, when the base and collector voltages are in phase, the collector will conduct current. The amount of current depends on the amplitude of the base voltage. The pulsating dc is then filtered, with the output being an average value of the combined collector current.

Millivolt-to-pressure (mV/P) converters serve as a link between electrically energized controller action and a pressure-actuated output. Typical converters of this type change varying dc signals into 3 to 15-lb/in² (20.7 to 103.4-kPa) air-pressure signals.

ACTIVITIES

CONVERTER ANALYSIS

1. Refer to the manufacturer's operational manual for the electronic converter to be used in this activity.

2. Identify the input transducer or sensor used with the instrument. In many cases, this can be used to identify the input function of the instrument. A thermocouple, for example, produces a millivolt input signal.

3. Identify the output function of the instrument. This can be determined by the load device being driven by the instrument. A low-resistance load generally indicates that the instrument has a current output function. The resistance of the load can be measured to determine this function.

4. What input/output function does the instrument used in this activity perform?

5. Remove the housing or cover from the converter.

6. With power applied to the instrument measure and record the input voltage produced by the sensor. This should represent a stable operating condition.

7. If possible, change the status of the input sensor so that it will indicate some value change in the input signal. Measure and record the value change in the input voltage.

8. If the instrument has a current output, connect a milliampere meter in series with the load. Measure and record the stable output current of the instrument.

9. Change the status of the input as before, while noting the change in output. Measure and record the output current change.

10. Disconnect the power or turn off the instrument.

QUESTIONS

1. Define the term *electronic converter*.

2. What are the functional blocks of an electronic converter diagram?

3. What is the function of the measuring circuit and power supply of an electronic converter?

4. What does the term *mV/I* of an electronic converter represent?

5. What is the function of the filter/chopper section of a discrete component electronic converter?

6. What is the function of the amplifier/demodulator section of a discrete component electronic converter?

7. Define the term *demodulation*.

8. What is the resulting output of the mV/I discrete component electronic converter of Fig. 15-5?

9. Describe the conversion procedure that takes place in the composite discrete component electronic converter of this chapter.

CHAPTER 16

MAGNETIC AND ELECTROMAGNETIC PRINCIPLES

OBJECTIVES

Upon completion of this chapter, you will be able to:

1. Explain the differences between permanent and electromagnetism.
2. Explain the operation of a magnetic circuit with respect to magnetomotive force, flux, and reluctance.
3. Identify the parts of a *B–H* curve.
4. Explain the response of a *B–H* hysteresis loop.
5. Identify the parts, electrical symbol, and physical construction of a saturable-core reactor.
6. Explain the operation of a saturable-core reactor.
7. Identify the components, electrical symbol, and circuitry of a magnetic amplifier.
8. Explain how a magnetic amplifier is used to achieve power control.
9. Define commonly used magnetic terms, such as *flux density*, *toroid*, *core saturation*, and *permeability*.

IMPORTANT TERMS

In the study of magnetic principles one frequently encounters a number of new and somewhat unusual terms. These terms play an important role in the presentation of this material. A few of these terms are singled out for study before proceeding with the chapter. A review of these terms will make the reading of the text more meaningful.

Abscissa: The horizontal coordinate of a point that is obtained by measurements that are parallel to the *x*-axis.

Alloy: A mixture of two or more metals that are combined in a molten state or intimately fused together during a manufacturing operation.

Ampere-turn: The unit of measurement of magnetic field strength; amperes of current times the number of turns of wire.

Coercive: A restraining or dominating force that nullifies an action.

Electromagnet: A coil of wire wound on an iron core so that as current flows through the coil it becomes magnetized.

Ferromagnetic: An iron material that has high magnetic

permeability, a definite saturation point, residual magnetism, and hysteresis.

Flux: Invisible lines of force that extend around a magnetic material.

Flux density: The number of lines of force per unit area of a magnetic material or circuit.

Hysteresis: The property of a magnetic material that causes actual magnetizing action to lag behind the force that produces it.

Left-hand rule: A rule for determining the direction of a magnetic field, the polarity of an electromagnet, or the direction of an induced current flow.

Magnet: A metallic material, usually made by alloying iron, nickel, or cobalt to produce magnetic properties.

Magnetic amplifier: A device similar in construction and appearance to a transformer but is a saturable-core reactor and its associated components that permits control of ac electrical power.

Magnetic field: Magnetic lines of force that extend from a north polarity and enter a south polarity to form a closed loop around the outside of a magnetic material.

Magnetic poles: Areas of concentrated lines of force on a magnet that exhibit magnetic properties.

Magnetomotive force: A force that produces magnetic flux around a device.

Permeability: The ability of a material to conduct magnetic lines of force as compared to air or the ability of a material to magnetize or demagnetize.

Reluctance: The opposition of a material to the flow of magnetic flux.

Residual magnetism: The magnetism that remains around a material after the magnetizing force has been removed.

Saturable-core reactor: An electromagnetic device that has a core control winding and two or more ac windings. Power control is achieved through this device by changing the value of dc applied to the core winding.

Silicon-controlled rectifier (SCR): A four-layer, three-terminal solid-state device that responds as an open circuit between the anode–cathode until gate current flows. After the device has been triggered into conduction, the gate loses its control capabilities.

Teslas (T): A unit of magnetic induction that is equal to 1 weber per square meter.

Toroid: A doughnut-shaped core piece of magnetic material that has a wire coil threaded through the center hole.

Weber (Wb): The unit of magnetic flux. The amount of magnetic flux applied to a single coil of wire that will induce 1 volt into the coil when it decreases uniformly to zero.

INTRODUCTION

In this and the following chapter, we study instruments that utilize magnetism and electromagnetism in their operation. This includes such things as magnetic amplifiers, magnetic flowmeters, transmitters, recorder pen mechanisms, and servomotors. A basic understanding of magnetic principles is essential in this study.

PERMANENT MAGNETISM

Iron, cobalt, and nickel are elements that are alloyed together in the construction of a permanent magnet. Each end of the formed magnet is called a *pole*. If a magnet were broken, each part would become a magnet. Each magnet would have two poles. Magnetic poles are always in pairs. When a magnet is suspended in air so that it can turn freely, one pole will point to the North Pole of the earth. The earth is like a large permanent magnet. This is why compasses can be used to determine direction. The north pole of a magnet will attract the south pole of another magnet. A north pole repels another north pole and a south pole repels another south pole. The two laws of magnetism are: (1) like poles repel, and (2) unlike poles attract.

The magnetic field patterns when two permanent magnets are placed end to end are shown in Fig. 16-1. When the magnets are farther apart, a smaller force of attraction or repulsion exists. This type of permanent magnet is called a *bar magnet*.

Some materials retain magnetism longer than others. Hard steel holds its magnetism much longer than soft steel. A magnetic field is set up around any magnetic material. The field is made up of lines of force or magnetic flux. These magnetic flux lines are invisible. They never cross one another, but they always form individual closed loops around a magnetic material. They have a definite direction from north to south pole along the outside of a magnet. When magnetic flux lines are close together, the magnetic field is strong. When magnetic flux lines are farther apart, the field is weaker. The magnetic field is strongest near the poles. Lines of force pass through all materials. It is easy for lines of force to pass through iron and steel. Magnetic flux passes through a piece of iron as shown in Fig. 16-2.

When magnetic flux passes through a piece of iron, the iron acts like a magnet. Magnetic poles are formed due to the influence of flux lines. These are called *induced poles*. The induced poles and the magnet's poles attract and repel each other. Magnets attract pieces of soft iron in this way. It is possible to magnetize pieces of metal temporarily by using a bar magnet. If a magnet is passed over the top of a piece of iron several times in the same direction, the soft iron becomes magnetized. It will stay magnetized for a short period of time.

When a compass is brought near the north pole of a magnet, the north-seeking pole of the compass is

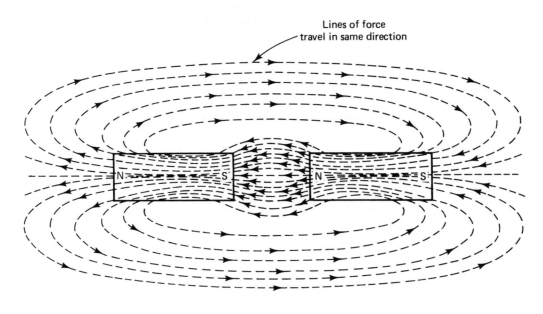

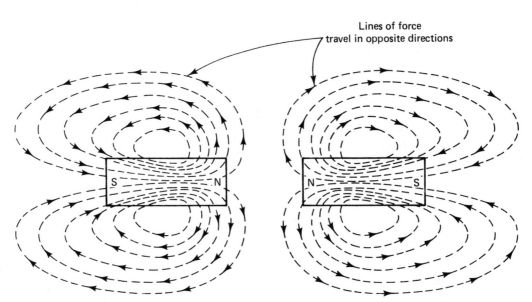

FIGURE 16-1   Magnetic field patterns when magnets are placed end to end.

attracted to it. The polarities of the magnet may be determined by observing a compass brought near each pole. Compasses detect the presence of a magnetic field.

Magnetic Field Around Conductors

When electrical current flows through a conductor it produces a magnetic field. It is possible to show the presence of a magnetic field around a current-carrying conductor. A compass is used to show that the magnetic flux lines are circular in shape. The conductor is in the center of the circular shape. The direction of the current flow and the magnetic flux lines can be shown by using the left-hand rule of magnetic flux. A conductor is held in the left hand as shown in Fig. 16-3(a). The thumb points in the direction of current flow from negative to positive. The fingers will then encircle the conductor in the direction of the magnetic flux lines.

A circular magnetic field is produced around a conductor. The field is stronger near the conductor and becomes weaker farther away from the conductor. A cross-sectional end view of a conductor with current flowing toward the observer is shown in Fig. 16-3(b). Current flow toward the observer is shown by a circle with a dot in the center. Notice that the direction of the magnetic flux lines is clockwise. This can be verified by using the left-hand rule.

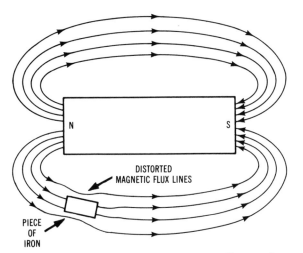

FIGURE 16-2 Magnetic flux lines distorted while passing through a piece of iron.

When the direction of current flow through a conductor is reversed, the direction of the magnetic lines of force is also reversed. The cross-sectional end view of a conductor in Fig. 16-3(c) shows current flow in a direction away from the observer. Notice that the direction of the magnetic lines of force is now counterclockwise.

The presence of magnetic lines of force around a current-carrying conductor can be observed by using a compass. When a compass is moved around the outside of a conductor, the needle will align itself tangent to the lines of force as shown in Fig. 16-3(d). The needle will not point toward the conductor when current flows in the opposite direction and the compass polarities reverse. The compass needle will again align itself tangent to the conductor.

Magnetic Field Around a Coil

The magnetic field around one loop of wire is shown in Fig. 16-4. Magnetic flux lines extend around the conductor as shown. Inside the loop, the magnetic flux is in one direction. When many loops are joined together to form a coil, the magnetic flux lines surround the coil as shown in Fig. 16-5. The field around a coil is much stronger than the field of one loop of wire. The field around the coil is the same shape as the field around a bar magnet. A coil that has an iron or steel core inside it is called an electromagnet. A core increases the magnetic flux density of a coil.

ELECTROMAGNETS

Electromagnets are produced when current flows through a coil of wire as shown in Fig. 16-6. The north pole of a coil of wire is the end where the lines of force come out. The south polarity is the end where the lines of force enter the coil. This is like the field of a bar magnet. To find the north pole of a coil, use the left-hand rule for polarity, as shown in Fig. 16-7. Grasp the coil with the left hand. Point the fingers in the direction of current flow through the coil. The thumb points to the north polarity of the coil.

When the polarity of the voltage source is reversed, the magnetic poles of the coil will also reverse. The poles of an electromagnet can be checked with a compass. The compass is placed near a pole of the electromagnet. If the north-seeking pole of the compass points to the coil, that side is the north pole.

Electromagnets have several turns of wire wound

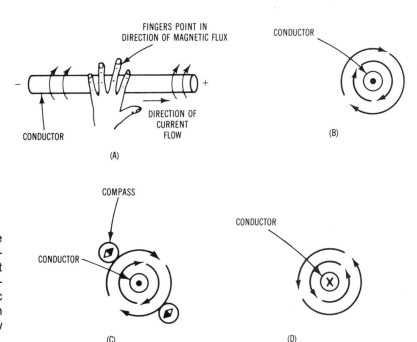

FIGURE 16-3 (a) Left-hand rule of magnetic flux; (b) cross section of conductor with current flow toward observer; (c) compass aligns tangent to magnetic lines of force; (d) cross section of conductor with current flow away from the observer.

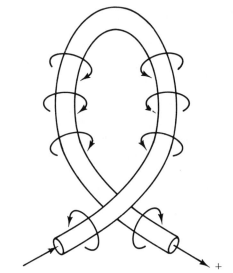

FIGURE 16-4 Magnetic field around a loop of wire.

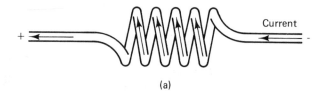

(a)

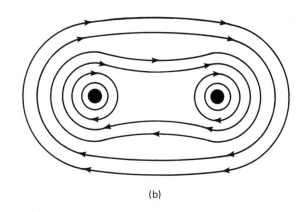

(b)

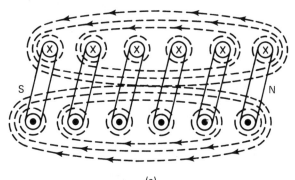

(c)

FIGURE 16-5 Magnetic field around a coil: (a) Coil of wire showing current flow; (b) lines of force around two loops that are parallel; (c) cross section of coil showing lines of force.

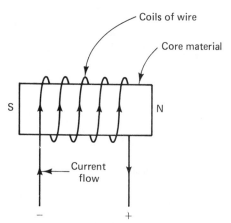

FIGURE 16-6 Electromagnet.

around a soft-iron core. An electrical power source is then connected to the ends of the wire coil. When current flows through the wire, magnetic polarities are produced at the ends of the soft-iron core. Three basic parts of an electromagnet are (1) an iron core, (2) wire coils, and (3) an electrical power source. Electromagnetism is made possible by electrical current flow, which produces a magnetic field. When electrical current flows through the coil, the properties of magnetic materials are developed.

Magnetic Strength of Electromagnets

The magnetic strength of an electromagnet depends on three factors: (1) the amount of current passing through the coil, (2) the number of turns of wire, and (3) the type of core material. The number of magnetic lines

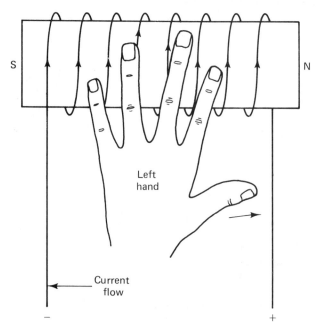

FIGURE 16-7 Left-hand rule for finding the polarities of an electromagnet.

of force is increased by increasing the current, by increasing the number of turns of wire, or by using a more desirable type of core material. The magnetic strength of an electromagnet is determined by the ampere-turns of each coil. The number of ampere-turns is equal to the current in amperes multiplied by the number of turns of wire, or $I \times N$. For example, 200 ampere-turns is produced by 2 A of current through a 100-turn coil. One ampere of current through a 200-turn coil would produce the same magnetic field strength. Figure 16-8 shows how the magnetic field strength of an electromagnet changes with the number of ampere-turns.

The magnetic field strength of an electromagnet also depends on the type of core material. Cores are usually made of soft iron or steel. These materials will transfer a magnetic field better than air or other nonmagnetic materials. Iron cores increase the flux density of an electromagnet. Figure 16-9 shows that an iron core causes the magnetic flux to be more dense.

An electromagnet loses its field strength when the current stops flowing. However, an electromagnet's core retains a small amount of magnetic strength after current stops flowing. This is called *residual magnetism* or "leftover" magnetism. It can be reduced by using soft-iron cores or increased by using hard-steel core material. Residual magnetism is very important in the operation of some electrical devices.

Electromagnetism is similar to permanent magnetism with respect to its magnetic field. However, the main advantage of electromagnetism is that it is easily controlled. The strength of an electromagnet can be increased by increasing the current flow through the coil. The second way to increase the strength of an electromagnet is to have more turns of wire around the core. A greater number of turns produces more mag-

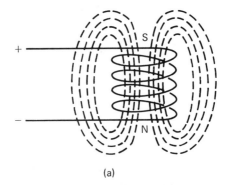

(a)

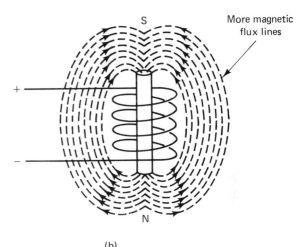

(b)

FIGURE 16-9 Effect of iron core on magnetic strength: (a) coil without core; (b) coil with core.

netic lines of force around the electromagnet. The strength of an electromagnet is also affected by the type of core material used. Different alloys of iron are used to make the cores of electromagnets. Some ma-

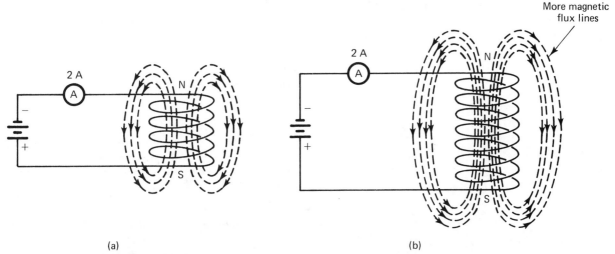

(a)

(b)

FIGURE 16-8 Effect of ampere-turns on magnetic field strength: (a) five turns, 2 amperes = 10 ampere-turns on magnetic field strength; (b) eight turns, 2 amperes = 16 ampere-turns.

terials aid in the development of magnetic lines of force to a greater extent. Other types of core materials offer greater resistance to the development of magnetic flux around an electromagnet.

SATURABLE-CORE REACTORS

Before we can readily understand the saturable-core reactor, we need to have firmly in mind some notions concerning the magnetic circuit. Consider a toroidal ring of ferromagnetic material with a coil of wire wound tightly and distributed evenly about it (Fig. 16-10). When there is a current through the coil, the resulting magnetic flux is confined almost entirely in the ring. Hence, this is the only magnetic path that needs to be considered. A magnetic-current problem then presents itself. To solve the problem two things must be considered—the flux-producing ability of the coil and the susceptibility of the ring material to magnetism. This problem is more exact for the ring than for any other magnetic circuit geometry. This is because of the almost complete uniformity of all significant variables throughout the ring. For example, the flux lines form concentric circles within the ring and the area of this field will be the same for any cross section. If the cross-sectional area of the ring is small compared to its overall diameter, then the flux path is approximately the same length anywhere. However, concepts developed from the ring are applicable to other magnetic-circuit geometries.

The flux-producing ability of the coil in Fig. 16-10 is proportional to the current through the coil and the number of turns. This is also true of any other magnetic circuit. The flux-producing ability of the coil is expressed as magnetomotive force ($\mathscr{F}$), and is given by the formula $\mathscr{F} = NI$ ampere-turns.

Besides the resulting flux being proportional to $\mathscr{F}$, it is also a function of the opposition of the core to carrying flux. This opposition is called the reluctance ($\mathscr{R}$) of the magnetic circuit. The flux Φ is then equal to $\Phi = \mathscr{F}/\mathscr{R}$ webers with $\mathscr{R}$ measured in ampere-turns per weber. As is true with resistance in the electrical circuit, reluctance is proportional to the length of the path, inversely proportional to the cross-sectional area, and dependent on the material. The magnetic susceptibility of the core material is called its permeability, expressed as μ. Expressed quantitatively, $\mathscr{R} = 1/\mu A$. The equation, $\Phi = \mathscr{F}/\mathscr{R}$, is sometimes called the Ohm's law of the magnetic circuit and illustrates the similarity between the magnetic and the dc electric circuits. Table 16-1 may help to apply previously learned facts about the dc circuit to the magnetic circuit.

Kirchhoff's laws may be written for the magnetic circuit. The first, analogous to the current law, is that the sum of the flux entering a junction in a magnetic circuit equals the sum of the flux leaving that junction. The second, analogous to the voltage law, states that around any closed path in the magnetic circuit the algebraic sum of the magnetic potentials ($\Phi \times \mathscr{R}$ drops, analogous to the $I \times R$ drops in electric circuits) must equal the net ampere-turns ($\mathscr{F}$) of excitation.

Magnetic circuits, however, differ from electric circuits in one very important way. The reluctance of a magnetic-circuit path containing iron, or other ferromagnetic material, is a function of the flux in that path. As the flux increases, it takes a larger change in F to produce the same change in flux. The circuit is said to *saturate*. The fact that $\mathscr{R}$ and μ are not constant makes it unusual to use the formula, $\Phi = \mathscr{F}/\mathscr{R}$, as we use it in electric circuits. That is, direct number substitution is not usually performed with the formula since $\mathscr{R}$ varies with Φ. The main advantage lies in the ability to use the electric-circuit analogy in the thought process of solving the magnetic-circuit problem. A graphical method of finding these values is usually used because of these nonlinearities.

The graphical approach to the magnetic circuit makes use of a *B–H* magnetization curve (Fig. 16-11), where *B* is the flux density of the toroid and H is the magnetizing force, or the intensity of the magnetic field. In the toroid of Fig. 16-10, the flux (Φ) is considered to be evenly distributed over the cross-sec-

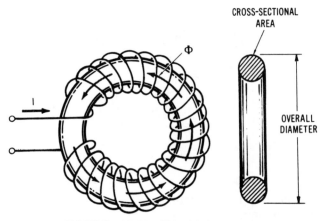

FIGURE 16-10 Toroidal magnet.

TABLE 16-1

Comparison of magnetic and electrical units

Magnetic-Circuit Quantity	Electric-Circuit Quantity
Magnetomotive force ($\mathscr{F}$)	Electromotive force (*V*)
Flux (Φ)	Current (*I*)
Reluctance ($\mathscr{R}$)	Resistance (*R*)
Flux density (*B*)	Current density (*J*)
Permeability (μ)	Conductivity (γ)

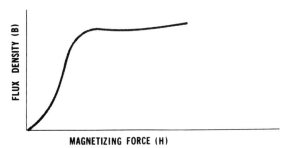

FIGURE 16-11 *B–H* magnetization curve.

tional area (A); therefore, the flux density (B) everywhere in the toroid is $B = \Phi/A$ teslas (webers per square meter). Because of the uniformity of the magnetic path of the toroid, the magnetomotive force ($\mathscr{F}$) expended per unit length (l) of the toroid is constant. This quantity is referred to as the magnetizing force (H) and is given by the formula $H = \mathscr{F}/l$ ampere-turns per meter. By substitution, we can find the relationship between B and H as follows:

$$B = \frac{\Phi}{A}$$

but since

$$\Phi = \frac{\mathscr{F}}{\mathscr{R}}$$

then

$$B = \frac{\mathscr{F}}{\mathscr{R}A}$$

Also,

$$\mathscr{R} = \frac{l}{\mu A}$$

and

$$\mathscr{F} = Hl$$

Therefore,

$$B = \frac{Hl}{(1/\mu A)A}$$

$$= \mu H$$

This equation, $B = \mu H$, is a basic relationship involving the property of the magnetic material. The B–H curve depends only on the material of the toroid and not on its dimensions. This makes it possible to develop a table of universal curves for different materials that can easily be used to find circuit parameters when a given core is involved.

In Fig. 16-12, when an alternating current is applied to the coil, a magnetizing force results which varies from $+H_{\max}$ to $-H_{\max}$. This causes the flux den-

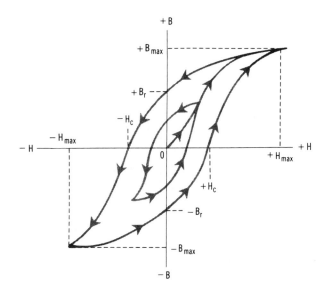

FIGURE 16-12 *B–H* hysteresis loops.

sity of the core material to vary from $+B_{\max}$ to $-B_{\max}$. At the start, however, when the current is first applied and the core is unmagnetized, the curve starts at the zero point and increases to the saturation point at $+B_{\max}$ as H values increase in a positive direction. H then decreases to zero, but B decreases only to the value of $+B_r$, not to zero, because of the circuit condition known as hysteresis. Hysteresis results from the fact that when the magnetizing force is removed, some residual magnetism remains in the core material. It takes a certain magnetizing force in the opposite direction to reduce the flux density to zero. This reverse magnetizing force ($-H_c$) is known as the coercive force of the core material. As H values then increase in a negative direction, B values also increase in a negative direction to the saturation point at $-B_{\max}$. The $-H$ values then decrease to zero, B decreases to $-B_r$, and the loop is finally completed with positive H values causing saturation at $+B_{\max}$ once again.

Thus B varies with H about a loop known as a hysteresis loop (Fig. 16-12). For any given core ma-

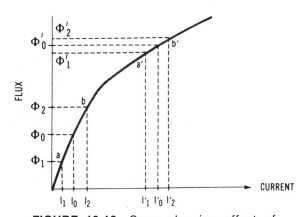

FIGURE 16-13 Curve showing effect of changing the dc excitation point.

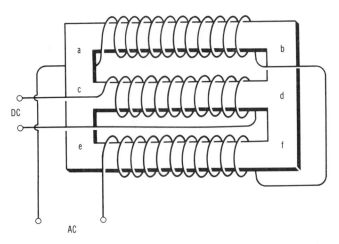

FIGURE 16-14 Schematic of a saturable-core reactor.

terial, the shape of this loop will always be the same. With a smaller ac current applied, however, a smaller loop results. Notice the similarity in shape of the two loops shown in Fig. 16-12.

Induction (L) depends on the ratio of the change in flux to the change in current. Therefore, the inductance of an iron-core coil can be varied by changing the portion of the magnetization curve or hysteresis loop being used. A variable reactor can be constructed using this principle. It employs a dc-excited coil to change the degree of saturation in the core on which the ac coils are wound.

In the magnetization curve shown in Fig. 16-13, if the dc excitation is constant at I_0, a superimposed ac excitation may cause variation from I_1 to I_2 and a flux change from Φ_1 to Φ_2. The apparent inductance

to alternating current is then proportional to the slope of the line ab. If, however, the dc excitation is changed to I_0', the flux change for the same ac excitation is only from Φ_1' to Φ_2'. In this region, the apparent inductance is proportional to the slope of line $a'b'$, a much smaller value than before. A change in dc excitation has effectively changed the inductance of the ac system. Hysteresis effects are not included in this discussion.

The schematic of a common form of saturable-core reactor is shown in Fig. 16-14. The ac magnetomotive force acts on the magnetic path abdfeca. The dc magnetomotive force acts on paths cabdc and cefdc in parallel. The dc coil then controls the reactance of the ac circuit by adding flux to one of the outer legs of the core while subtracting flux from the other outer leg. The amounts being added and subtracted are un-

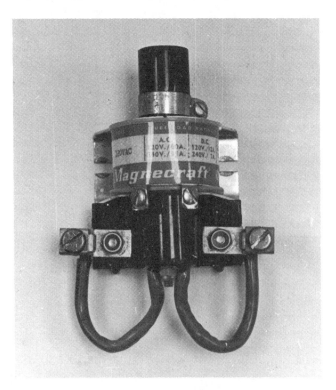

FIGURE 16-15 Commercial saturable-core reactor. (Courtesy of Magnecraft Electronic Co.)

equal because of saturation. In other reactors, the ac and dc magnetomotive forces may be supplied by the same coil that carries an alternating current superimposed on a direct current. Saturable-core reactors have many applications in control circuits. Special types will be used in magnetic amplifiers. Waveform distortion due to hysteresis and other core losses should be kept in mind as they are discussed. Figure 16-15 shows a saturable-core reactor.

MAGNETIC AMPLIFIERS

A magnetic amplifier is a device that uses saturable-core reactors either alone or in combination with other circuit elements to secure amplification or control. The devices are amplifiers because the expenditure of relatively small amounts of power in the control winding permits control over relatively large amounts of power in the output windings. Magnetic amplifiers are also known by various trade names, such as Magamp and Amplistat.

A very common form of magnetic amplifier consists of saturable reactors in association with diode rectifiers. A specific example is given by the circuit diagram of Fig. 16-16. This is a single-phase, full-wave, center-tapped rectifier circuit. The magnitude of the direct current in the load is controlled by adjusting the relatively small current in the control windings. The principal source of power is the center-tapped transformer to the ac supply. A simplified explanation of the operation may be given by comparing the amplifier of Fig. 16-16 with the full-wave rectifier in Fig. 16-17A. Figure 16-17A is simply a redrawing of the full-wave rectifier. The idealized load-voltage waveform is shown in Fig. 16-17B.

The magnetic amplifier of Fig. 16-16 differs from

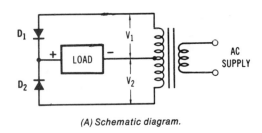

(A) Schematic diagram.

(B) Load-voltage waveform.

FIGURE 16-17   Rectifier circuit.

the simple rectifier circuit of Fig. 16-17 only by the addition of the saturable reactors in series with each rectifier element. The object of the reactors is to permit conduction by a rectifier element to be delayed after the positive half-cycle for that element starts. The delay in the magnetic amplifier is determined by the properties of the magnetic core in conjunction with the value of the control magnetomotive force.

An idealized hysteresis loop for a typical core material is shown in Fig. 16-18. Until the core flux on the positive half-cycle of voltage reaches a value corresponding to the almost horizontal line def, the reactor presents such a high impedance that practically all the voltage is dropped across it. Only a small amount of current is then present in the rectifier element and, hence, in the load. When the core becomes saturated, however, the slope of line def is so small that the impedance of the reactor is practically negligible. Practically all of the voltage appears across the load, and the rectifier is conducting a fairly large amount of current. On the negative half-cycle, diode D_1 blocks current through it, and diode D_2 undergoes the preceding process.

The point in the positive half-cycle at which sat-

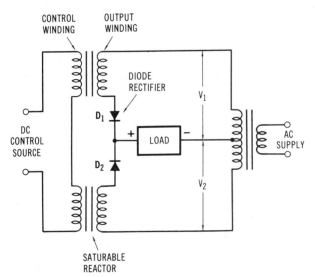

FIGURE 16-16   Magnetic amplifier circuit.

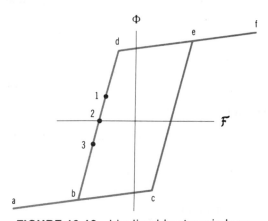

FIGURE 16-18   Idealized hysteresis loop.

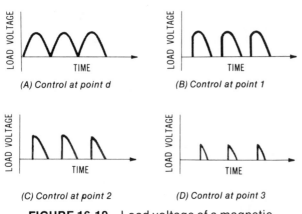

(A) Control at point d

(B) Control at point 1

(C) Control at point 2

(D) Control at point 3

FIGURE 16-19   Load voltage of a magnetic amplifier.

uration is reached can be controlled by the dc magnetomotive force of the control windings. Thus, if the control magnetomotive force corresponds to the abscissa of point d (Fig. 16-18) or any point to the right of it, the core is already saturated. Under these conditions, the load-voltage waveform is that of complete full-wave rectification as shown in Fig. 16-19A. If, however, the control magnetomotive force corresponds to the abscissa at point 1, time must elapse until the flux builds up to the saturation value before effective conduction can start. The load-voltage waveform becomes like that of Fig. 16-19B. For points farther to the left, such as 2 and 3, still further delay is occasioned. The resulting waveforms are approximately as shown in Fig. 16-19C and D. Finally, a value of control magnetomotive force will be reached when no effective conduction takes place and the load voltage becomes practically zero.

A magnetic amplifier gives the same general type of control over the output current as is obtained by a solid-state device known as a silicon-controlled rectifier (SCR).

The saturable reactors and diode rectifiers are quite rugged and require less maintenance than many other types of amplifiers. Additional windings may be added to provide for feedback to give better amplification or better control. Magnetic amplifiers are available in power sizes from a few milliwatts up to hundreds of watts. The development of magnetic amplifiers has to a considerable extent been dependent upon the development of suitable diode rectifiers and core materials with properties approaching those in Fig. 16-18.

SUMMARY

A permanent magnet is made of an alloy of iron, cobalt, and nickel. A magnet has a north (N) pole and a south (S) pole in its construction. Magnetic flux lines travel from N to S outside a magnet. When like poles are brought together they repel and unlike poles attract.

When electric current flows through a conductor it produces a circular magnetic field around the conductor. When conductors are wound into coils, each turn of the coil produces a magnetic flux. A large number of turns produces a strong magnetic field. The magnetic field of an electromagnet can be turned on or off by the applied current. The strength of an electromagnet depends on the amount of current passing through the coil, the number of wire coils, and the material type and structure of the core.

A magnetic amplifier is primarily a saturable-core reactor employing diodes that permits control of large amounts of electrical power.

A saturable-core reactor has two or more ac windings connected in a series arrangement. Applying dc to an independent core winding increases or decreases the magnetomotive force of the core to a level such that large amounts of ac can be controlled by small values of dc. When saturation of the core occurs, coil inductance is changed or controlled as a result of this action.

B–H curves are used to show the relationship between flux density and the magnetizing force of a coil. Inductance depends on the ratio of the change in flux to the change in current. The inductance of an iron core can be altered by changing the portion of the magnetization curve or hysteresis loop being used. Through this action, coil reactance can be altered, which permits control of large amounts of ac power without the use of moving parts.

ACTIVITIES

PERMANENT MAGNETS

1. Obtain two permanent magnets, a sheet of paper, and some iron filings.

2. Place one magnet on a flat surface and cover it with a piece of paper.

3. Sprinkle iron filings on the sheet of paper to see the resulting magnetic field.

4. Remove the paper from the magnet and make a sketch of the observed magnetic field.

5. Place two magnets on a flat surface with like poles facing each other. Cover the magnets with a sheet of paper and sprinkle iron filings over the paper.

6. Make a sketch of the resulting field display.

7. Repeat steps 5 and 6 for unlike poles facing each other.

ELECTROMAGNETS

1. Construct an electromagnet with approximately 20 ft of copper wire on a large bolt or metal rod.

2. Cover the coil with a piece of stiff paper or card board.

3. Energize the coil with a dry cell or 2-V source.

4. Sprinkle iron filings over the top of the coil and make a sketch of the field.

5. Disconnect the source of energy while observing the field.

6. Determine the polarity of the coil using the left-hand rule.

7. Determine the polarity of the coil with a compass.

8. Reverse the polarity of the energy source.

9. Turn on the electromagnet and repeat steps 6 and 7.

10. Disconnect the energy source.

SATURABLE-CORE REACTORS

1. Identify the windings of the saturable-core reactor.

2. With an ohmmeter, measure the resistance of the windings.

3. Make a symbol sketch of the SX and indicate the resistance value of each coil.

4. Depending on the rating of the SX, connect the magnetic amplifier of Fig. 16-16. The diodes should have a 1 A or better current rating. The load should be a low-wattage lamp.

5. Apply ac to the circuit while observing the intensity of the lamp.

6. According to the rating of the SX, apply a suitable value of dc to the control winding. How does this action alter the intensity of the lamp?

7. Alter the value of the control winding dc while observing the intensity of the lamp. Does this have any effect on lamp intensity?

8. Turn off the ac and dc power and disconnect the circuit components.

QUESTIONS

1. What is meant by the term *permanent magnet*?

2. What is meant by the term *electromagnet*?

3. How can the strength of an electromagnet be increased?

4. How can the polarity of an electromagnet be reversed?

5. Discuss the relationship of magnetomotive force, magnetic flux, and reluctance in a magnetic circuit.

6. What are the basic parts of an electromagnet?

7. Discuss the term *magnetic amplifier*.

8. What is a toroidal magnet?

9. What is hysteresis?

10. How does the magnetic amplifier of Fig. 16-16 achieve control of the load device?

CHAPTER 17

APPLICATIONS OF MAGNETISM IN INSTRUMENTS

OBJECTIVES

Upon completion of this chapter, you will be able to:

1. Describe some applications of the magnetic principle in electronic instruments.
2. Analyze the operation of a magnetic differential-pressure transmitter.
3. Explain how the magnetic principle is used in the operation of a permanent-magnet recorder pen mechanism.
4. Explain how the magnetic principle is used in the operation of an electrodynamometer recorder pen mechanism.
5. Describe how electromagnetism is used to produce a letter with a dot matrix print head.
6. Identify the components of a synchro system.
7. Explain the operation of an ac synchronous motor.
8. Explain the operation of a dc stepping motor.

IMPORTANT TERMS

In this chapter we investigate applications of the magnetic principle in electronic process control instruments. A number of new and somewhat unusual terms will be encountered. These terms play an important role in the presentation of this material. A few of these terms are singled out for study before proceeding with the chapter. A review of these terms will make the reading of the text more meaningful.

Absolute pressure: The pressure of a liquid or gas measured in relation to a vacuum or zero pressure.

Actuator: A piece of hardware that is designed to accept a signal at its input and convert this signal to an appropriate mechanical motion.

Alnico: An alloy of aluminum, nickel, iron, and cobalt used to make permanent magnets.

Armature: A part of the moving member of an instrument on which the magnetic flux reacts to provide deflecting torque.

Atmospheric pressure: The pressure exerted on a body by the air, equal at sea level to about 14.7 lb/in^2.

Brushes: A sliding contact made of carbon and graphite which touches the commutator of a generator or motor.

Commutator: An assembly of copper segments that provide a method of connecting rotating coils to the brushes of a dc rotating machine.

Diaphragm: A thin, flexible partition used to transmit pressure from one substance to another while keeping them from direct contact.

Differential pressure: The difference in pressure between two pressure sources, measured relative to one another.

Direct writing recorder: An instrument that prints or marks on a chart by energy received directly from the signal being measured.

Electrodynamometer: An electromagnetic instrument that has a set of stationary electromagnetic field coils and a movable coil that interact to produce movement.

Flow rate: The weight or volume of flow per unit of time.

Laminated core: Thin strips of sheet metal connected together into a single unit that serves as the center or foundation on which a coil is wound.

Laminations: Thin pieces of sheet metal used to construct the metal parts of a motor, generator, or transformer.

Pen mechanism: An assembly of components that move or drive the marking pen of a recorder.

Power interface: A device that interconnects a delicate instrument to a high power load.

Pressure: Force per unit area. Measured in pounds per square inch (psi), or by the height of a column of water or mercury that it will support in inches, feet, or centimeters.

Print head: A mechanism that graphically reproduces a mark or character on paper in the operation of a printer.

Recorded value: A value recorded by a marking device on a chart, with reference to the division lines marked on the chart.

Recording pen: The marking device of a recording instrument.

Slip rings: Solid metal rings mounted on the end of a rotor shaft and connected to the brushes and the rotor windings; used with some types of ac motors.

Static pressure: Pressure or force that is at rest or the result of weight that is not in motion.

Stepping motor: A motor that rotates in small uniform steps that is controlled by pulses.

Stylus: A pen-shaped or needle-like instrument for marking, writing, or indicating measured values on a chart or paper.

Synchronous motor: An ac motor that operates at a constant speed regardless of the load applied.

Voltage divider: A network consisting of impedance elements connected in series to which a voltage is applied and from which one or more voltages can be obtained across any portion of the network.

INTRODUCTION

In this chapter we study some instruments and devices that utilize magnetism or electromagnetism in their operation. This includes such things as transmitters, converters, recorders, indicators, actuators, and transducers. These instruments and devices use magnetism to generate a signal, to produce torque, move a recording pen, energize an actuator, or detect a change in the position of an object. In general, magnetism is used to control automated manufacturing processes in an operational system.

In automated manufacturing systems magnetics is often thought of as a control element. It essentially generates a signal for processing or employs a signal to do some form of work. Magnetic sensors or transducers generate a signal from some physical change that takes place at the input of the system. At the output of a system, magnetism is used to actuate the final control element. This response is used to turn a valve, rotate a shaft, energize a contactor, move the arm of a robot, or deflect the pen of a recorder. Magnetism plays a very important role in the operation of automated manufacturing system instrumentation.

DIFFERENTIAL-PRESSURE TRANSMITTERS

The differential-pressure transmitter of Fig. 17-1 is considered to be a magnetic instrument because it employs a feedback motor and a laminated-core armature in its operation. This transmitter can be adjusted to measure differential pressures of 0 to 850 in. of water or 0 to 210 k Pascal. Modifications can be made so that

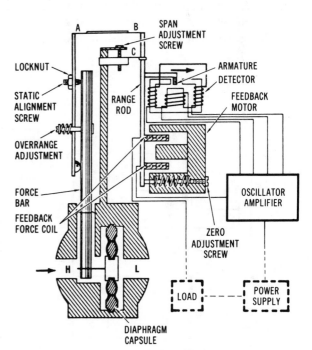

FIGURE 17-1 Foxboro differential-pressure transmitter. (Courtesy of The Foxboro Co.)

this instrument will measure flow rate up to 30 ft³ per minute or 14 liters per second. It can also measure liquid level, static pressure, and absolute pressure. The transmitter is composed of three basic sections: the body, the force-balance unit, and the oscillator/amplifier. The body consists of the component parts of the transmitter that come in contact with the fluids being measured. This represents the sensing element of the instrument. It is composed of two metal diaphragms separated by a silicone fluid. This type of construction prevents overrange damage and provides for noise filtering at the point of measurement.

The force-balance system is composed of a detector and a feedback motor. Detection is provided by a movable laminated core or armature in the inductor of the coil assembly. The position of the armature is varied by a lever mechanism connected to the sensing element. The feedback motor provides a mechanical null on the lever connected to the armature, as shown in Fig. 17-1. The detector is excited by the amplifier and it also amplifies its output. The oscillator/amplifier employs transistors or operational amplifiers similar to those used in equipment discussed previously. The 10- to 50-mA output signal passes through the feedback motor as well as the load. From the diagram, it can be seen that a two-wire transmitting system is used. The oscillator/amplifier assembly can be obtained both as a remote unit or as the integrally mounted unit shown here.

High- and low-pressure inputs are applied to the diaphragm capsule of Fig. 17-1. A change in the differential pressure across the diaphragm will cause a slight movement of the armature. As shown in the diagram, this movement is accomplished through a system of levers. The amount of movement of the armature is set by three adjustment screws: the zero adjust, range adjust, and overrange adjust. An increase in high pressure decreases the air gap between the primary and secondary cores in the detector transformer. This movement increases the coupling between the two windings, and as a result the amplifier input increases.

The oscillator/amplifier used with the differential pressure transmitter is shown in Fig. 17-2. Transistor Q_1 acts as the oscillator; Q_2 and Q_3 form a two-stage dc-coupled Darlington amplifier. The detector signal is applied to the base of Q_1. The amplified output is taken from the collector. As described in the section on transistors, the kind of amplifier hookup provides an output that is in phase with the input as long as the load is resistive. The load of Q_1 is composed of the primary coil of the output transformer (T_1) for this stage, with a parallel capacitor (C_3). These components form a resonant circuit at about 1 kHz. It is, therefore, a resistive load at that frequency. Feedback coupling is provided by a capacitor (C_1) in series with the primary detector coil. These two components form a series-resonant circuit at 1 kHz. Feedback is in phase at a frequency of 1 kHz, so the circuit of Q_1 will oscillate at this frequency. The amplitude of this oscillation depends on the amount of coupling in the input detector transformer. This, of course, is determined by the position of the armature.

The output of Q_1 is applied to the base of Q_2 by transformer coupling. The output of the coupling transformer is rectified by a full-wave, solid-state rectifier and filtered by an RC pi-section filter. The dc voltage thus obtained is the input to Q_2. The greater the am-

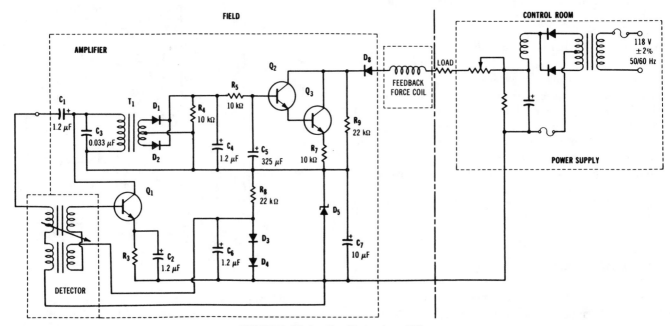

FIGURE 17-2 Oscillator/amplifier.

plitude of the oscillator output, the greater will be this dc voltage. Q_2 and Q_3 amplify this dc voltage and provide a current output proportional to the input. The current passes through the load, power supply, and feedback force coil, and is adjusted to the range 10 to 50 mAmpere by a series load-adjust potentiometer. The power supply contains a full-wave solid-state rectifier and a choke input filter. As shown, the power supply is normally located in the control room with the load or receiver, and the two-wire transmission system is used. Since the feedback force coil is in series with the load, the same current passes through it as through the load. This current produces a magnetic field that moves the detector armature to keep the force-balance system in equilibrium.

You will notice that a zener diode (D_5) is used to maintain a constant emitter-to-collector voltage in Q_1. A portion of this voltage is taken from a voltage divider formed of a resistor (R_8) and two diodes (D_3 and D_4) to provide a base voltage to bias Q_1. This bias voltage is applied through the secondary of the detector coil. Since the bias voltage is applied through this coil, it will also bias the transformer. This bias will not change since it is zener controlled and filtered by a parallel capacitor (C_6).

MAGNETIC AMPLIFIER APPLICATIONS

Magnetic amplifiers are used in industrial instruments to control large amounts of electrical power. As a rule, this applies to the final control element or the load device that performs the work function of the system. Industrial loads may require hundreds of amperes at hundreds of volts of ac. If this power were simply to be applied, or removed, an electromechanical relay would work adequately. However, automated controller output often calls for continuous or proportional control of large amounts of power. The positioning of a conveyor line, the speed control of a large horsepower motor, or the temperature control of an electric furnace are examples of high-power adjustments.

A device that is used to interconnect a delicate electronic instrument to a high-power load is called a *power interface*. This type of device must accept the low current or voltage output of a controller and use it to alter large amounts of power supplied to the final control element. A power interface device must be capable of tracking the input signal, isolating the input from the output, minimizing electrical noise, and efficiently controlling large amounts of power. A magnetic amplifier is a common power interface device that achieves these control features.

A general-purpose magnetic amplifier furnace control circuit is shown in Fig. 17-3. This type of circuit ranges from a simple self-contained 120-V single-phase unit to large three-phase assemblies that employ feedback control circuitry. This general-purpose unit has a saturable reactor that controls ac applied to the furnace element control transformer. The control winding is energized by rectified dc that is developed from the output of a controller. The dc control winding changes the saturation level of the core. This, in turn, deter-

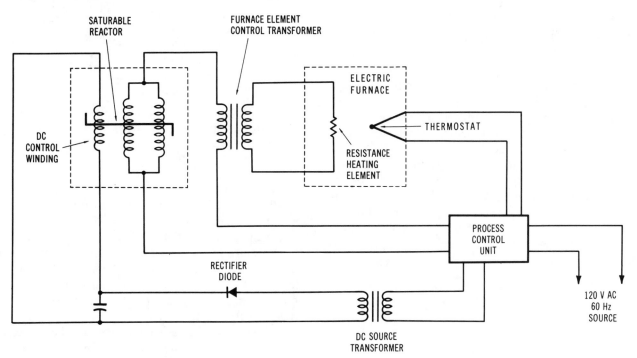

FIGURE 17-3 General-purpose magnetic amplifier circuit application.

mines the amount of reactance offered by the ac coils. This type of control uses the saturable reactor as a variable inductor. A high dc voltage value causes the core to saturate, which lowers the reactance of the inductor. This, in turn, causes more ac power to be applied to the element control transformer. Reducing the dc control voltage lowers the saturation level of the core, which causes a higher value of reactance. This action lowers the value of ac supplied to the element control transformer. Variable power control is achieved by simply altering the value of the dc voltage going to the control winding. Since this type of control employs no moving parts, it serves as a very efficient and reliable method of controlling large amounts of ac power.

Large amounts of dc power can be effectively controlled by a magnetic amplifier similiar to the one shown in Fig. 17-3. The ac coils of the saturable reactor would each employ a high current diode to rectify the line voltage being delivered to the load device. The resistance heating element in this case, would be driven directly by the rectified output. The element control transformer would be removed from the output and the heating element would be directly energized by dc. A magnetic amplifier with controlled dc output can also be used to regulate power to motors, lighting circuits, high-wattage heating assemblies, electric welders, and serve as a source of dc power for electrochemical operations. The primary role of a saturable reactor or magnetic amplifier today is in the control of large amounts of electrical power.

RECORDER MECHANISMS

The recording mechanism of an electronic instrument is responsible for producing some form of physical action or movement. This term applies to the pen mechanism, which is responsible for making an impression on paper or the chart drive assembly which moves the paper being marked. Recorder mechanism operation is based on an interaction of magnetic and electromagnetic fields. Its operation is similiar to that of an electric motor. The actual mechanism varies a great deal between different instrument manufacturers. Electronic pen recording mechanisms are classified as permanent magnet/moving coil units or electrodynamometer devices.

Permanent Magnet/Moving Coil Mechanisms

A pictorial representation of the D'Arsonval meter movement is shown in Fig. 17-4A. The similarity between this and the permanent magnet/moving coil re-

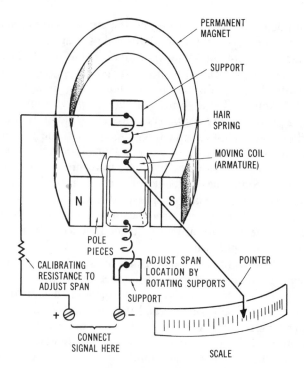

(A) D'Arsonval movement.

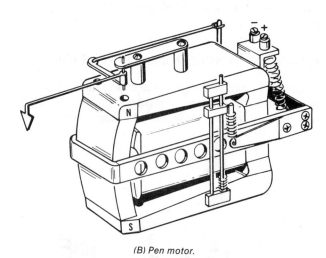

(B) Pen motor.

FIGURE 17-4 Comparison of D'Arsonval meter movement and pen motor.

corder mechanism can be seen by comparing parts A and B of Fig. 17-4. In the D'Arsonval movement the primary components are a moving coil or armature and a permanent-magnet field. When current is applied to the armature in the direction shown, the left side of the armature becomes a north magnetic pole. This induced north pole is opposed by the north magnetic pole of the permanent-magnet field. There is also some attraction between the north pole of the armature and the south pole of the permanent magnet. The attractive force adds very little force to the armature compared to the opposing force because of the difference in distance. The right end of the armature is affected by forces similar to those acting on the left end. These forces result in the armature rotating in a clockwise direction. The direction of rotation depends on the resting point or zero position of the armature. As a result, the armature is placed at a slight angle with the magnetic lines of force of the field. The rotating forces acting on the armature are opposed by a spring. Since the amount of tension on an ideal spring is proportional to the force applied, the spring produces no nonlinearity in armature deflection. An increase in current increases the pole strength on the armature, and a greater deflection of the pointer results. The mechanical movement of the pointer is proportional to the current through the meter.

A pictorial representation of the recorder pen assembly is shown in Fig. 17-4B. A strong alnico magnet is used as the permanent magnetic field in this device. A small moving coil or armature is placed near one pole of this magnet. The motion or movement of the coil is transferred by means of levers into the pen assembly. A calibrating spring is used as the opposing force to the armature movement. By adjusting the tension of this spring, the zero position of the pen can be set. Since the resistance of the armature is very small, several units can be connected in series without serious error.

A permanent magnet/moving coil pen assembly is generally used as a direct writing recorder. This means that the pen deflection power is obtained directly from the signal being measured. Pen units can be driven by a number of different input values. The versatility of this device is accomplished by adding amplifiers, resistors, and rectifiers to the moving coil or armature circuit.

Electrodynamometer Mechanisms

The electrodynamometer pen mechanism operates on the same principle as the permanent magnet/moving coil assembly. Deflection of the moving element is obtained by an interaction of two magnetic fields. However, in the electrodynamometer, two stationary coils are used in place of the permanent magnet. The movable coil is attached to the central shaft of the mechanism so that it is deflected inside the two stationary coils. Electrical connections for the movable coil are made through two spiral springs attached to the top of the mechanism. The springs also provide the restoring force to return the coil to its original or starting position.

When an electrodynamometer is used to respond to small voltage or current values, all of the internal coils are connected in a series configuration. Since the polarity of an electromagnet depends on the direction of current passing through it, each coil of the series circuit will change polarity at the same time. This means that the moving coil will deflect in the same direction regardless of the direction of current flowing through the coils. For this reason, the electrodynamometer may be used with equal precision to measure dc or ac values. The electrodynamometer mechanism seems to be more favorably suited for measuring small values of ac. For instruments that need to respond to large values of ac a current transformer may be connected to the input of the mechanism.

Thermal Writing Recorders

Thermal writing recorders use the magnetic pen mechanism to move the writing stylus as a chart moves with respect to time. In this case, the stylus is heated instead of making an ink mark on the chart paper. Heat causes a permanent mark to be scribed into the temperature-sensitive chart paper. In practice, this recording principle is better than the ink distribution principle because it eliminates problems in ink smudges, clogged ink lines, and fountain filling operations. Figure 17-5 shows a representative thermal-writing strip-chart recorder. Note the fine detail of the chart pattern reproduced.

A variation of the thermal-writing recorder is the thermal print head recorder. The print head mechanism is a solid-state thermal printing device that utilizes a matrix of individual semiconductor heating devices in its structure. The thermal print head matrix is made up of four vertical columns and five horizontal rows. Figure 17-6 shows a representative print head matrix displaying the number eight. Each semiconductor heating element can be selectively energized to form a dot. Dots are combined to form numbers and data points. The print head matrix elements are energized by a decoder/driver logic circuit.

The print head mechanism of this instrument is moved or driven by a special linear servomotor. The printing element and moving wiper contact are integral parts of the servomotor armature. The armature assembly of the servomotor is the only moving part of the mechanism. It rides on ball-bearing knife-edge rollers. This reduces sliding friction and backlash when the assembly moves linearly. A simplified drawing of the print head mechanism is shown in Fig. 17-7. The response time of this print head is approximately 0.3 s for its full-scale span. The servomotor rides back and forth on the guide armature with low inertia for a nearly frictionless operation.

Impact Printing Recorders

Impact printing mechanisms include the ball-type element, daisy wheel head, cup elements, engraved belts, rotating cylinders, and type blocks. All of these printing mechanisms respond by making an impression of

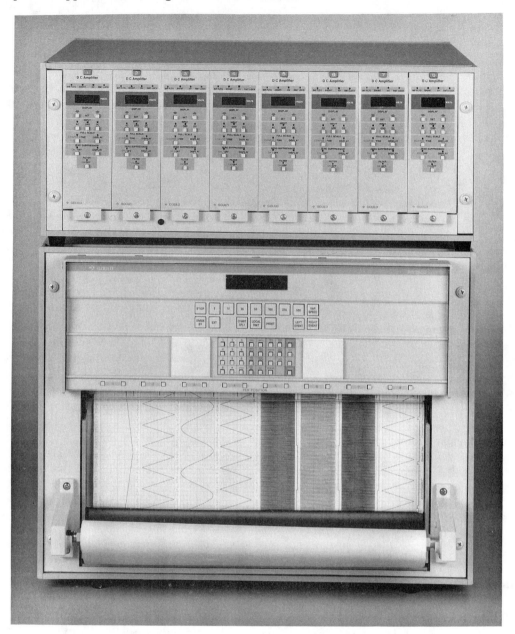

FIGURE 17-5 Thermal-writing strip-chart recorder. (Courtesy of Gould, Inc.)

a preformed character on a paper chart. No matter how the type element is mounted, it must be positioned in front of a platen and struck against an inked ribbon to produce an image on paper. This type of mechanism has a limited operational speed and creates some noise while printing. Impact printing recorders can supply a great deal more information to the chart. A time-of-event mode of operation permits the tabulation of increased information. This type of printing mechanism is also well suited for multiple-channel event displays. As a rule, impact character printing mechanisms are found in microprocessor-based data logging instruments and large multiple-channel printing applications.

Figure 17-8 shows an example of a printout from a column impact character printing mechanism.

The speed limitation of the impact character printer is responsible for the development of an impact printing process called the *dot matrix mechanism*. The printed image or character developed by this type of printer is formed by a rectangular matrix of dots. This pattern is typically nine dots high and seven dots wide. Printing is performed by moving a print head containing a column of nine pins across the paper and selectively actuating the necessary configuration of pins in seven discrete columns to form a specific character. The print head is made of nine individual electromag-

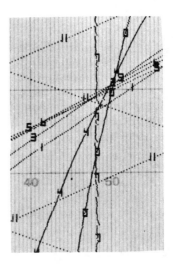

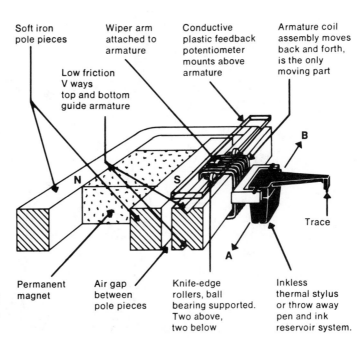

Thermal matrix printhead energizes to form analog value dot and point number 8 printout. Typical sample of characters shown above.

ROW 1
ROW 2
ROW 3
ROW 4
ROW 5

COL. 1
COL. 2
COL. 3
COL. 4

FIGURE 17-6 Thermal matrix printhead energizes to form analog value dot and point number 8 printout. Typical sample of characters shown above.

Soft iron pole pieces
Wiper arm attached to armature
Conductive plastic feedback potentiometer mounts above armature
Armature coil assembly moves back and forth, is the only moving part
Low friction V ways top and bottom guide armature
B
N   S
A
Trace
Permanent magnet
Air gap between pole pieces
Knife-edge rollers, ball bearing supported. Two above, two below
Inkless thermal stylus or throw away pen and ink reservoir system.

FIGURE 17-7 Simplification of a print head mechanism. (Courtesy of Esterline Angus.)

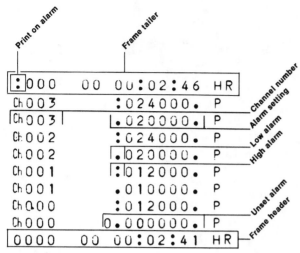

FIGURE 17-8 Printout from a column impact character printer. (Courtesy of Esterline Angus.)

characters per second and 85 lines per minute being considered as typical operating speeds. However, the chart speed, number of channels being recorded, and logging interval dictate the actual printing speed on the instrument.

A matrix nine-wire print head generates all alphanumeric and graphic characters as well as time and trend lines. The main components of the printing assembly are the print head, print wheel, and chart drive. Multiple-color printers usually employ an electromagnetically actuated mechanism that presses the print head against ink-filled color wheels to produce a display. Figure 17-9 shows an example of a chart printout from a dot matrix impact printer that shows alphanumeric characters, time designations, and trend lines.

netic coils with a movable wire pin core in each coil. The wire pin is made of high-strength tungsten carbide steel for durability. When a particular coil is energized it pulls the spring loaded core into the coil. This causes the pin to impact an inked ribbon. The impact operation causes ink to be transferred to the paper or chart of the printer. Operation is exceedingly rapid, with 167

SYNCHRO AND SERVO SYSTEMS

A number of new and unique machines have been developed to control manufacturing operations automatically. Particular emphasis is now being placed on the design of machines controlled by computer systems, as actuators in robots, and being used for automated industrial process control applications. In this regard we have synchro and servo systems. These devices

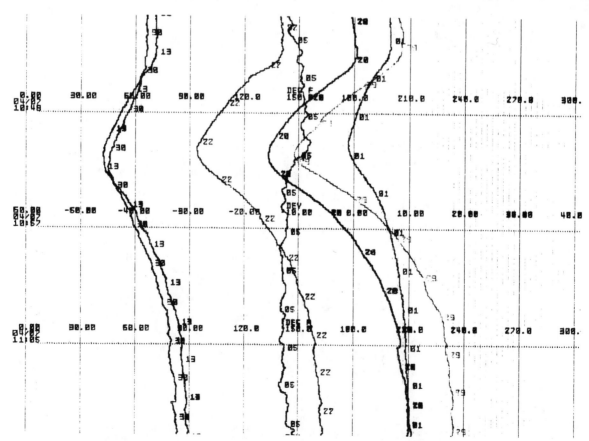

FIGURE 17-9 Chart printout from a dot matrix impact printer.

rely very heavily on the principles of magnetism and electromagnetism in their operation.

Synchro Systems

A synchro system contains two or more electromagnetic devices that are similar in appearance to small electric motors. These devices are connected together in such a way that the angular position of the generator shaft can easily be transmitted to the motor or receiver unit. Figure 17-10 shows a schematic diagram and symbols of a synchro unit. As a rule, the generator and motor units are identical electrically. Physically, the motor unit has a flywheel attached to its shaft to prevent shaft oscillations or vibrations when it is powered. The letters G and M inside the electrical symbol denote the generator and motor functions.

Figure 17-11 shows the circuit diagram of a basic synchro system. Single-phase ac voltage is used to power this particular system. The line voltage is applied to the rotors of both the generator and motor. The stationary coils, or stator windings, are connected together as indicated.

When power is initially applied to the system, the motor positions itself according to the location of the generator shaft. No physical change takes place after the motor unit aligns itself with the generator position. Both units remain in a stationary condition until further

action takes place. Turning the generator shaft a certain number of degrees in a clockwise direction causes a corresponding change in the motor unit. If calibrated dials are attached to the shaft of each unit, they would show the same angular displacement.

The stator coils of a synchro are wound inside a cylindrical laminated metal housing. The coils are uniformly placed in slots and connected to provide three poles spaced 120° apart. These coils serve functionally like the secondary windings of a transformer.

The rotor coil of the synchro unit is also wound on a laminated core. This type of construction causes north and south magnetic poles to extend from the laminated area of the rotor. Insulated slip rings on the shaft are used to supply ac power to the rotor. The rotor coil responds functionally like the primary windings of a transformer.

When ac is applied to the rotor coil of the synchro unit, it produces an alternating magnetic field. By transformer action, this field cuts across the stator coils and induces a voltage in each winding. The physical position of the rotor coil determines the amount of voltage induced in each stator coil. If the rotor coil is parallel with a stator coil, maximum voltage is induced. The induced voltage is minimum when the rotor coil is at right angles to a stator coil set.

The stator coils of the generator and motor of a synchro system are connected as indicated in the cir-

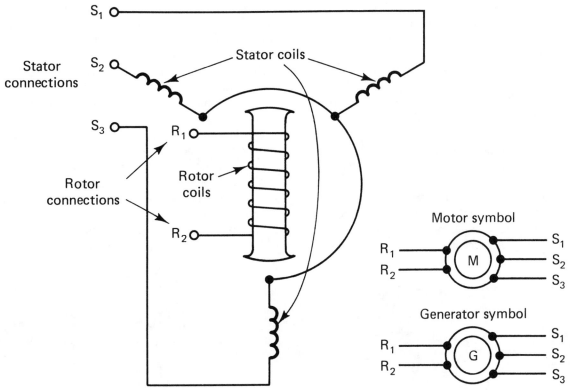

FIGURE 17-10 Schematic diagram and symbols of a synchro system.

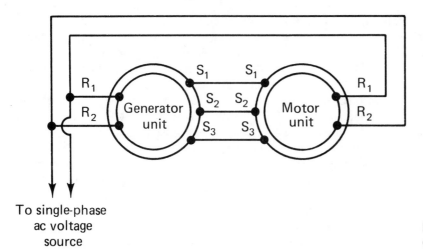

To single-phase
ac voltage
source

FIGURE 17-11   Circuit diagram of a synchro system.

cuit diagram of Fig. 17-11. Voltage induced in the stator coils of the generator therefore causes a resulting current flow in the stator coils of the motor. This in turn causes a corresponding magnetic field to be established in the motor stator. Line voltage applied to the rotor of the motor unit causes it to align itself with the magnetic field of the stator coils.

Any change in rotor position of the generator unit is translated into an induced voltage and applied to the motor stator coils. Through this action, linear displacement changes can be effectively transmitted to the motor through three rather small stator coil wires. Systems of this type are becoming very important in recorder and transmitter applications that involve position-control and motion-detection applications.

Servo Systems

Servo systems are rotating machines that accomplish a variety of functions in the operation of an electronic instrument. This includes such things as changing the mechanical position of the recording head or controlling the speed of a paper-chart mechanism feeding a recorder. A servo system is a fairly complex assembly of components that has electronic circuitry, a feedback loop, and servomotor. The circuitry is responsible for amplifying the input signal, comparing it with the feedback signal, and developing an error signal that drives the servomotor. The servomotor is ultimately used to achieve a very precise degree of rotary motion in the operation of the instrument.

Two distinct types of servomotors are used today in the operation of process control instruments. An ac type of motor, called a *synchronous motor,* is commonly used in lower-power applications. Excessive amounts of heat developed during the starting operation tends to limit this motor to rather low output power applications. Dc stepping motors represent the other servomotor.

AC Synchronous Motors.   The construction of an ac synchronous motor is quite simple. It contains no brushes, commutators, slip rings, or centrifugal switches. As shown in Fig. 17-12, it is simply made up of a rotor and a stator assembly. There is no direct physical contact between the rotor and stator. A carefully maintained air gap is always present between these two components. As a result of this construction, the motor has a long operating life and is highly reliable.

The speed of a synchronous motor is directly proportional to the frequency of the applied ac and the number of stator poles. Since the number of stator poles cannot be effectively altered after the motor has been manufactured, frequency is the most significant speed factor. Speeds of 28, 72, and 200 r/min are typical, with 72 r/min being a common industrial standard.

Figure 17-13 shows the stator layout of a two-

FIGURE 17-12   Exploded view of an ac synchronous motor. (Courtesy of Superior Electric Co.)

FIGURE 17-13 Stator layout of a two-phase ac synchronous motor. (Courtesy of Superior Electric Co.)

phase synchronous motor with four poles per phase. Poles N_1–S_3 and N_5–S_7 represent one phase while poles N_2–S_4 and N_6–S_8 represent the second phase. There are places for 48 teeth around the inside of the stator. One tooth per pole, however, has been eliminated to provide a space for the windings. Five teeth per pole, or a total of 40 teeth, are formed on the stator. The four coils of each phase are connected in series to achieve the correct polarity.

The rotor of the synchronous motor in Fig. 17-12 is an axially magnetized permanent magnet. There are 50 teeth cast into its form. The front section of the rotor has one polarity, while the back section has the opposite polarity. The physical difference in the number of stator teeth (40) and rotor teeth (50) means that only two teeth of each part can be properly aligned at the same time. With one section of the rotor being a north pole and the other section being a south pole, the rotor has the ability to stop very quickly. The rotor can also accomplish direction reversals very rapidly because of this gearlike construction.

A circuit diagram of a single-phase synchronous motor is shown in Fig. 17-14. The resistor and capacitor of this circuit are used to produce a 90° phase shift in one winding. As a result, the two windings are always out of phase regardless of whether the switch is

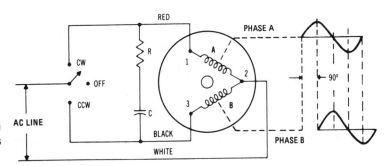

FIGURE 17-14 Circuit diagram of a single-phase synchronous motor.

in the clockwise (CW) or counterclockwise (CCW) position. When power is applied, the four coils of one phase produce an electromagnetic field. The rotor is attracted to and aligns itself with these stator coils. Then, 90° later, the four coils of the second phase produce a corresponding field. The stator is again attracted to this position. As a result of this action, the rotor sees a moving force across first one phase and then the other. This force gives the rotor the needed torque that causes it to start and continue rotation when power is applied.

The synchronous motor just described has the capability of starting in one and one-half cycles of the applied ac source frequency. In addition to this, it can be stopped in five mechanical degrees of rotation. These two characteristics are attributed primarily to the geared rotor and stator construction. Synchronous motors of this type have another important characteristic. They draw the same amount of line current when stalled as they do when operating. This characteristic is very important in automatic machine operations, where overloads occur frequently.

DC Stepping Motors. Dc stepping motors are specialized electrical machines that are used to control instruments automatically. They convert electrical pulses into rotary motion. The shaft of a dc stepping motor rotates a specific number of degrees when a pulse is applied to it. The amount of movement caused by each pulse can be repeated. The shaft of a dc stepping motor is used to position a machine accurately to a specific location. The speed, distance, and direction of a machine can be controlled by a dc stepping motor. The stator construction and coil placement of a stepping motor are similar to those of the synchronous motor shown in Fig. 17-12. The rotor is made of permanent-magnet material that is placed between two series-connected stator coils as shown in Fig. 17-15. With power applied to the stator, the rotor would be repelled in either direction. The direction of rotation in this case is unpredictable. Adding two more stator coils to this simple motor, as indicated in Fig. 17-16, would make the direction of rotation predictable. With the stator polarities indicated, the rotor would align itself midway between the two pairs of stator coils. The direction of rotation can be predicted and is determined by the polarities of the stator coil sets. Adding more stator coil pairs to a motor of this type improves its rotation and makes the stepping action very accurate.

Operation of the stepping motor of Fig. 17-17 may be achieved in a four-step switching sequence. Any of the four combinations of switches 1 or 2 will produce a corresponding rotor position location. After the four switch combinations have been achieved, the switch-

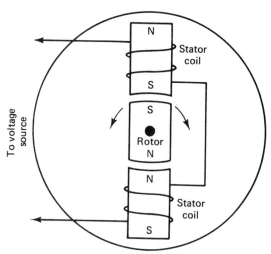

FIGURE 17-15   One set of stator coils of a stepping motor.

ing cycle repeats itself. Each switch combination causes the motor to move one-fourth of a step.

A rotor similar to the one shown in Fig. 17-13 normally has 50 teeth. Using a 50-tooth rotor in the circuit of Fig. 17-17 would permit four steps per tooth, or 200 steps per revolution. The amount of displacement or step angle of this motor is therefore determined by the number of teeth on the rotor and the switching sequence.

A stepping motor that takes 200 steps to produce one revolution will move 360° each 200 steps or 1.8° per step. It is not unusual for stepping motors to use

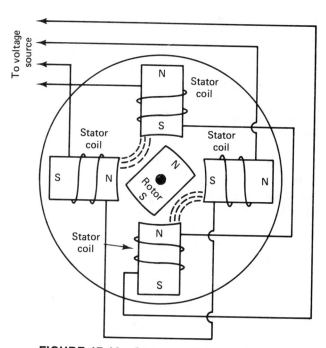

FIGURE 17-16   Stator coil placement of a dc stepping motor.

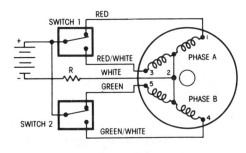

SWITCHING SEQUENCE *

STEP	SWITCH 1	SWITCH 2
1	1	5
2	1	4
3	3	4
4	3	5
1	1	5

* To reverse direction, read chart up from bottom.

FIGURE 17-17 Circuit diagram and switching sequence of a bifilar-wound dc stepping motor. (Courtesy of Superior Electric Co.)

eight switching combinations to achieve one step. In this case, each switching combination could be used to produce 0.9° of displacement. Motors and switching circuits of this type permit a very precise controlled movement. An eight-step switching sequence called *half-stepping* is also used. During this type of operation, the motor shaft moves half of its normal step angle for each input pulse applied to the stator. This allows more precise movement control and greater speed capability. Dc stepping motors provide a very accurate control procedure for electronic instruments.

SUMMARY

A differential transmitter is designed to measure pressures of 0 to 850 in. of water or 0 to 210 kiloPascal of pressure. It contains a body, a force-balance system, and an oscillator/amplifier. The body deals with those things that come in contact with the fluids being measured. The force-balance unit employs a detector and a feedback motor. The oscillator/amplifier is responsible for electronic signal conversion.

The pen mechanism of an electronic instrument operates on the same principle as the D'Arsonval meter movement. This mechanism employs a permanent magnet and a movable coil attached to an indictor hand or marking pen. When the polarity of the movable coil is the same as the permanent magnet poles, repulsion of the movable coil occurs. Through this action electrical energy is changed into mechanical movement. Pen mechanisms respond to an interaction of permanent magnet and electromagnetic fields or two electromagnetic fields.

A thermal writing pen mechanism is moved as a result of magnetic field interaction. The writing stylus of this mechanism is heated instead of making an ink mark on the chart paper. The heat printing element may be a stylus or a solid-state print head device that produces a dot matrix display.

In impact printing an impression is made as the result of paper being positioned in front of a platen and being struck against and inked ribbon. A nine-wire matrix print head is an impact printer that uses electromagnetism in its operation. The print head of this mechanism generates alphanumeric, graphic, time, and trend-line characters in its operation.

A synchro system contains a generator that is interconnected to a motor. When ac is applied to the rotor of the generator it produces a magnetic field. By transformer action, this field cuts across the stator coils and induces voltage in each winding. The stator coils of the generator and motor are connected together by three wires. A magnetic field change in the generator is transferred electrically to the motor. This action couples a magnetic field change in the position of the generator rotor to the rotor of the motor. In effect, a change in rotor position of the generator produces an equivalent positional change in the rotor of the motor unit.

An ac synchronous motor can be used as a servo system. This type of servomotor is used to achieve very precise control of rotary motion in the operation of an electronic instrument. The rotational speed of a synchronous motor depends on the frequency and the number of stator poles; 72 r/min is a common industrial standard. A synchronous motor contains an axially magnetized permanent magnet rotor and four poles per phase in the stator. The construction of this motor permits it to start, stop, and reverse direction very quickly.

Another type of servo system employs the dc stepping motor to produce rotary motion. This type of motor converts electrical pulses into rotary motion or positional changes. A stepping motor has the same construction as an ac synchronous motor. Operation of the stepping motor is achieved by a four- or eight-step switching process. Each combination of the switching sequence will produce a rotor position change. After the switching sequence has been achieved, the process repeats itself. Each switch combination produces a step in the operation of this motor. A stepping motor that takes 200 steps to produce one 360° revolution moves 1.8° per step.

ACTIVITIES

DIFFERENTIAL PRESSURE INSTRUMENTS

1. Refer to the manufacturer's operational manual for the differential pressure transmitter to be used in this activity.

2. Identify the inputs for this instrument. They may be identified as high/low or P_1/P_2. Note the identification procedure for the instrument.

3. Identify the output of the instrument. What does the output of this instrument represent?

4. With the electrical power off or disconnected and the pressure source off, remove the housing from the instrument.

5. Identify the main body of the instrument. This has the pressure inlets attached to it.

6. Note the location of the force-balance mechanism. Does this assembly have any adjusting mechanisms? If so, point out where they are located.

7. Locate the electronics of the instrument. Does this part of the instrument have any adjustments? If so, describe what is accomplished by the adjustment.

8. Return the housing to the instrument. Connect the pressure lines and apply power to the instrument.

9. Describe the function of this instrument in the operation of a system.

RECORDER PEN MECHANISM

1. Refer to the manufacturer's operation manual of the instrument being used in this activity.

2. Turn off or disconnect the electrical power.

3. Open the front door or remove the housing from the instrument.

4. Locate the pen mechanism of the instrument. Is this a single- or multiple-point system?

5. Identify the type of pen mechanism used in this instrument.

6. Explain how the mechanism produces a mark on the chart paper.

7. Explain how electromagnetism is used in this mechanism.

8. Reassemble the instrument.

SYNCHRO SYSTEMS

1. Connect the synchro system of Fig. 17-11.

2. Apply 120 V/60 Hz to the rotors of the system.

3. Turn the shaft of the generator unit while observing the motor shaft.

4. Describe the response of the motor shaft.

5. Try several other moving operations on the generator shaft while observing the response of the motor shaft. Turn the motor shaft while observing the generator shaft. Does this procedure change the operation of the system?

6. Describe the operation of a synchro system.

7. Disconnect the system.

AC SYNCHRONOUS MOTORS

1. Connect the single-phase synchronous motor circuit of Fig. 17-14. Representative capacitor and resistor values are 20 μF and 5 kΩ.

2. Apply 120 V ac to the input.

3. Place the switch in the clockwise position. Note the direction of rotation.

4. Change the switch to the counterclockwise position. Note the direction of rotation.

5. Explain why the switch position determines the direction of rotation.

6. Turn off the ac source voltage and disconnect the circuit.

DC STEPPING MOTORS

1. Connect the dc stepping motor circuit of Fig. 17-17.

2. The value of dc applied to the circuit depends on the stepping motor used. Typically, 10 V dc is used with a resistor value of 50 Ω at 20 W.

3. Position the two switches to represent an off–off or 0–0 condition. We will assume that this occurs when the switches are both in the down or bottom position.

4. Step the two switches through an operational sequence of 0–1, 1–0, and 1–1 while observing the motor shaft.

5. Does this stepping sequence change the position of the shaft?

6. Repeat the stepping sequence several more times. How does each sequence alter the shaft position?

7. Turn off the dc source and disconnect the circuit.

QUESTIONS

1. What are three basic sections of the differential pressure transmitter?

2. How does a differential pressure transmitter detect a change in pressure?

3. How does a magnetic amplifier control ac power?

4. How does a magnetic amplifier control dc power?

5. How does a D'Arsonval movement respond when dc is applied to it?

6. What is the difference between a D'Arsonval movement and a permanent-magnet pen mechanism?

7. What is an electrodynamometer?

8. How does the thermal-writing mechanism differ from an ink pen writing mechanism?

9. Describe the operating principle of the impact printing mechanism.

10. How does a dot matrix mechanism produce a character or number?

11. What is a synchro system?

12. How is an ac synchronous motor used as a servo system?

13. What makes a dc stepping motor move positionally?

14. What is the stepping sequence for a four-input stepping motor?

ELECTRICAL AND ELECTRONIC SYMBOLS

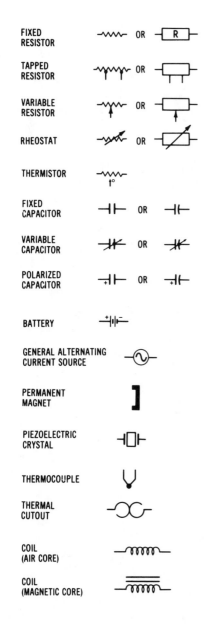

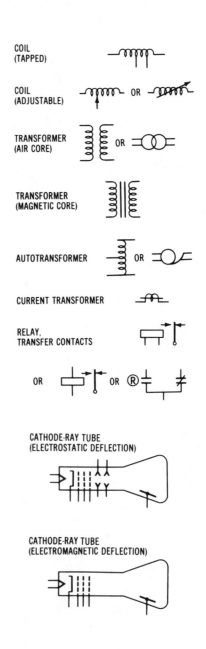

SEMICONDUCTOR SYMBOLS:

DIODE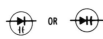

CAPACITIVE DIODE (VARACTOR) OR

TEMPERATURE-DEPENDENT DIODE

PHOTODIODE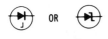

LIGHT-EMITTING DIODE (LED)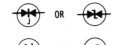

ZENER DIODE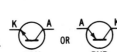

THYRECTOR DIODE

TUNNEL DIODE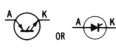

TRIGGER DIAC, UNIDIRECTIONAL NPN ... PNP

TRIGGER DIAC, BIDIRECTIONAL NPN PNP

THYRISTOR REVERSE BLOCKING-DIODE TYPE

THYRISTOR, REVERSE-BLOCKING-TRIODE TYPE OR SCR

THYRISTOR BIDIRECTIONAL DIODE TYPE (DIAC)

THYRISTOR BIDIRECTIONAL TRIODE TYPE (TRIAC)

BIPOLAR TRANSISTOR PNP NPN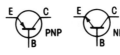

PHOTOTRANSISTOR PNP NPN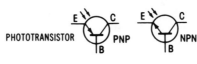

DARLINGTON TRANSISTOR NPN

UNIJUNCTION TRANSISTOR N-TYPE BASE P-TYPE BASE

N-CHANNEL JFET

P-CHANNEL JFET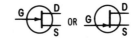

N-CHANNEL MOSFET, DEPLETION TYPE

N-CHANNEL MOSFET, ENHANCEMENT TYPE

P-CHANNEL MOSFET, DEPLETION TYPE

P-CHANNEL MOSFET, ENHANCEMENT TYPE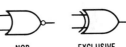

PHOTOVOLTAIC CELL (SOLAR CELL)

PHOTON-COUPLED ISOLATOR (PHOTOEMISSIVE DIODE & PHOTOTRANSISTOR

GENERATOR (GENERAL)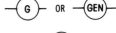

DC GENERATOR

AC GENERATOR

SYNCHRONOUS GENERATOR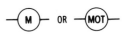

MOTOR (GENERAL)

DC MOTOR

AC MOTOR

SYNCHRONOUS MOTOR

ANALOG AND DIGITAL LOGIC SYMBOLS:

OPERATIONAL AMPLIFIER

AND FUNCTION   OR FUNCTION   NAND FUNCTION

NOR FUNCTION   EXCLUSIVE OR FUNCTION   INVERTER (NOT) FUNCTION

FLIP-FLOP (GENERAL)

INDEX